AF335719

THE CLOCKWORK UNIVERSE

The publication of this catalog and the production of the exhibition described herein were made possible by the support of the NCR Corporation, Dayton, Ohio

THE CLOCKWORK UNIVERSE

German Clocks and Automata
1550–1650

Edited by
Klaus Maurice and Otto Mayr

Smithsonian Institution, Washington, D.C.

1980
Neale Watson Academic Publications, New York

English-language edition published by Neale Watson Academic Publications, Inc., 156 Fifth Avenue, New York. Copyright © 1980 by the Smithsonian Institution. All rights reserved. Except for use in reviews, no portion of this book may be reproduced in any form without written permission.

A German-language edition of this book, *Die Welt als Uhr,* is published by Deutscher Kunstverlag, Munich, 1980, ISBN 3-422-00709-1.

Editor for English edition: Diana Menkes
Editor for German edition: Georg Himmelheber
Translator for English edition: H. Bartlett Wells
Translator for German edition: Winfried Petri

Graphics: Elisabeth Zanzen
Composition: ComCom, Allentown, Pennsylvania
Printing: aprinta Druck-KG, Wemding, Federal Republic of Germany
Binding: The Haddon Craftsmen, Scranton, Pennsylvania

ISBN 0-88202-188-5

Library of Congress cataloging in publication data will be found on the last page of this book

Contents

Introduction

Once in a while it happens in history that a civilization produces a surpassing technological achievement, like the Pharaonic pyramids, the Gothic cathedrals, or the American space program. By channeling all resources, by focusing all talents of the society, such achievements far exceed all other technological efforts of the age. It also happens that such singular achievements of technology—again like the pyramids, the cathedrals, the space ships—may lack utility. They may solve no practical problems, save no labor, produce no commercial profit, nor feed the hungry. They are merely flights of the creative spirit, materialized fantasies, projections from the realm of ideas into the real world.

The mechanical clock of early modern Europe represented such an achievement. One could hardly call it practical or useful: unreliable, imprecise, and overloaded with such extraneous capabilities as astronomical prediction, mechanical music, and automatic theater, it was a problematic timekeeper. Such shortcomings, however, did not count, for the clock was a wonder of inventiveness, a triumph of craftsmanship, an example of the particular beauty of machinery. From the point of view of engineering design, it represented the vanguard of European technology for centuries. Its design elements solved complex problems with a mechanical sophistication that has yet to be adequately appreciated by modern scholarship. Its parts were structured, spatially and logically, in strict order. In the logic of their action, the components of the clock formed firmly linked causal chains that received their commands from one central point where the actions of the whole mechanism were pre-programmed from the beginning onward through all eternity.

The impact of the mechanical clock was astonishing, all the more so considering its limited utility. It was produced in large quantities and in all sizes; as a conceptual image it took possession of the minds and spirits of an entire civilization, in a way no machine had ever done before.

As the technical and artistic expression of an era the clock reached its apex in the hundred years between approximately 1550 and 1560 and geographically in the area of German-speaking central Europe which coincided approximately with the boundaries of the Holy Roman Empire. The mechanical clock did not originate in this area. Its invention, a little before 1300, more likely took place in northern Italy or in England. But economic and social factors created a favorable environment for its development in Germany. At that time Germany was Europe's leader in the extraction of metals, which provided the raw material for clocks, and also in metal-working skills. The self-governing craft guilds in the city republics advanced the training of craftsmen and insisted on high standards of quality. In other countries, therefore, the opinion was widespread that technical skill and mechanical inventiveness were innate to the Germans. For example, in *Bilancia politica* Traiano Boccalini (1556–1613) mentions the gift of the Germans for setting up city republics and for making instruments: "cosi sottili e eccelenti istitutori di republiche, come inventori e fabricatori di varii instromenti." Helpful external socio-economic circumstances coincided with favorable internal technical conditions. By the middle of the 16th century the mechanical clock had reached maturity as a machine; the repertoire of available design elements was ample enough to deal with all occurring tasks, and the play of artistic fancy was not hampered by technical impossibilities.

And yet, the factors suggested here do not suffice to account for a technological flowering that was sustained for an entire century. Why were clocks produced in such large numbers, on an almost industrial scale, for so extensive a market? What caused so great a demand?

The appeal of the clock did not lie in its utility but was of an abstract, almost mystical character whose roots the connoisseurs and collectors of these machines hardly understood. The clock represented the sharpest conceivable contrast to the prevailing reality, with the collapsing political and social order of the Middle Ages, with the wars of religion arising out of the Reformation, with the multitude of revolutionary new ideas and the social unrest which these unleashed. The clock exemplified what was lacking in the real world: a centrally organized, unalterably functioning, rational order. People started to formulate their idea of the universe on the model of the clock and to conceive of the three essential systems within which mankind exists—the cosmos, the state, and the body—as being clock mechanisms. The relationship between God and Creation became analagous to that between the clockmaker and the clock; the harmony of the universe was explained through the regularity of the clock. The animal body was understood as an automaton directed by clockwork, and the technology of automata promised realization of the ancient dream of the creation of artificial life. Clocks and automata became the fundamental analogies of medicine. The same analogy held true for the kind of state toward which people were aspiring: a structure with a central authority, whose parts worked together with the same inevitability, predictability, and rapidity as the wheels in clockwork. Rule by absolute monarchy, which became increasingly extended in Europe at the start of the 17th century, closely reflected this aspiration.

In Germany the flowering of clockmaking came to an end in the middle of the 17th century. The decline of the craft can be understood through the interaction of many factors. The great princely courts, which had been concerned with the advance of technology, found themselves weakened after the Thirty Years' War; after the Peace of Westphalia an extreme federalism consisting of numerous small states no longer either challenged or supported the skills of craftsmen. The guilds, always preoccupied with conserving traditions and unsympathetic to innovation, were incapable of developing new initiatives. Leadership in clockmaking moved westward, where the new centrally administered national states, notably France and England, promoted technology and science through the foundation of national academies in their capitals. It was here that the pendulum clock, invented in 1657, was promptly introduced into practical use. Within a few decades the clock was transformed from the mechanical expression of a worldview into a functional instrument of precision measurement.

The spiritual spell of the clock was broken only at a later date. The concept of order embodied in it, for all its virtues, was decidely authoritarian: order, peace, equilibrium were to be achieved through central planning, steering, decision-making. All the members of the system outside the sole central authority remained mere little wheels in a great train, and they possessed neither freedom nor individuality. But now an alternative concept of order arose in England as a consequence of its political and religious revolutions of the 17th century. This new liberal order insisted on the maintenance of equilibrium and peace without a central authority. What was discovered was the concept of the self-regulating system, or the idea of dynamic equilibrium. In such a system equilibrium is restored only by forces that arise spontaneously from deviations from equilibrium, without the intervention of any external authority. This liberal conception of order was converted into practice promptly and in many ways, for example in the new state form of constitutional monarchy with its checks and balances and in the capitalist notion of an economy regulated by the law of supply and demand.

While the ideas of liberalism gained acceptance, the deterministic metaphor of the clock lost its power. As a model for the universe, the state, and the organism, the clock was now deliberately and explicitly rejected. To the Romantic movement the clock and the automaton had turned into objects of horror, personifications of a demonic intervention into organic nature and human freedom.

The task of the exhibition described herein has been to portray an extraordinary technological and cultural achievement through its works, and to illuminate it through its historical circumstances. The available evidence suggests that the origins and causes of the phenomenon examined here lay preponderantly on the level of intellectual history and that the German clocks and automata of 1550–1650 will have to be understood as an expression of the thoughts, feelings, and hopes of that period. Rarely in history has a machine so directly expressed, and in turn affected, the intellectual climate of its time. Just as the concrete mechanism—the clock as a universe—served to depict the cosmic motions on a small, humanly comprehensible scale, so it furnished methods and structures of thought by which man could explore, on a larger scale, the mysteries of the universe in which he existed and could interpret this universe as clockwork.

OTTO MAYR

1 A Mechanical Symbol for an Authoritarian World

1. A symbolic clock appears in the portrait of Duke Philip of Bavaria, Bishop of Regensburg. Engelhard de Pee (?), Munich, 1596/1598. Munich, Residenzmuseum.

Up to the time of the mature steam engine, late in the 18th century, the mechanical clock was the culmination of European technology. It represented a refinement of workmanship and a sophistication of mechanical design never seen before. That it was also immensely popular is attested to by the large numbers in which early clocks were produced and have survived to the present.

What was the basis of this fascination? Why did the best technical minds of the time devote their lives to the making of ingenious clocks? Why did clocks become favorite objects of princely ostentation and the pride of towns and private owners? The answer which immediately suggests itself to the modern observer with his utilitarian cast of mind—that the mechanical clock filled a need for reliable and accurate timekeeping—misses the point. Admitted the clock was a timekeeper, and it did introduce a measure of regularity into the public life of urban communities of the late Middle Ages. But it was only the pendulum clock, devised by Christiaan Huygens in 1657 and introduced into practical use in the late 17th century, that represented an accurate and reliable timekeeper. Before that, clocks were elaborate objects of decorative art, capable of performing astronomical predictions, mechanical music, and automatic theater, but they were unreliable and inaccurate timekeepers, with hour dials sometimes hard to locate among the ornamentation. Their successors, the pendulum clocks, were scientific precision instruments of functional appearance—often quite literally "black boxes"—designed for only one task, the accurate measurement of time.

In short, the reasons for the extraordinary fascination with the mechanical clock in early modern Europe lie deeper than the level of a material need like that of timekeeping; they must be sought on the levels of basic intellectual attitudes and aesthetic preferences. How does one find access to historical factors as vague and elusive as these? A key to the problem may present itself in the following. Almost from the moment of its invention the mechanical clock was used frequently in comparisons, figures of speech, metaphors. When a writer refers to clocks to make a point in some entirely different context, he may reveal his true feelings about clocks, although his mind is not on that subject. The history of the clock metaphor shows a profusion of applications that seems bewildering at first glance but which on some level of abstraction reveals a unifying theme.[1]

1 The present essay is a very brief condensation from a book-length study that is still in progress.

The first authors to refer metaphorically to the mechanical clock, Dante in the *Divine Comedy* (c. 1320), Henricus Suso in his *Horologium sapientiae* (c. 1340), and Berthold of Freiburg in the *Horologium devotionis circa vitam Christi* (c. 1350) invoke it rather loosely to illustrate a divine order and harmony.[2] Much more specific is Jean Froissart in his long poem *The Clock of Love* (c. 1380), an elaborate allegory comparing various aspects of chivalrous love with the parts of the mechanical clock.[3] Significantly, he connects the central element of the clock, the verge-and-foliot escapement, with the virtue of moderation (*mesure* = temperance, self-control), the highest in the canon of virtues of the medieval knight. This identification of the clock with *temperantia,* and occasionally *sapientia,* has recurred in subsequent literature and, more widely, in the iconography of art.[4] Nicole Oresme (c. 1370) employs the clock image in another context: discussing the possibility of irrational numbers in the working of the universe, he characterizes a proposed hypothesis as follows: "if someone should construct a material clock, would he not make all the motions and wheels as nearly commensurable as possible? How much more [then] ought we to think [in this way] about that architect who, it is said, has made all things in number, weight, and measure?"[5] This comparison of the clockmaker with the creator of the universe, which Oresme repeats once more elsewhere,[6] is important because it became the basis for most subsequent clock metaphors and later developed into a formal proof of the existence of God, the "argument from design," which played such a prominent role in theological discussions of the 17th and 18th centuries.

Subsequently, until well into the 18th century, the clockwork metaphor was employed in steadily increasing frequency and over a wide range of applications. Significantly, it grew into the chief metaphor for three central human concerns: the world, the body, and the state.

The first of these had already been foreshadowed in the 14th century. It remained alive and active beyond the time of Newton and Leibniz. Joachim Rheticus, for example, in his *First Account* of Copernicus' new system (1540) asked rhetorically: "why then should we not give God, the Creator of Nature, credit for the kind of skill that we see in common clockmakers?"[7] Philippe de Mornay, a Huguenot theologian, stated in 1581: "Sure, the sky is as the great Wheele of a Clocke," and John Norden, an Elizabethan poet, wrote in 1614: "This moving world, may

2   Dante Alighieri, *La divina commedia,* Paradiso, 10:139–148; 24:13–15; 33:144. Henricus Suso, *Horologium aeternae sapientiae,* Joseph Strange, ed. (Cologne, 1850). Berthold, *Das andächtige Zeitglöcklein des Lebens und Leidens unseres Herrn Jesus Christus* (Einsiedeln, 1874).
3   Jean Froissart, "Li orloge amoureus," in *Oeuvres: Poésies,* Auguste Scheler, ed., 3 vols. (Paris, 1870), Vol. I, pp. 53–86.
4   For a definitive treatment, see Lynn White, jr., "The Iconography of *Temperantia* and the Virtuousness of Technology," in Theodore K. Rabb and Jerrold E. Seigel, eds., *Action and Conviction in Early Modern Europe* (Princeton, 1969), pp. 197–219.
5   Edward Grant, ed., *Nicole Oresme and the Kinematics of Circular Motion: Tractatus de commensurabilitate vel incommensurabilitate motuum celi* (Madison, 1971), pp. 293–295.
6   Nicolas Oresme, *Le livre du ciel et du monde,* Albert D. Menut, trans. (Madison, 1968), p. 289.
7   Georg Joachim Rheticus, *Erster Bericht über die sechs Bücher des Kopernikus von den Kreisbewegungen der Himmelsbahnen,* Karl Zeller, trans. (Munich, 1944), p. 56.

well resembled be, T'a Jacke, or Watch, or Clock, or to all three."[8] The astronomer Johannes Kepler completely shared these views: "My goal is to show that the celestial machine is not like a divine being but like a clock."[9] Johannes Geyger, a German Lutheran pastor, offered the following version of the design argument:

> . . . how much wiser, more intelligent and ingenious must this master be who has created the master of the clockwork itself, and who gave him understanding and skill to make the clockwork, and who indeed has created by his omniscience the whole heavenly firmament and clockwork, sun, moon, planets, constellations, stars, etc., and who, by his omnipotence and wisdom maintains these in proper running order; who can he be but the Lord . . . ?[10]

While to authors such as these the clockwork metaphor was perhaps no more than a casual literary ornament, Descartes made the clock, or rather the automaton (which he regarded as a larger category that included clocks), the subject of an analogy that was basic for his mechanistic philosophy of nature. It was such an automaton that he had in mind when he postulated: "There is a material world machine; or, to put it more forcefully, the world is composed like a machine of matter; or, in material things all causes of motion are the same as in machines built by artifice."[11] Clockwork metaphors and analogies were also frequent in the writing of Robert Boyle; for example,

> . . . when . . . I see a curious clock, how orderly every wheel and other parts perform its own motions . . .; I do not imagine, that any of the wheels, etc. or the engine itself is endowed with reason, but commend that of the workman, who framed it so artificially. So when I contemplate the actions of those several creatures, that make up the world, I do not conclude . . . the vast engine itself, to act with reason or design, but admire and praise the most wise Author. . . .[12]

The clock metaphor was given its most central role in the philosophy of Christian Wolff, whose *Cosmologia generalis* (1731) was based on the simple premise "Mundus propemodum se habet ut horologium automaton" the world behaves like a clockwork automaton.[13] Proceeding in this vein, he employs the metaphor dozens of times; the reason for his fascination with the clock is the manner in which it functioned, a manner which he believed illustrated accurately the deterministic character of the

8 Philip of Mornay, *A Worke Concerning the Trunesse of Christian Religion,* Philip Sydney and Arthur Golding, trans. (London, 1617), p. 95. John Norden, *The Labyrinth of Man's Life* (London, 1614), p. D-2v.

9 Johannes Kepler, *Opera omnia,* Christian Frisch, ed., 8 vols. (Frankfurt, 1858–1871), Vol. II, p. 84.

10 Johannes Geyger, *Horologium politicum* (Nuremberg, 1621), p. 69.

11 René Descartes, *Oeuvres,* C. Adam and P. Tannery, eds., 12 vols. (Paris, 1897–1913), Vol. V, p. 546.

12 Robert Boyle, *Works,* 6 vols. (London, 1772), Vol. II, p. 40.

13 Christian Wolff, *Cosmologia generalis,* 1731 (Frankfurt, 1737), p. 103.

world. "If the only common characteristic [between world and clock] is a predetermined manner of functioning, in contrast to the freedom of man, then the world is like a clock."[14]

For representatives of the French Enlightenment the view of the world as a clock was similarly linked with a belief in determinism. Voltaire, who held that "Free-will is a world absolutely void of sense,"[15] was fond of reciting the design argument: "The universe embarrasses me, and I cannot imagine that this clockwork exists and has no clockmaker."[16] Diderot, whose views on free will were similar ("Study it closely and you will see that the word liberty is meaningless; there is none and there cannot be a free agent"),[17] likewise held that the world is simply "a machine, with its wheels, its ropes, its pulleys, its springs, and its weights."[18] Thus, over the centuries, the clockwork image developed into a symbol of determinism. It fell into disuse when determinism was rejected.

If the creator of the universe was a clockmaker, then his second most fascinating creation, living organisms—especially human beings—had to be clockworks as well. This conclusion followed from logic, and it was drawn in due course. At first, comparisons between the living body and the clock were loose and poetic. John Donne derived comfort from the image in a funeral elegy:

> But must we say she's dead? may't not be said
> That as a sundered clock is piecemeal laid,
> Not to be lost, but by the maker's hand
> Repolished, without error then to stand, . . .[19]

Others compared sleep to the repair of clocks:

> . . . in thy ebony box [night]
> Thou dost inclose us, till the day
> Put our amendment in our way,
> And give new wheels to our disorder'd clocks.[20]

or to their winding: "the body of man requiring to be a sundry times relieved, even as a Clocke winding up, to keepe it in motion."[21] Somewhat more specific is the obvious comparison of the pulse with the ticking of a clock. "The pulse is in man the escapement, and as the clock cannot run without one, so does man's life end with his pulse."[22]

14 *Ibid.,* p. 107.
15 Voltaire, *A Philosophical Dictionary,* 6 vols. (London, 1824), Vol. III, p. 256.
16 Voltaire, "Les Cabales" (1772), *Oeuvres complètes* (Paris, 1877), Vol. X, p. 182.
17 Denis Diderot, *Oeuvres complètes,* J. Assézat and M. Tourneux, eds., 20 vols. (Paris, 1875–1877), Vol. IX, p. 434.
18 *Ibid.,* Vol. I, pp. 132–133.
19 John Donne, "A Funerall Elegie," in *The Poems of John Donne,* H. J. C. Grierson, ed., 2 vols. (Oxford, 1912), Vol. II, pp. 37–40.
20 George Herbert, "Even Song," in *The Works of George Herbert,* F. E. Hutchinson, ed. (Oxford, 1941), p. 64.
21 Sir William Pelham, *Meditations upon the Gospell by Saint John* (London, 1625), p. 98. For this quotation I am indebted to Prof. Francis C. Haber, University of Maryland.
22 Georg Philip Harsdörffer, *Delitiae mathematicae et physicae* (Nuremberg, 1651), p. 347.

It was Descartes who advanced the clock metaphor, as applied to matters of physiology, from a literary ornament to a scientific analogy. His frequent references to "the machine of our body" were an indication of a serious belief: "There are certainly no rules in Mechanics that do not also belong to Physics [i.e., physiology], of which it is a part or special case: it is no less natural for a clock composed of wheels to tell the time than for a tree grown out of a given seed to produce the corresponding fruit."[23] This conviction that the laws of mechanics apply to physiology is reaffirmed after a discussion of the functioning of the heart: "this . . . follows from the mere disposition of the organs of the heart which we can see with the eye, and from the heat which we can feel with the fingers, and from the nature of the blood, which we can understand through experiment, just as necessarily as does the motion of a clock from the force, situation and shape of its counterweights and wheels."[24]

Inevitably, the analogy led to determinism, although Descartes, who did not wish to offend orthodoxy and share Galileo's fate, took care to limit his determinism to animals: "From the extreme perfection of certain actions we suspect that they do not have a free will." He explained his reasoning repeatedly, for example as follows: "I know quite well that animals do many things better than we, but that does not astonish me; for precisely that serves to prove that they act naturally and by such spring forces as a clock which indicates what time it is far better than our judgment tells us. And when the swallows come in Spring, they doubtless act like clocks."[25]

Descartes' influence was enormous, and the mechanical interpretation of physiology was quickly and widely adopted. In the process, his distinction between men and animals did not long survive. When LaMettrie's manifesto *L'homme machine* (1747) placed man and animal unambiguously on the same plane and declared them both equally subject to determinism, this was heresy only on the level of theology, not of physiology.

After the living organism succumbed to the clockwork imagery, the state was bound to follow, because the state had traditionally been viewed in analogy to the human body, with its different branches of society corresponding to the various organs. The earliest manifestations of the shift were not sharply focused. Antonio de Guevara's widely read *Horologium principum* (*Diall of Princes*, 1529) was merely an instruction manual for young princes in morals and manners. Perhaps inspired by its title, John Webster announced that

> The lives of princes should like dyals move,
> Whose regular example is so strong,
> They make the times by them go right or wrong,[26]

---

23 Descartes, *Oeuvres*, Vol. VIII A, p. 326.
24 *Ibid.*, Vol. VI, p. 52.
25 *Ibid.*, Vol. IV, p. 575.
26 John Webster, "White Devil," in *The Complete Works*, F. L. Lucas, ed., 4 vols. (New York, 1959), Vol. I, p. 120.

and similarly, Christoph Lehmann,

> A Prince and Ruler is the Country's clock,
> Everyone directs himself after the same in his actions,
> As though after the clock in business.[27]

2. Cardinal Richelieu. Philippe de Champaigne (1602–1674). Chantilly, Le Musée Condé.

The comparison here is between the clock and the ruler as a unique, commanding individual. Rulers and other dignitaries were commonly portrayed next to clocks, as seen in Figs. 1, 2, and 3.

A more modern view at the same time regarded the entire state as a complex interacting clockwork. The Spanish state, for example, was a "huge machine composed of so many parts . . . encumbered by its own weight that moves by its secret spring."[28] And as more and more countries on the European Continent committed themselves to the absolutist form of monarchy, the clockwork image became increasingly the controlling analogy for the state. A Spanish political thinker wrote:

> The wheels of a clockwork move in such secrecy that one can neither see nor hear them; . . . such harmony should likewise prevail between a prince and his councillors. . . . A monarchy is distinguished from other forms of government in that only one commands, others however obey. . . . Therefore in the clockwork of government the prince should be not only a hand but also the escapement that tells all other wheels the time to move.[29]

The German dramatist Daniel Casper von Lohenstein (1635–1683) repeated this thought in various forms: "The weight that drives the clock of commonwealth should be the force of a lively intellect"; ". . . Augustus alone was the escapement in the clock of the [Roman] commonwealth"; "In the clock of government the councillors are only wheels, the prince, however, must be not the hand but rather the weight."[30]

A radical mechanist among political philosophers was Thomas Hobbes, as is manifested in the prefaces of *De cive* (1642) and *Leviathan* (1651). In the former he recommends the watchmaker's approach to political analysis. "For as in a watch, or some such small engine, the matter, figure, and motion of the wheels cannot be known, except it be taken insunder and viewed in parts; so to make a more curious search into the rights of states and duties of subjects, it is necessary, I say, not to take them insunder, but yet that they be so considered as if they were dissolved."[31] *Leviathan* begins with the famous triple analogy that insists that the state, as an artificial construction, must be compared to an automaton, which in turn is a mechanical imitation of the living body.

27 Christoph Lehmann, *Florilegium politicum* (n.p., 1630), p. 693.
28 Henri Duc de Rohan, *Trutina statuum Europae,* 1638 (Leyden, 1644), pp. 10–11.
29 Diego de Saavedra Fajardo, *Idea de un principe politico christiano,* 1640 (Amsterdam, 1659), quoted from A. Henkel and A. Schöne, eds., *Emblemata* (Stuttgart, 1976), cols. 1341–1342.
30 Daniel Casper von Lohenstein, *Arminius,* E. M. Szarota, ed., 2 vols. (Bern, 1973), Vol. I, p. 229a; Vol. II, p. 962b; and *Dem . . . Fürsten . . . Georg Wilhelm . . . Gewidmete Lobschrift* (Breslau, 1679), p. 1F2b.
31 Thomas Hobbes, *De cive,* in *The English Works,* William Molesworth, ed. (Aalen, 1962), Vol. II, p. xiv.

3. Queen Maria Anna of Spain. School of Velázquez. Madrid, c. 1653. Vienna, Kunsthistorisches Museum.

The purest manifestation of the absolutist state was probably the Prussia of Frederick II. Not surprisingly, Frederick saw the state in consistently mechanical terms:

> A body of perfect laws should be the crowning achievement of the human spirit as regards the policy of government: one would observe a unity of design and of so exact and so well proportioned rules that a state conducted by such laws would resemble a watch all of whose springs had been made for the same purpose;—everything would be anticipated, everything would be coordinated, and nothing would be subject to mishap.[32]
>
> As all the springs of a watch conspire to achieve a common purpose, namely that of measuring time, so the springs of government should be so regulated, that all the different branches of the administration may likewise work together for the greatest good of the state.[33]

Similar quotations from his followers and disciples could be added at will. The state that was compared to clockwork was always an absolutist state based on a central authority that denied its citizens any significant freedom. In contrast, the movement of political liberalism arising out of the English revolutions of the 17th century, and advocating constitutional government, had no use for the clockwork metaphor except in the completely transformed version of "checks and balances."

The clockwork metaphors for body, state, and universe presented here are a small sampling from a vast supply of similar ones which in turn are surrounded by an abundance of clockwork metaphors from the same period applied to other themes. What do they all have in common? First, an obvious but not trivial observation: the clock is uniformly cited affirmatively, employed to illustrate and to support positive values. Second, they all express dynamic systems composed of parts that are connected by positive, direct linkages. Third, there is the underlying, and sometimes explicit, commitment to a central controlling element, an authority (king, head, heart) to which all other elements of the system are directly subject and for which the clock provided an obvious model (in the form of escapement, spring, or weight). Together, these aspects define an ideal method of organization, a conception of order which was applicable to every facet of life and which was shared, tacitly and without reflection, by most members of society. This authoritarian conception of order was perfectly expressed in the clock. The extraordinary rise of this machine, from its invention to the advent of the pendulum clock, is best understood as an interaction between a technological development and a social value. The latent, unarticulated preference for an authoritarian conception of order furnished an eager audience for an intriguing but otherwise not overly useful new invention. The mature clock in turn provided the model by which this conception of order could define itself.

32 Frédéric le Grand, *Oeuvres,* J. D. E. Preuss, ed., 30 vols. (Berlin, 1846–1857), Vol. IX, p. 24.
33 *Ibid.,* p. 200.

Francis C. Haber

2 The Clock as Intellectual Artifact

The credentials of the mechanical clock in the area of practical utility are impressive, whether it is considered as a timekeeper, a machine, or an instrument of regulation in industrial society. Given the importance of these functions in the development of modern science, technology, and economic life, it is natural that practical utility would be the primary focus in historical writing about the mechanical clock. However, the mechanical clock has had another role as a symbol and a model for the expression and illustration of thought and values apart from the practical. No other machine was employed so extensively in the pre-industrial-revolution period as an intellectual artifact by the literate classes, and this use was well developed in the late Middle Ages and Renaissance when the economic potential of the mechanical clock was only partially realized.

In the 19th century many writers saw the machine as the enemy of spiritual values, but in the 14th century the religious establishment was actively engaged in the promulgation of the machine through the construction of monumental clocks in cathedrals and chapels. Theologians also used the analogy of clockwork to illustrate some of the principles of the divine creation, thereby helping to legitimate a mechanical philosophy and to accommodate the pursuit of mechanics with religious values. The need for such an accommodation was still sufficiently felt in the early 17th century for writers on mechanics to engage in a kind of technological exegesis of the Bible in order to justify their activities,[1] but the process of integrating the machine into Western values had begun with the earlier uses of the mechanical clock as an intellectual artifact. Some of these uses will be suggested here.

There were at least four different, though often interrelated, aspects of the mechanical clock that can be singled out as sources for analogy: the simple timekeeper, the astronomical clock, clockwork automatons, and the making of the clock. As a timekeeper, it is usual to define the mechanical clock in terms of the use of an escapement device that made possible the conversion of the force of a descending weight or spring into a source of power to drive clockwork. By this definition, the mechanical clock began its career in the late 13th century with the invention of the escapement and a new class of machine based upon it. The widespread dissemination of clocks and watches domesticated the machine and made it available as a familiar reference for analogies and sentiments about temporality. A few instances will suffice here. Many of the public clocks were inscribed with *memento mori*, warning of the shortness of life, the inevitability of

---

1 Many examples are given in Ansgar Stöcklein, *Leitbilder der Technik: Biblische Tradition und technischer Fortschritt* (Munich, 1969).

damnation for the sinner, and the urgency of preparing now for the afterlife.[2] The beat of the clock also became a popular symbol for the passage of time, regularity, and balance.

By the early 15th century, a small portable clock had been added to the symbols carried by the female figure of Temperance in the iconography of the cardinal virtues.[3] Lynn White has found in a 15th-century poem the clock among the symbols used with Temperance to signify regularity, promptitude, reliability, restraint, perspicacity, and industriousness. He has also pointed out that a principal consumer of this iconography, with elements of the so-called bourgeois ethic, was the nobility.[4] The clock used as a symbol of the disciplined life was still very much in vogue at the court of Louis XIV, as Klaus Maurice has shown.[5]

The three other aspects of the clock—the astronomical clockwork, the automatons, and the idea of a maker—all have their origins in antiquity but flourished in the early centuries of the mechanical clock. In the literary tradition passed down from antiquity, the astronomical clock began with the two celestial globes made by Archimedes. Cicero, who had seen both, "concluded that the famous Sicilian had been endowed with greater genius than one would imagine for a human being to possess."[6] What Archimedes had done was to think out a way to represent accurately with a machine, operated by a single turning device, the various and divergent movements and different rates of speed of the sun and moon and the five planets. Cicero compared this achievement with the god who built the world in Plato's *Timaeus,* an achievement Archimedes could not have done without "divine genius." This led Cicero to raise the question whether things of this world were not also the product of divine reason and intelligence, for the perfection of the original shows a craftsmanship many times as great as does the counterfeit of Archimedes.[7] Lactantius developed this thought in a Christian context by asking whether God could not do as well in making the original as Archimedes did in making the imitation of the heavens, and would not a Stoic be forced to say that the stars were moved "by the skill of the designer" rather than by their own purpose.[8]

Derek Price has convincingly argued that the mechanical clock had its

2 Examples can be found in Alfred Ungerer, *Les horloges astronomiques et monumentales le plus remarquables de l'antiquité jusqu'à nos jours* (Strasbourg, 1931), pp. 44, 142, 197–198, 227, 242, 400.

3 Klaus Maurice, *Die französische Pendule des 18. Jahrhunderts: Ein Beitrag zu ihrer Ikonologie* (Berlin, 1967), pp. 4, 14. Lynn White, jr., "The Iconography of *Temperantia* and the Virtuousness of Technology," in Theodore K. Rabb and Jerrold E. Seigel, eds., *Action and Conviction in Early Modern Europe* (Princeton, 1969), pp. 197–219.

4 White, "Iconography of *Temperantia,*" pp. 214, 218–219.

5 Maurice, *Die französische Pendule,* pp. 4–5.

6 *De re publica,* I, xiv (21–22), cited in Derek de Solla Price, *Gears from the Greeks: The Antikythera Mechanism—a Calendar Computer from ca. 80 B.C.* (Philadelphia, Transations of the American Philosophical Society, 1974, Vol. 64, Pt. 7), p. 56.

7 *Tusculan disputations,* I, 63, and *De natura deorum,* II, xxxiv–xxxv (87–88), cited in Price, *Gears from the Greeks,* p. 57.

8 *The Divine Institutes,* II, 5, 18, cited in Price, *Gears from the Greeks,* p. 57.

origins in the attempts to simulate with mechanisms the motions of the heavenly bodies, and that the simple daily timekeeper is merely a by-product of the astronomical clockwork.[9] This would have been true of the water clocks of antiquity as well as the mechanical clock. It is a matter of record that by 1271 efforts were being made by clockmakers to devise a mechanical clock that would be more accurate than an astrolabe or other astronomical instrument.[10] Since the time of daily use was based on the unequal hours, any clock that had a predetermined indication of the hours of sunrise and sunset would entail astronomical knowledge, but the concern with accuracy appears to have been much more directly related to the science of astronomy than to daily timekeeping. The two earliest well-documented clocks were astronomical, those of Richard of Wallingford, Abbot of St. Albans, made about 1327–1330, and Giovanni de' Dondi, completed in 1364 after sixteen years of labor.[11] These clocks were the most complex, precise machines to be made for about two centuries, and their makers gained the kind of fame and admiration given to Archimedes in the literary tradition. The direction toward which these mechanical representations of the universe would develop was through the orrery to the modern planetarium. From an intellectual point of view, it should be no surprise that the universe would be modeled after this type of mechanical clock, for the model had attempted to counterfeit the original: it was a machine designed to imitate and illustrate nature through a tradition dating from Archimedes.

The adaptation of clockwork to cosmology was one of the many amalgamations of Greco-Roman and Judeo-Christian traditions that carried inherent difficulties. The clockwork analogy could also be used to support the idea of the world as a perpetual machine, an eternal world of natural laws, instead of a created world. One of the problems faced by the Church with the revival of Aristotelianism was its eternalism. In the Condemnation of 1277 by Bishop Tempier of Paris there were many propositions on eternalism.[12] If all the 219 propositions condemned were seriously being taught, the omnipotent Christian Creator based on Genesis was being threatened by naturalistic philosophy and by the Aristotelian prime mover. This threat had been countered by Saint Augustine, and Augustinians would continue to argue for the idea of God the Creator who made the world and time together. With the advent of the mechanical clock in the aftermath of this 13th-century struggle, the act of making the clock was used as an analogy to support the idea of the creation of the world

---

9 Derek de Solla Price, "On the Origin of Clockwork, Perpetual Motion Devices and the Compass," *Contributions from the Museum of History and Technology* (Washington, D.C., Smithsonian Institution, 1959, Bull. 218, Paper 6), pp. 81–112.

10 Lynn Thorndike, *The Sphere of Sacrobosco and Its Commentators* (Chicago, 1949), pp. 229–231.

11 See John D. North, *Richard of Wallingford: An Edition of His Writings with Introductions, English Translation and Commentary*, 3 vols. (Oxford, 1976); Silvio A. Bedini and Francis R. Maddison, *Mechanical Universe: The Astrarium of Giovanni de' Dondi* (Philadelphia, Transactions of the American Philosophical Society, 1966, Vol. 56, Pt. 5).

12 P. Mandonnet, *Siger de Brabant et l'averroïsme latin au XIIIme siècle*, 2me partie (2nd ed., Louvain, 1908), pp. 175–191.

machine. Again, the precedents were drawn from antiquity, as well as the Bible.

We have seen that Cicero maintained that Archimedes must have had "divine genius" in order to make his celestial globes and that this opened the possibility that divine reason had made the world, using the analogy between the human artisan and the Platonic Divine Artisan. In the Greek aristocratic tradition there were expressions of contempt for the artisans, but Aristotle at least discriminated between the masterworker who knew the causes of things and the manual laborer who did not. The artist could be admired, according to Aristotle, as long as he did not invent things directed to the necessities of life. The idea of the artist or architect was developed further by the Roman architect Vitruvius, but without the strictures against utility.

The ideal architect of Vitruvius was trained in theory and in craftsmanship. He "should be a man of letters, a skilful draughtsman, a mathematician, familiar with historical studies, a diligent student of philosophy, acquainted with music, not ignorant of medicine, learned in the responses of juriconsults, familiar with astronomy and astronomical Calculations."[13] This ideal architect was rare, and Vitruvius could point only to Aristarchus, Philolaus, Archytas, Apollonius, Eratosthenes, Archimedes, and Scopinas, men who had "left to after times many treatises on machinery and clocks, in which mathematics and natural laws are used to discover and explain."[14] His ideal architect was also inspired by a divine genius. By happy circumstance Vitruvius included clockwork in his ten books *On Architecture,* along with other machines.[15]

The Christian scholar of the late Middle Ages and the humanist scholar of the Renaissance found it easy to conflate the divine genius of classical thought with biblical texts, particularly the text from Genesis: "Let us make man in our own image, after our likeness." This was interpreted as a divine spark, among other things, and was considered to be a source of original creativity in man that imitated that of the Creator.[16] The work of Charles Trinkaus, *In Our Image and Likeness,* leaves no doubt about the degree to which this idea permeated Renaissance humanism.[17]

The idea was also adapted to the artist or architect who made ingenious clockwork. In the early phase of clockmaking, it was associated with the work of Richard of Wallingford and de' Dondi. Of Richard it was said, "he wished to show a miracle by some great work not only of inventiveness, but also of learning and excellent craftsmanship" and of de' Dondi, "Using his keen mind, he built a perfect machine . . . so that it seems to

---

13 Vitruvius, *On Architecture,* ed. and trans. Frank Granger, 2 vols. (Cambridge, Mass., 1962), Bk. I, Ch. 1, par. 3.
14 *Ibid.,* par. 17.
15 *Ibid.,* Bk. IX: Dials and Clocks.
16 On imitation, see Hans Blumenberg, " 'Nachahmung der Natur': Zur Vorgeschichte der Idee des schöpferischen Menschen," *Studium Generale,* 1957, *10:* 266–283.
17 Charles Trinkaus, *In Our Image and Likeness: Humanity and Divinity in Italian Humanist Thought,* 2 vols. (London, 1970).

be a divine rather than a human work."[18] The idea of divine imitation
survived the Reformation and was re-expressed, for instance, by Calvin,
who offered as a proof of our having divine souls our ability to do things
not dictated by the necessity of satisfying bodily needs. The nimbleness
of the soul surveys heaven and earth, joins past to future, and retains things
in memory. "Manifold also is the skill with which it devises things incred-
ible, and which is the mother of so many marvelous devices. These are
unfailing signs of the divinity of man."[19] Clearly, for Calvin the mother
of invention was, not necessity, but the divine human soul.

The genius who invented machines in imitation of the Divine Artisan
was deserving of fame, but in its early formulation at least it was not a fame
of vanity but of spiritual exercise. Saint Bonaventure, for example, writing
in the 13th century, explained that "the illumination of mechanical knowledge is the path to the illumination of Sacred Scripture."[20] It was his
conviction that Divine Wisdom lies hidden in sense perception and all the
works of God. In the illumination of the mechanical arts we can see the
Incarnation of the Word ("the Word was made flesh"), the pattern of
human life, and the union of the soul with God. The artificer aims to
produce a work that is beautiful, useful, and enduring so that he may
derive praise, benefit, or delight therefrom, a threefold purpose that corresponds to the threefold reason why God made the soul rational: that it
might praise Him, serve Him, and find delight in Him.[21]

The sole purpose of the mechanical arts, in the view of Saint Bonaventure, was the production of works of art. In considering the manner of
their production, he expressed another dimension in the act of making that
had been present in ancient thought, was clearly expounded by Saint
Augustine, was favored in Platonic idealist thought, but was not incompat-
ible with the matter-form philosophy of Aristotelians. He wrote: "we shall
see that the work of art proceeds from the artificer according to a model
existing in his mind; this pattern or model the artificer studies carefully
before he produces and then he produces as he has predetermined."[22]
This was a fundamental concept in the "argument from design" of natural
theology, and its association with clockwork extended from Lactantius in
the early 4th century to William Paley in the early 19th century.[23]

One of the reasons why the act of making in the case of Archimedes was
so compelling in Greek thought, and so closely associated with the Demi-
urge in Plato's *Timaeus,* was its resemblance to the activity of a geometer.

18 John Leland (c. 1540), and de' Dondi family document, cited in Bedini and Maddison,
Mechanical Universe, pp. 7, 5.
19 John Calvin, *Institutes of the Christian Religion* (1560), Bk. I, Ch. 5, par. 5. See also F. C.
Haber, "The Cathedral Clock and the Cosmological Metaphor," in J. T. Fraser and N.
Lawrence, eds., *The Study of Time, II* (New York, 1975), p. 408.
20 Saint Bonaventure, "Retracing the Arts to Theology, or Theology the Mistress among the
Sciences" (1253?), in Arthur Hyman and James J. Walsh, eds., *Philosophy in the Middle Ages*
(Indianapolis, 1973), p. 427.
21 *Ibid.*
22 *Ibid.,* p. 426.
23 William Paley, *Natural Theology: or, Evidences of the Existence and Attributes of the Deity, Collected
from the Appearances of Nature* (London, 1802).

4. The Creator measuring the universe with a compass, 13th century. Vienna, Nationalbibliothek, Codex 2.554, fol. 1.

The celestial globes of Archimedes reduced the heavenly motions to geometry and mathematics by means of material substances, rather than lines on paper, confirming that the architecture of the world was in reality like the mental concept based on theorems in the head of the geometer. Natural law and rationally deduced theorems coincided in the actual world. Also the process of conceptualizing in geometry and mathematics made the process of designing by an artificer similar and in sharp contrast to a technology of accidental discovery and modification through adaptation in the course of producing the necessities of life. The latter view had found expression in Lucretius and influenced Vitruvius, but it was the former that seemed most suitable to Christian concepts of the Creation.

A world that was preplanned and then brought into existence was more congenial to the description in Genesis of a world finished in all its essentials when Creation was completed. The idea of God as the Geometer was buttressed by the text from Wisdom 11:21, where it is stated that God had ordered all things by measure, number, and weight. This text was widely used in late scholastic philosophy, and God was sometimes portrayed as a Geometer (see Fig. 4). It was also one of the most often used texts in the technological exegesis of Scripture in the 17th century.[24] In the very influential books of Vitruvius, it was his picture of the architect who was trained in mathematics, geometry, and astronomy, and who engaged in designing a work prior to construction that was favored over his Lucretian evolution of technology.

The conflation of Genesis and mechanics is well illustrated in the 14th century by the lectures on Genesis of Henry of Langenstein. According to Nicholas Steneck, Henry was an Aristotelian-trained scientist with an Augustinian bias. He taught science by way of an extended commentary on Genesis in which he treated Creation as a *machina mundi,* frequently invoking the clock and the clockmaker as models. He suggested that knowing the cleverness of the craftsman and the subtlety of an artistic work enhanced the understanding and appreciation of the craftsmanship of God.[25] He assumed that the Master Artisan would have the design of Creation potentially present even in the first instant of actualizing form and matter, as well as in the construction of the world machine. The existence of a harmoniously ordered world system was evidence for Henry that it was made from a design, while the presumption of a design no doubt influenced him in finding that the world was an orderly system. Like Saint Bonaventure, he advocated the study of Creation to illuminate the ways of God, not for its material benefits.[26]

The analogy of the making of clockwork was used in 1453 by the mathematically inclined Augustinian Nicolaus Cusanus to expound the way that eternity and temporality are in opposition but also coincide. Comparing God's design with the concept of the clockmaker, having the

24 Stöcklein, *Leitbilder der Technik,* pp. 68–72.
25 Nicholas H. Steneck, *Science and Creation in the Middle Ages: Henry of Langenstein (d. 1397) on Genesis* (Notre Dame, 1976), e.g., p. 141.
26 *Ibid.,* p. 138.

plan of the whole clock in mind from the outset, the concept was timeless
or eternal, but once the clock was made, its running was temporal succession. With the turning of the hands, events would be brought forth as
planned by the Eternal Clockmaker, but we, as temporal beings, could
only await the turning of the hands to discover what would be brought
forth. The concept enfolds everything, but things are unfolded with the
running of the clock.[27]

The world-clock metaphor of Cusanus represented not only the temporal course of nature but the course of human history as well, for it too was
part of God's design in the Christian concept of Creation. As formulated
by Saint Augustine, for example, time was the linear series of present
instants, but it was also much more. The instant of the present in our
consciousness was capable of extension into the past by memory and into
the future by anticipation, but we were part of the temporal condition.
From the position of the Designer, all time, from its creation with the
creation of the world, to its end with the end of the world, was eternally
present. The unfolding of sacred history was preplanned around certain
marked events and persons with a direct linkage to God, and therefore
they had an atemporal and archetypal significance. Although they were
incarnated in actual history, and might reappear in different historical
actuality in successive times, what the events and persons "figured" was
outside of the time of history as well as in it. All secular history and sacred
history were united by Saint Augustine in his *City of God* under this
figuralist or typological view of reality. Adam, Moses, and Joshua, for
instance, existed as actual historical persons, but at the same time they
were prefigurations of Christ. The sacrifice of Isaac prefigured the sacrifice
of Christ. Figural interpretation, Erich Auerbach has written, with particular reference to the *City of God,* "establishes a connection between two
events or persons in such a way that the first signifies not only itself but
also the second, while the second involves or fulfills the first. The two
poles of a figure are separated in time, but both, being real events or
persons, are within temporality."[28]

This figuralist or typological view of history flourished down through
the middle of the 17th century in theology and literature, but it had a
special significance for early monumental clocks with automatons, for they
portrayed these two dimensions of history. The simple timekeeper ticked
away the temporal passage of instants, but the sacred scenes and person-
ages portrayed by complicated trains of automatons were figuralist, and
their daily re-enactment in the clockwork kept them symbolically in the

---

27 Nicolaus Cusanus, *The Vision of God,* trans. by E. M. Salter of *De visione Dei,* 1458 (London,
    1928), p. 52. See also F. C. Haber, "The Darwinian Revolution in the Concept of Time,"
    Studium Generale, 1971, *24:* 296–297, and *The Study of Time,* J. T. Fraser, F. C. Haber, and
    G. H. Müller, eds. (New York/Berlin, 1972), pp. 391–392.
28 Erich Auerbach, *Mimesis: The Representation of Reality in Western Literature* (Princeton, 1953),
    p. 73. See also his "Figura," *Archivum Romanicum.* 1938, *22:* 436–489, and *Typologische Motive
    in der Mittelalterlichen Literatur* (Krefeld, 1953); Joseph Anthony Mazzeo, *Renaissance and
    Seventeenth-Century Studies* (New York, 1964), pp. 183–208; and C. A. Patrides, *The Grand
    Design of God: The Literary Form of the Christian View of History* (Toronto, 1972).

5. Mechanical cock, installed on the clock of the Strasbourg Cathedral, 1354. Strasbourg, Musée des Beaux Arts, Château des Rohan.

present of the viewer in accord with their genuine timelessness. The automated sacred scene thus served as a continuing reminder that temporality was enfolded in the eternal plan of salvation. The central figure that revealed the significance of all earlier ones and prefigured all future ones was, of course, that of Christ.

One of the most famous, and most often imitated, of the early monumental clocks with automatons was that built in the Strasbourg Cathedral in the years 1352 to 1354. In addition to a large automated astrolabe and a perpetual calendar, it had automatons of the Virgin holding the Christchild, before whom the three Magi presented themselves, while a carillon mechanically played hymns. The most remarkable of the automatons was a mechanical cock, also driven by the clockwork (Fig. 5), that flapped its wings and crowed. The architect of the second clock of the cathedral, Conrad Dasypodius, said of this cock that it was "made two hundred years ago and placed in the old clock, and since it was customary to commemorate the Passion of Christ in the Christian church, this cock by its crowing warned men of the denial of Peter."[29]

In the second clock of the Strasbourg Cathedral, finished in 1574 (Fig. 6), many of the expressive and illustrative functions of the early history of the clock found their culmination and grandest fulfillment. It was a temple within a temple standing some 18 meters high in the south transept of the cathedral, fitted with a celestial globe, an astrolabe, and other astronomical mechanisms driven by the clockwork to represent the heavenly motions and also to present all divisions of time from centuries to minutes. It was also furnished with elegant automatons and richly decorated with paintings and sculptures, to exhibit eternity and everything from history and sacred and profane writings that can delineate time, according to its architect.[30] A pelican, a well-known figural symbol for Christ, supported the celestial globe in place of Atlas to represent eternity. An angel turned a sandglass at the quarter hours, the four ages of life passed before death during the hour, and at the last hour Christ appeared. The cock from the first clock was restored and mounted atop the weight tower, and it flapped its wings and crowed to the accompaniment of mechanically played hymns. The paintings included scenes of the creation of the world, the resurrection of the dead, Christ judging the world, the Last Judgment; representations from the Apocalypse of the Four Monarchies; and the three Fates disposing of the thread of life. The clockwork also drove a public clock outside the cathedral, suggesting that the expensive and magnificent clockwork within the cathedral was funded for reasons other than economic utility.[31]

29 See Haber, "The Cathedral Clock," p. 400, and Conrad Dasypodius, *Heron mechanicus: seu De mechanicis . . . eiusdem horologii astronomici* (Strasbourg, 1580).

30 Dasypodius, *Heron mechanicus* and *Warhafftige Ausslegung und Beschreybung des Astronomischen Uhrwercks zu Strassburg* (Strasbourg, 1580).

31 The curators of the cathedral in 1571 contracted for a mathematician-astronomer, clockmakers, a painter, a constructor, and a sculptor at an expenditure of 700 gulden for all work, to be completed in one year. The work took three years and probably involved cost overruns. Wealthy citizens seem to have furnished the money. A copy of the MS is in Archives de la Ville de Strasbourg, Administration de la Fondation de l'Oeuvre Notre Dame.

16

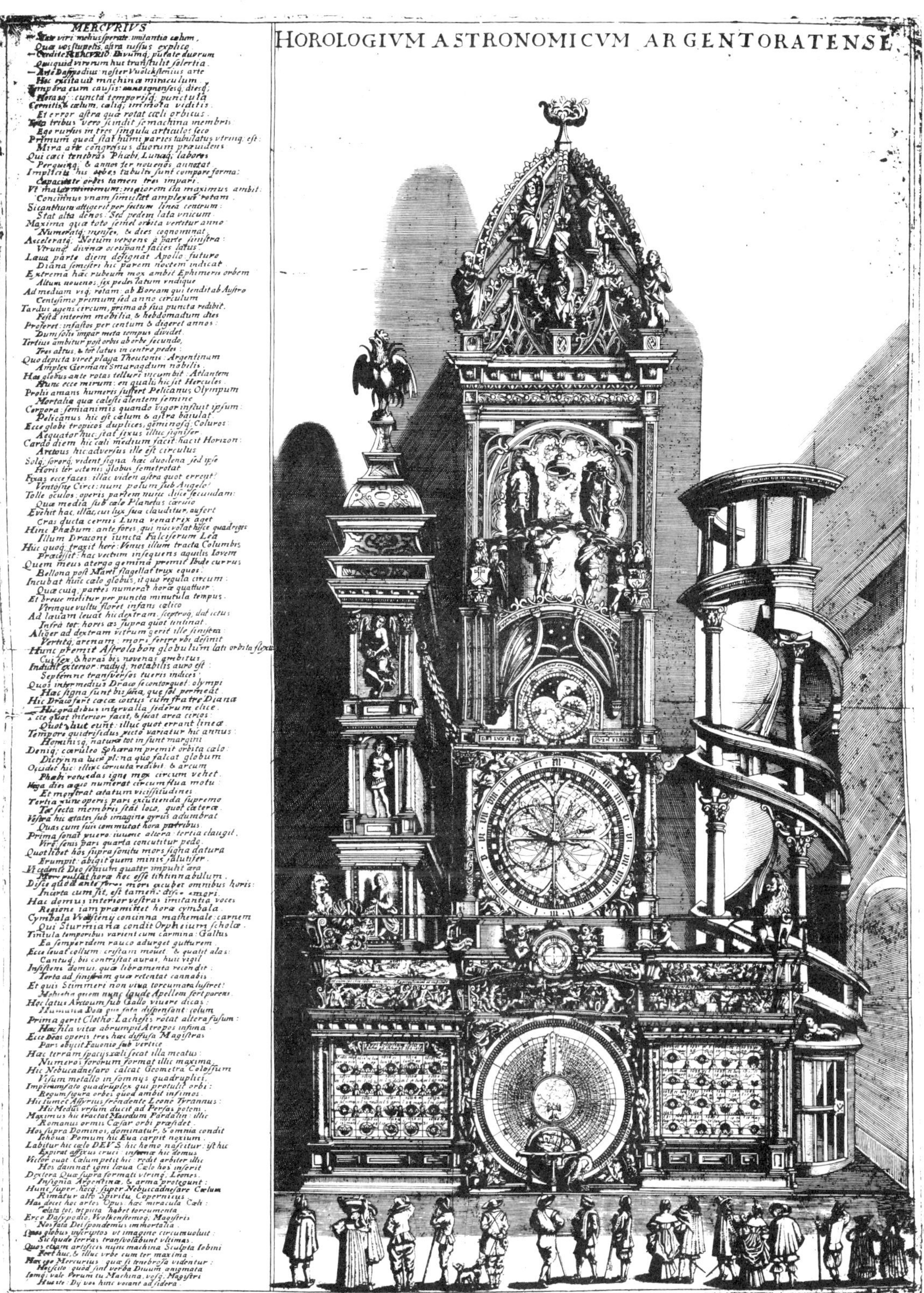

6. Astronomical clock of the Strasbourg Cathedral, 1574. Copperplate engraving after the woodcut by Tobias Stimmer, 17th century.

Dasypodius, the architect of the clock, wrote two descriptions of it, one in German, and one in Latin, with different prefatory material. The Latin introduction was a small treatise on the history of mechanics in antiquity, with mention of some of his near predecessors. Part of the treatise was a restatement of the idea of Vitruvius about the training and function of the architect, with a sharp distinction between rational theory and practice and the necessity of both in mechanics. Wise and ingenious architects conceive ideas, make them into models, and then, as far as possible, simulate the model in the actual work. He made a strong plea for the application of logic, geometry, and mathematics to mechanics in the practical areas; for a work such as his clock they were absolutely indispensable. In stating the goals for the clock, "splendor," "magnificence," and "fame" were words reminiscent of Saint Bonaventure, but they were now directed to the glory of the moderns, Germany, the city of Strasbourg, its cathedral, and posterity.

In a contemporary poem memorializing the Strasbourg clock and its makers by Nicodemus Frischlin, an accomplished neo-Latin humanist, the clock was hailed as one of the wonders of all time. Frischlin echoed familiar themes: "Oh divine inventions of the human hand! What work does either God or Nature do anywhere which we do not imitate with our thumb, a people rivaling our Father?" But the signification had shifted somewhat more strongly toward the powers of man.[32] He also emphasized how the clock simulated the motions of the heavens and taught what it was essential to know about temporality and eternity.

There were hundreds of writings about the Strasbourg clock, and its fame was spread far and wide down through the 17th century. It was the exemplar clock in the writings on the "new mechanical philosophy," and notably so in the work of Robert Boyle. There was an enthusiasm for building monumental clocks for about half a century after the Strasbourg clock, and monumental clocks continued to be built through the 19th century, but by the middle of the 17th they had begun rapidly to lose their appeal as a concrete manifestation of the ideas with which they were associated. The ideas of clockwork and clockmaker became abstract models for a new science and secular cosmology on the one hand, and for a natural theology on the other. Yet while they flourished, they helped legitimate a mechanical philosophy in the high culture as well as the low. And through the clockwork analogies it can be seen that not only was the early mechanical philosophy not alien to Christian culture, it was an expression of it.

---

32 Nicodemus Frischlin, *Carmen de astronomico horologio Argentoratensi,* 1575, in *Operum poeticarum* (Strasbourg, 1598).

Silvio A. Bedini

3 The Mechanical Clock and the Scientific Revolution

The scientific revolution, which was sparked in the mid-16th century by the publication of *De revolutionibus* of Nicolaus Copernicus, generated new activity in many of the sciences in addition to astronomy. Its impact was manifested as well on some of the crafts which were to contribute to scientific development, not the least of which was the craft of the clockmaker.

Before the middle of the 16th century, clockmaking as a profession was virtually nonexistent. Until then the clock had served two major functions. Public clocks, installed in churches and city halls, acted as community timekeepers; domestic clocks, produced in small numbers for royal patrons and wealthy members of the courts and merchant class, served as curiosities or as models for the study of astronomy. The timetelling function was a secondary consideration, for there was as yet no need for precision timekeeping. The demand for timepieces was consequently limited, but even more restricting to the growth of the craft was the scarcity of skilled workers. These factors effectively delayed the establishment of clockmaking as a specialized profession.

The making of clocks was a craft which derived originally from that of the blacksmith and was directly related to metalworking skills. As a separate craft it eventually emerged from the established professions of the gunfounder and the locksmith. Confirmation of this evolution is found in numerous community records of the 15th century in which clockmaking was designated as a secondary skill. Among the earliest examples is a gunfounder of Lille who also made clocks, and in the same period there was in Fribourg a *magister bombardum et horologiorum.* A clockmaker working at Caffa in 1455 was designated *bombardarius et magister orologii comunis,* while the keeper of the clock in the church of Saint Gothardus in Milan in 1474 was also a *bombardarius.* [1]

The workers in metal in the first half of the 16th century were limited in number, and those having mechanical skills as well were even fewer. Because the demand for clocks was at best occasional, these craftsmen who also made clocks had to move from one community to another as their talents were required. The primary employment available to them was the repair and maintenance of public clocks. For the most part these workers

---

1 P. Henrard, "Les fondeurs d'artillerie. Documents per servir à l'histoire de l'artillerie en Belgique," *Annales de l'Académie d'Archéologie de Belgique,* 1889, *45:* 169. A. Fillet, "Les horloges publiques dans le Sud-Est de la France," *Bulletin Archéologique du Comité des Travaux Historiques et Scientifiques,* 1902, *20:* 102–119. L. T. Belgrano, "Degli antichi orologi pubblici d'Italia con aggiunte e notizie della Posta in Genova," *Archivio Storico Italiano,* 1868, 3rd Ser., 9 (Pt. 1):30–65.

were of German origin and travelled into France, the Low Countries, Italy, and England to pursue their skills.

It was not until the demand for the production of domestic clocks increased substantially that the craft developed. This demand came primarily from the royal courts of Europe, where timepieces became popular additions to cabinets of curiosities for princes and prelates and members of the court. Gradually the market was extended to the slowly rising class of wealthy merchants and the bourgeois. In time the fashion called for domestic clocks of increasingly smaller size and greater elaboration, requiring other skills which were borrowed from the talents of the goldsmith, the silversmith, and the engraver of printed illustrations. The records frequently mention these new combinations of occupations. Little by little clockmakers achieved an identity of their own and in time commanded an enhanced social position, although they were not to reach the higher levels of the craft status until much later.[2]

As the market for clocks expanded, the specialized craftsmen required to produce them increased in number and gradually formed into groups which became centers of clockmaking. These centers developed in cities able to serve the royal courts of Europe, which had established metalworking skills and which were commercial centers for export trade. The earliest were formed at Blois and Paris in France, Augsburg and Nuremberg in Germany, and Geneva and London.[3]

Although beset by a multitude of political and religious troubles throughout the second half of the 16th century, Paris emerged as a major clockmaking center, and by the middle of the next century it had expanded sufficiently to include the formation of one of the earliest guilds of clockmakers. The second half of the 17th century was to witness its abrupt decline as a center, however, due primarily to increasing competition from Geneva and to the restrictions imposed by the guild system.[4]

The formation of groups of clockmakers into centers was followed by a new phenomenon, the division of labor. Until the mid-17th century the clockmaker generally produced a complete timepiece in his own shop. Then gradually craftsmen began to specialize in the production of components, which were purchased by clockmakers and incorporated in their products. The first specialists to emerge were the makers of clock springs; then followed the makers of tools for the clockmakers. In time the fusee cutters, escapement makers, turners, and founders also assumed individual identities.[5]

This division of skills had the effect of increasing production, and in time it had another effect on the craft as well. Before the beginning of the

2　A. Babel, "Histoire corporative de l'horlogerie, de l'orfèvrerie, et des industries annexes," *Mémoires et Documents publiés par la Société d'Histoire et d'Archéologie de Genève.* 1916, *33:*445.
3　E. Develle, *Les horlogers Blésois au XVIe et au XVIIe siècle* (Blois, 1917), pp. 20–22. R. de Lespinasse, *Les métiers et corporations de la Ville de Paris* (Paris, 1886–1897), Vol. III, pp. 554–556. Alfred Franklin, *La vie privée d'autrefois. Arts et métiers, modes, moeurs, usages de Parisiens du XIIe au XVIIIe siècles. La mesure de temps* (Paris, 1888), pp. 78–83.
4　Franklin, *La vie privée d'autrefois.* pp. 141–147. Lespinasse, *Les métiers et corporations.* Vol. III, p. 555.
5　Babel, *Histoire corporative.* pp. 91–99. Develle, *Les horlogers Blésois.* p. 93.

17th century, clockmakers generally sold their own products, clocks large and small, as complete units, usually made to order one or two at a time. The new trend toward specialization brought with it a trade in clock parts, making it possible to increase the rate of production. In time this led to the creation of a new group, the clock merchants, who purchased finished products from the clockmakers in some numbers and then sold them to other markets. To the corporative structure of the clockmaker's craft, with its rigid distinctions between master and apprentice, was now added the class of the clock merchant. These merchants first appeared in Geneva and shortly thereafter in other clockmaking centers.[6]

By the 16th century other centers of metalworking skills developed in the Low Countries and in Italy, and a migration of craftsmen began, often encouraged by the prospect of royal patronage or by changing economic conditions. This occasional dispersion of craftsmen was substantially increased by the advent of the Thirty Years War in 1618.[7] The migrant craftsmen had a strong impact on the industry of England, for example, which had lagged far behind that of other countries until the middle of the 17th century. With the infusion of new skills, clockmaking in England began to develop with fresh inspiration and slowly moved forward to assume supremacy over the industry in other countries. This development in England was prompted as much by the needs of men of science as by the market created by a steadily rising middle class.[8]

Just as England assumed the lead following the decline of clockmaking in Germany, Paris lost its importance as a major center. The competition with Geneva and the constrictions of the guild system brought a sharp decline, followed soon after by the dispersal of Huguenot craftsmen after the repeal of the Edict of Nantes. The decline was to continue until the beginning of the 18th century, when the importation of English craftsmen restored French clockmaking to new heights.[9]

Until the mid-17th century the clock remained an inaccurate timekeeper. It was regulated first by the relatively crude crown wheel and foliot escapement, which was subsequently replaced by the balance wheel. Its power source was improved by the replacement of the falling weight with the spring drive, but the clock remained at best a poor measurer of time. For the practical purpose of timetelling, the sundial—produced in a great variety of forms and elaboration by numerous makers in Nuremberg and Augsburg and other centers—continued to be the most popular

6 Babel, *Histoire corporative,* pp. 495–502.
7 Silvio A. Bedini, *Johann Philipp Treffler, Clockmaker of Augsburg* (Columbia, Pa., 1957), p. 4. August Fink, "Das Augsburger Kunsthandwerk und der Dreissigjährige Krieg," *Forschungen und Studien zur Kultur und Wirtschaftgeschichte Augsburgs* (Augsburg, 1955), p. 323. Friedrich Biendinger, "Augsburger Handel im Dreissigjährigen nach Konzepten von Fedi di Sanita, Politen, Attesten, u.ä.," *Wirtschaftskräfte und Wirtschaftswege. II. Wirtschaftskräfte in der europäischen Expansion. Festschrift für Hermann Kellenbenz,* 1978. G. Frischholz, "Nürnberg in der Geschichte der Uhren," *Deutsche Uhrmacher-Zeitung,* 1897, *21:* 252–260.
8 Franklin, *La vie privée d'autrefois,* p. 140. F. J. Britten, *Old Clocks and Watches and Their Makers* (7th ed., London, 1956), pp. 280–281.
9 J. Savary, *Dictionnaire universel du commerce* (Copenhagen, 1761), Vol. III, col. 331. Britten, *Old Clocks,* p. 285.

device long after the clock had been improved.[10] It was not until the scientific revolution had enlisted the clock as a potential tool to serve the new astronomical sciences that it achieved the improvements necessary to convert it into a precise time measurer. During the same period another important use was visualized for it—determining the longitude at sea.

The clock as a mechanical device for timekeeping in Europe had a relatively slow period of evolution. Still unknown as late as 1270, it manifested itself first in the tower clock on churches and public buildings from the beginning of the 14th century. Whether these were weight-driven mechanical clocks or combinations incorporating the clepsydra cannot be determined.

Its next development was in the form of extremely complicated and sophisticated planetary models operated by clockwork which first appeared in the 14th century and went through a series of refinements for the next two centuries. The earliest example was the great planetary clock constructed by the English mathematician and astronomer Abbot Richard of Wallingford for the Benedictine monastery of Saint Albans in the first quarter of the 14th century.[11] This was succeeded by the independent production of the astrarium between 1348 and 1364 by a physician and professor of astronomy, medicine, astrology, and philosophy at the University of Padua, Giovanni de' Dondi.[12] Curiously enough, these 14th-century masterworks were not derived primarily from the craft of the clockmaker; they were designed mainly as models for the study of astronomy, with timetelling a secondary function. Nonetheless they demonstrated that the metalworking arts of the clockmaker had made rapid advances during the preceding century.

The tradition of these clockwork planetary models continued through the 16th century with examples of increasing sophistication. Juanelo Turriano, mechanician in the service of Emperor Charles V, developed a great astronomical clock which he spent twenty years in designing and three and a half years in constructing. It was not yet completed at the time of the Emperor's death in 1558, and Turriano subsequently modified it in accordance with the new calendar reform.[13] In 1561 Eberhard Baldewein, clockmaker to the Landgrave Wilhelm IV of Hesse, produced an astronomical timepiece based on the new astronomical tables of Peter Apian and the astronomical calculations made by the Landgrave and his astrono-

10 The range and variety of these instruments are described and illustrated in the following works: Alfred Rohde, *Die Geschichte der wissenschaftlichen Instrumente* (Leipzig, 1923); Maximilian Bobinger, *Alt-Augsburger Kompassmacher* (Augsburg, 1966); Ernst Zinner, *Deutsche und niederländische astronomische Instrumente des 11.–18. Jahrhunderts* (Munich, 1956).

11 The most comprehensive study of the achievements of Richard is the 3-volume work by John D. North, *Richard of Wallingford: An Edition of His Writings with Introductions, English Translation and Commentary* (Oxford, 1976).

12 Silvio Bedini and Francis R. Maddison, *Mechanical Universe: The Astrarium of Giovanni de' Dondi* (Philadelphia, Transactions of the American Philosophical Society, 1966, N.S. Vol. 56, Pt. 5).

13 Ambrosio de Morales, *Las antiguedades de las ciudades de España* (Madrid, 1573), fols. 92r and v, 93r and v. Casto Maria del Rivero, "Nuevos documentos de Juanelo Turriano," *Revista Española de Arte.* 1936, primer trimestre, pp. 17–21.

mer, Andreas Schöner.[14] Of a similar nature was a masterwork of Jost Bürgi (see Catalog No. 53) made for Emperor Rudolph II in 1604.[15] Thus the clockmaker may be said to have been in the service of the astronomer from the early 14th century, although it was not until new levels in the sciences were achieved in the early decades of the 17th century that the clock assumed a more significant role as a timekeeper.

The scientific preoccupations of the 17th century, ranging from astronomy, navigation, and microscopy to mechanics and scientific demonstration, brought men of science for the first time into a close relationship with the crafts and the craftsmen needed to design the apparatus of science. Among the greatest needs were tools for measurement. The clock was readily adaptable for the purposes of science, particularly astronomy, but it was still greatly in need of improved regulation.

A solution seemed to be readily at hand in the form of the pendulum. It had been used for centuries as a rocker for providing propulsion in mills and in other machines. Its application to clockwork was not an entirely new concept. It was claimed that an Arab astronomer named Ibn Junis had employed a pendulum for his astronomical studies, and its application to clockwork had already been anticipated in the 15th century. More recently clockwork regulated by a pendulum had appeared in the drawings of Leonardo da Vinci and the master clockmaker of Florence, Benvenuto della Volpaia, among others.[16]

The realization of the pendulum-regulated clockwork came in the mid-17th century following a long period of experimentation by Galileo Galilei. While a student at the University of Pisa in about 1582, he had discovered the isochronism of the pendulum and visualized its potential application to clockwork. His invention of the pulsilogium a short time later to measure a patient's pulse rate was followed by his attempts to discover an acceptable solution to the problem of determining longitude at sea.[17] The great voyages of discovery undertaken for Spain and Portugal in the 16th century in the wake of the discovery of the American continent by Columbus made the need for determining longitude increasingly urgent. As early as 1530 Gemma Frisius proposed that this could be accomplished by some form of timekeeper, and numerous attempts were made thereafter to achieve such an invention.[18] In 1598 the King of Spain offered a reward of 1,000 crowns for a practical solution to the problem, an offer that was soon supplemented by a prize of an additional 10,000 gulden offered by the States General of the Netherlands. The rewards went unclaimed, but not for lack of efforts to win them.

14 H. Alan Lloyd, *Some Outstanding Clocks over Seven Hundred Years, 1250–1950* (London, 1958), pp. 46–60.
15 *Ibid.,* pp. 61–69.
16 Antonio Simoni, *Orologi Italiani dal cinquecento all' ottocento* (Milan, 1965), pp. 31–36.
17 Silvio A. Bedini, "Galileo Galilei and the Measure of Time," pp. 3–6 in *Saggi su Galileo Galilei* (Florence, 1967).
18 Gemma Frisius, *De principiis astronomiae et cosmographiae* (Antwerp, 1530). Charles H. Cotter, *A History of Nautical Astronomy* (New York, 1968), p. 24.

In 1612 Galileo began his first serious studies of the subject and continued to struggle with the problem for the next several decades. In 1636 he proposed that the longitude could be determined by plotting the motions of the satellites of Jupiter, which he had discovered in his astronomical observations. He specified, however, that a sufficiently accurate telescope would be required that could be used conveniently on shipboard, as well as an accurate timekeeper. The latter, he said, "must be a timepiece such that, if 4 or 6 alike were made at the same time, they would prove their accuracy by showing the difference of only one second, not only from hour to hour, but also from day to day and from month to month, so uniformly would they operate."[19] Such a timepiece had yet to be produced, however. In a letter to Admiral Lorenze Real of the Dutch East India Company in the following year, Galileo described a vibrations counter which he had devised in an effort to produce such a timepiece. It required manual manipulation to keep it vibrating and was not applicable to timetelling. It was in that year that Galileo was afflicted with blindness, but he nonetheless continued his negotiations with the States General until two years before his death.[20]

Despite his blindness, he next visualized a clock regulated by a pendulum for making astronomical observations, and he published a small work on the subject in 1639.[21] Two years later he conveyed to his son, Vincenzio Galilei, his concept for a pendulum-regulated clock. Vincenzio had a model made by a blacksmith, but it was not completed before his own death in 1649. It remained for Johann Philipp Treffler to make it operable by the addition of a pendulum for Prince Leopold de' Medici a few years later.[22] Had Galileo retained his health, it is probable that he would have perfected a pendulum-regulated clock that would not only have been practical but would have been far superior to the one devised and patented by Christiaan Huygens in 1657.[23] What Galileo did achieve, however, was the invention of the frictional or pinwheel escapement for clocks, which was the most accurate escapement to be produced before the mid-18th century. It was applied successfully to the tower clock of the Palazzo Vecchio in Florence in 1666 but remained forgotten until it was independently reinvented in France in about 1741.[24]

Experimentation with a method for determining longitude at sea continued through the 17th century. Huygens developed a marine clock which utilized his new invention of the balance spring, but since it had to rely nonetheless on a weight drive, it was impractical for marine use.[25] The Campani brothers, clockmakers of Rome, proposed several intriguing

19 *Le opere di Galileo Galilei,* Edizione Nazionale (Florence, 1909), Vol. XVI, pp. 469–471.
20 *Ibid.,* Vol. XVII, pp. 96–105.
21 Galileo Galilei, *L'usage du cadran où de l'horloge physique universel* (Paris, 1639). This work was reprinted in Dom J. Allexandre, *Trâité general des horloges* (Paris, 1734), pp. 296–299.
22 Bedini, *Treffler,* pp. 6–12. Bedini, "Galileo," pp. 25–30.
23 Lloyd, *Some Outstanding Clocks,* pp. 70–72.
24 Bedini, "Galileo," pp. 33–39.
25 A. Wolf, *A History of Science, Technology and Philosophy in the 16th and 17th Centuries* (1950; reprint: New York, 1961), Vol. I, pp. 115–117.

solutions to both King Louis XIV and Archduke Ferdinand de' Medici, but none of them came to realization with the production of a practical timepiece.[26] The prizes offered by the King of Spain and the States General of the Netherlands were supplanted at the beginning of the 18th century by a reward offered by the English Parliament. Although much effort was directed to finding a suitable solution, it was not achieved until 1764 with the invention of the chronometer by John Harrison.[27]

The introduction of the pendulum-regulated clock in the mid-17th century was followed by several other improvements in rapid succession. Among those who experimented with perfecting the pendulum clock were Robert Hooke, first curator of the Royal Society, and the clockmakers William Clement and Joseph Knibb. Each claimed the invention of the anchor escapement, which greatly improved the accuracy of the pendulum clock, but it is probable that it was the achievement of Clement, who first applied it in about 1670.[28] Little time was lost in adapting this invention to timekeepers for astronomical purposes.

Next came experimentation with the suspension of the pendulum and with its length. A spring suspension claimed by Hooke as his invention became widely used, and a longer, 39-inch pendulum beating seconds proved to be a major improvement. These achievements, coupled with the invention of the balance spring, claimed by both Hooke and Huygens, made possible even greater precision of the timepieces necessary for astronomical observations.[29] After the Royal Observatory at Greenwich was established in 1675, the quest for improved accuracy continued with even greater eagerness. Among the products of this effort was the year clock produced by the dean of English clockmakers, Thomas Tompion.[30]

It was not until the beginning of the 18th century that the next major achievement in timekeeping was made. This was the invention of the dead-beat escapement by a London clockmaker, George Graham, in 1715. A major improvement over the anchor escapement, it became the standard escapement for observatory clocks. It was coupled with further improvements in the pendulum—the mercury pendulum devised by Graham and the gridiron pendulum of James and John Harrison—and was not surpassed until the appearance of Reifler's escapement in 1893.[31]

The art of the clockmaker also had a strong influence in the 17th century on the development of tools and apparatus for the scientist. Prior to this time scientific instruments, which had been produced in great numbers primarily in Germany and the Low Countries, were rarely useful for

26 Matteo Campani degli Alimeni, "Novum horologium ad omni erroris et inaequalitatis periculo liberum ad mentem Galilaei primi indagatoris et quasi indicis," A. Fabroni, ed., *Lettere inedite di uomini illustri per servire d'appendice all'opera Vitae Italorum* (Florence, 1773), Vol. I, pp. 227–228. [Matteo Campani degli Alimeni], *L'oriuolo giusto d'Antimo Tempera utilissimo a' naviganti* (Rome, 1668), pp. 3–19.

27 Humphrey Quill, *John Harrison: The Man Who Found Longitude* (New York, 1966). E. G. R. Taylor, *The Mathematical Practitioners of Hanoverian England 1714–1840* (Cambridge, 1966), pp. 45–46, 68–69, 125–127.

28 Lloyd, *Some Outstanding Clocks,* pp. 72–80.

29 *Ibid.,* pp. 75–76, 85–87.

30 R. W. Symonds, *Thomas Tompion, His Life and Work* (London, 1951), pp. 129–130.

31 Britten, *Old Clocks,* pp. 114–117, 147–149.

precise measurement. Generally they served as models to explain theories or were examples of finely decorated metalwork for the cabinets of curiosities of the wealthy. For the most part they evolved from the tradition of the sundial makers, who frequently also produced books on the practical sciences of surveying and navigating, but few of their products were actually practical for field use.

A major figure in the production of this literature on the practical sciences and related instruments was Regiomontanus of Nuremberg. He had established a studio for the production of instruments, with a printing press and an observatory.[32] From this center the movement spread north to the Low Countries and south to Italy along the trade routes, with the consequent rise of the mathematical practitioners—the cartographers, surveyors, navigators, and in time the science teachers—who gradually formed into groups and movements in response to the need for the practical sciences. The strongest of these developed in England with the need for the division of lands and definition of boundaries following the redistribution of monastic lands. This brought about the professionalization of the art of the surveyor and the production of practical instruments for that purpose. In Germany and the Low Countries the emphasis was on instruments for military field use. The skills of the clockmaker, coupled with those of the engraver, were brought into service for the development of the new craft of the instrument maker.[33]

As the market for clocks developed, and as the requirement for more accurate clocks to serve the sciences increased, clockmakers were forced to devise more and better tools with which to perform the more sensitive aspects of their trade. This led them to investigate the properties of metals, such as their thermal aspects, for example, required for the improvement of the elasticity of springs. As a consequence, clockmakers were able eventually to make available to the scientist a range of tools and equipment which were applicable to the production of scientific instruments and apparatus for a variety of uses.

In this manner the 17th century that witnessed the advent of precision timekeeping was also responsible for the birth of the industry of precision scientific instrumentation. Just as precision timekeepers and instruments assisted the progress of science, science contributed to the improvement of timekeepers and scientific instruments. Lewis Mumford may have exaggerated when he wrote that "the clock, not the steam engine, is the key-machine of the modern industrial world," but there can be no doubt that the clock played a central role in the advancement of science during the scientific revolution. As Mumford went on to say, the clock "by its essential nature . . . disassociated time from human events and helped create the belief in an independent world of mathematically measurable sequences: the special world of science."[34]

32 Zinner, *Astronomische Instrumente,* pp. 65–67, 479–483.
33 E. G. R. Taylor, *The Mathematical Practitioners of Tudor and Stuart England* (Cambridge, 1954), pp. 27–48.
34 Lewis Mumford, *Technics and Civilization* (New York, 1934), p. 14.

26

4 Propagatio fidei per scientias:

Jesuit Gifts of Clocks to the Chinese Court

We have constantly been concerned to achieve access to the king through presents not of costly articles (for such we have not), but instead by unfamiliar articles that have never been seen in this country.
—Diego Pantoja, *Historie und eigentliche Beschreibung, was Gestalt des Evangelium in China eingeführt* (Munich, 1608).

A clock striking was at first regarded as a marvel among the Chinese: they set guards over it to see whether it was striking all by itself.
—*Dictionnaire de Trévoux* (Paris 1721), Vol. IV, s.v. *Merveille.*

The Chinese have done without geometry because they have had no taste for demonstrations of what one sees one can prove through *a priori* reasoning: the origin of things which are revealed through induction in terms of numbers. The art of revealing something through induction runs the same course as the art of putting forth hypotheses and the art of arriving at definitions.
—Gottfried Wilhelm Leibniz, *Relation de l'état présent de la République des Lettres* (1675).

The concept of "technological transfer" between continents is used today without emotional overtones. Natural science is taught at present in the same way in the universities of all nations, and widely differing social systems acknowledge the validity of its laws. This was not so in early times. The quotations presented at the left show us the novelty, for another culture, of the sciences which had developed in the West and of the array of instruments which had been devised here. Science and instruments both accompanied, in a most effective fashion, the Christian missionary endeavor in China. Before we speak of the first Jesuit presentations of clocks to the Chinese court, let us begin by citing two instances of later clock presentations by the British and the French courts and in this way recall a tradition which the Jesuit order had founded: Western technique accompanied Christian missionary work, just as European policy did later on.

In 1793 Lord Macartney, Ambassador of his Britannic Majesty, delivered to the Emperor Ch'ien-lung the presents offered by George III. In the diplomat's accompanying note, composed "somewhat in oriental style," there is an exposition of the gifts, which were not there just for their own sake but were intended to convince the Chinese court of the scientific and technological advancement of the British nation:

> The King of Great Britain, willing to testify his high esteem and veneration for his Imperial Majesty of China, by sending an Embassy to him at such a distance, and by choosing an Embassador among the most distinguished characters of the British dominions, wished also that whatever presents he should send, might be worthy of such a wise and discerning monarch. Neither their quantity nor their cost could be of any consideration before the Imperial throne, abounding with wealth and treasures of every kind. Nor would it be becoming to offer trifles of momentary curiosity, but little use. His Britannic Majesty had been, therefore, careful to select only such articles as might denote the progress of science, and of the arts in Europe, and which might convey some kind of information to the exalted mind of his Imperial Majesty, or such other articles as might be practically useful. The intent and spirit accompanying presents, not the presents themselves, are chiefly of value between sovereigns.[1]

The first of these gifts was a planetarium, a heliocentric model of the skies with all its celestial motions. The clockwork train for the planetary

---

1 George Staunton, *An Authentic Account of an Embassy from the King of Great Britain to the Emperor of China . . . taken chiefly from the Papers of his Excellency the Earl of Macartney* (London, 1797), Vol. II, pp. 491 ff.

7. Planetarium by Ole Roemer. Paris, c. 1680.
Copenhagen, Rosenborg Castle.

gear had been so painstakingly calculated that it was able to run synchronously with the heavenly constellations even after a thousand years, and —as Macartney concluded the description of this machinery which was at the time unknown in China—it would "long be a monument to the respect in which the virtues of his Imperial Majesty are held in some of the remotest parts of the world."[2] After the planetarium followed a laudatory description of the other presents: a telescope, a celestial and a terrestrial globe, clocks, a barometer, air pump, firearms, and minor gifts.

A hundred years previously Louis XIV had sent two planetaria along with French Jesuits for their missionary activity in China. The Frenchmen added to their array of instruments pendulum clocks, magnets, burning glasses, and quadrants, just as the Englishmen delivered such items later on. The two models, which could represent on a reduced time scale the motion of the planets or eclipses of the sun and moon, had been designed by Ole Roemer; his paired replicas, for Louis XIV and for the Danish King Christian IV, are still preserved at Paris and Copenhagen (Fig.7) P. Bouvet, S.J., one of the French missionaries, reported in a letter (which Leibniz also published in 1699)[3] that thanks to the Jesuits the Emperor Cam-Hy had put aside his prejudices against foreigners and had acknowledged that science and art flourished even outside his own realm. He then had the use of the two machines explained to him, displayed them with the greatest appreciation, and deduced from their incomparable precision and accompanying beauty that in France not only were mathematical instruments contrived to perfection, but all other arts flourished mightily. He had two of the seconds-beating pendulum clocks set up in his chamber and had their use for observating heavenly bodies explained to him (the missionaries delivered the third seconds-pendulum clock to the heir to the throne). The Emperor then founded an academy on the French model, in which clocks and instruments were also to be manufactured. Europe, particularly Paris, was to be the pattern. Thus through his support for the Jesuit mission, Louis XIV also promoted French trade.

During his stay in Paris, Leibniz addressed Colbert, the minister charged with the organization of French economy and science, and advocated sending an array of instruments along with political missions and Christian missionary activity—a set of instruments which would be destined to lead to a scientific comprehension of the world and thus to political domination of the earth by the Christian West:

> A king of Persia will exclaim at the effect of the telescope, and a Chinese mandarin will be astonished and delighted upon realizing the infallibility of a geometer-missionary. What will these people say when they come to

2 *Ibid.* The planetarium was by the Württemberg pastor Philipp Matthäus Hahn, who had worked thirty years on its construction; cf. I. L. Cranmer-Byng, *An Embassy to China, being the Journal kept by Lord Macartney during his Embassy to the Emperor Ch'ien-lung, 1793–1794* (Hamden, Conn., 1963), p. 361.
3 G. W. Leibniz, *Novissima sinica* (2nd ed., London, 1699), pp. 67–69, 96. On the French Jesuit mission see P. Bornet, "La mission Français à Pékin 1688–1775," *Bulletin Catholique à Pékin,* Oct.–Nov. 1936, pp. 551–563, 609–616.

28

see this marvelous machine you have had made, a machine which represents accurately the state of the heavens at any given time? I believe that they will recognize that the spirit of man derives from the divinity, and that this divinity communicates itself most particularly to Christians. The secret of the heavens, the greatness of the earth, and the measure of time are all of this nature.[4]

The delivery of the planetaria designed in England and France shows —although we should not reduce the actual history of the shifting political influence of the two nations to a sequence of state presents—that the Western countries knew well the effect of their physical instruments and mechanisms upon the Chinese court and its dignitaries. It was natural science and instruments that had opened up entry into China at the end of the 16th century. *Propagatio fidei per scientias* as a missionary method was used for the first time by the order of the Societas Jesu in the Central Kingdom.

All religious orders took part in the dispatch of missions to the lands newly discovered by Spain and Portugal in the 16th century: Franciscans, Dominicans, Carmelites, Benedictines, and Augustinians. All of them had disseminated the new faith among the natives through the Christian virtue of compassion, by building hospitals and orphanages. But in contrast to the indigenous peoples of America and Africa who lived divided up into many tribes, China confronted the Asian missionaries as an enormously large, politically unified, well-administered empire. The Chinese people possessed a highly developed philosophy and an intricately articulated ethical and social system, but it shut itself off from all that was alien.

The individual orders had long since delimited territories for their missionary activity in order to insure uniformity of missionary practice, and it was thus that Pope Gregory XIII in 1585 relegated Japan and China exclusively to the Jesuits. Because of the enormous size of the two Asiatic countries, however, the mendicant orders were dissatisfied with this decree, and after 1600 the preaching and teaching of other orders was again permitted in China and Japan. But soon thereafter the divergent missionary methods of the various orders led to a conflict of rites which impaired the work of all the orders.[5]

Francis Xavier, S.J. (1506–1552), the founder of the Jesuit missions in Asia, had been the first to initiate systematic exploration of the new missionary area. In addition to the unfamiliar customs of the Chinese, he primarily studied language, reported regularly to the order, and built up the peninsula of Macao—leased by the Portuguese since 1557 and lying

4 Onno Klopp, *Die Werke von Leibniz gemäss seinem handschriftlichen Nachlasse in der königlichen Bibliothek zu Hannover* (Hannover, 1864), Vol. III, p. 211.
5 The literature on the Jesuit mission to China has been covered bibliographically by Robert Streit, O.M.I., *Bibliotheca missionum*. Vol. IV (Aachen, 1928), Vol. V (1929). For biographies of individual missionaries see Louis Pfister, *Notices biographiques et bibliographiques sur les Jesuites des anciens missions de Chine. 1553–1773* (Shanghai, 1932–1934; reprint 1972). For the conflict of rites see Georg Schurhammer, S.J., *Das kirchliche Sprachproblem in der japanischen Jesuitenmission des 16. und 17. Jahrhunderts* (Tokyo, 1928); Anton Huonder, S.J., *Der chinesische Ritenstreit* (Aachen, 1921).

across from Hong Kong—as the center for the mission. In the *annuae,* the accounting reports sent annually to Rome, the missionaries told of their successes or their persecutions, their expulsion and the martyrdom of the first Christians, and also of political events. Copies of the reports were distributed to the various Jesuit houses, but for the most part they were printed and in that form widely disseminated. Extensive historical, ethnographic, and geographic descriptions of the countries were written by individual missionaries after longer sojourns. These travel accounts, frequently reprinted in other languages, made the culture of China familiar in Europe from the beginning of the 17th century, just as did the *annuae.*

Following Xavier, Matteo Ricci (1552–1610; see Fig. 8), together with his Jesuit companion Michele Ruggiero, first succeeded in entering the Chinese empire and in building up mission stations in four of the thirteen provinces beginning in 1583. Almost two decades later, in 1601, he entered the imperial city of Peking. He had developed for his order missionary methods that were particularly fruitful for China: first of all, he adapted the Christian message to Chinese thought and feeling; next, always before attempting religious conversion, he explained Western knowledge through the natural sciences, expounding by means of accompanying sets of instruments; and he frequently bestowed gifts upon the Chinese, mainly mechanical clocks which were unknown to them. This method of accommodation[6] (Ricci adopted Chinese names for gods in his translation of the catechism published in 1595, he said mass in Chinese, and he allowed converts to continue honoring Confucius and their ancestors) led to the conflict of rites to which we have referred: a conflict over the question whether Christianity, bearing the stamp of the history and culture of Europe, ought to be adapted to another culture.

Ricci's second method, the *propagatio fidei per scientias,* put scientific teaching in astronomy, mathematics, and geometry in first place. Ricci had himself been a pupil of Clavius at Rome. It was precisely the Jesuits who had, more than other orders, stressed the training of their missionaries in the natural sciences and had thereby achieved scientific preeminence over the Chinese scholars, whose knowledge and training had fallen into decline in the Ming period. Still, scientific argument was effective for only a few and could not be used in addressing a congregation. Nicholas Trigault, S.J. (1577–1628), who came to China in 1607, reported in the *annua* of 1610 that while the number of converts might be low, the Chinese nevertheless perceived that it was not greed nor temporal advantage that brought the missionaries there, but the propagation of a code of law that did no harm to the well-being of the rich. Accordingly the missionaries shunned mass gatherings and popular assemblies; they sought to turn officials in their favor, for the latter asserted that the people would follow their lead. Concerning the baptism of an official of high rank in 1611 Trigault reported that his conversion weighed as much in the balance as a thousand such among the *bassa gente;* and immediately there-

6 J. Bettray, *Die Akkommodationsmethode des P. M. Ricci S.J. in China* (Rome, 1955).

8. The Jesuit fathers Ricci, Schall, and Verbiest. Copperplate engraving from P. I. B. Duhalde, S.J., *Description de la Chinoise* (The Hague, 1786).

after he remarked how "little stability there was among the plebeian folk without the favor of the great."[7]

First the talk is of the sciences, but thereafter always of virtue—so Trigault reported on the Jesuit method. Taking the sciences as a springboard, one might then exhibit some curious items, a clock, a clavicembalo, a picture of the Savior; then one might explain the difference between our images and their idols, our God and their gods. Trigault then told of Hsü, the Chancellor General of the Nanking residency, who, despite long discussions with the Jesuits, did not wish to convert, but who was finally brought to the faith after he had received an armillary sphere, a terrestrial globe, a map of the world, and a work on geometry, and after he had studied the writings of the missionaries concerning God, the immortality of the soul, original sin and the incarnation, and also Ricci's catechism.

In 1595 in his *Storia dell'introduzione del christianesimo in Cina* Ricci reported about the Chinese water clocks and fire clocks and remarked in jest that because of his own knowledge of the construction of clocks he was taken there for a Ptolemy.[8] Just as Ricci had done, other Jesuits in the

7 Streit, *Bibliotheca missionum*. Vol. V, Nos. 2088, 2087; both letters are included in Nicolai Trigault, *Historie von Einführung der Christlichen Religion in das grosse Königreich China durch die Societas Jesu* (Augsburg, 1617; Latin edition, 1615).

8 Pascuale M. D'Elia, S.J., *Storia dell'introduzione del Christianesimo in Cina. Scritta da Matteo Ricci S.J.*. 3 vols. (Rome, 1942–1949), Lib. I, cap. IV, p. 33, n. 44. Ricci had conceived the plan for this work in 1594 but had not carried it out himself. Trigault brought his manuscript to Europe and used it for his account in the *Christiana expeditio* (Augsburg, 1615).
The origin and the functioning of all the clocks that Ricci delivered to the various Chinese dignitaries is found in detail in Silvio Bedini, "Oriental Concepts of the Measure of Time," in J. T. Fraser, N. Lawrence, eds., *The Study of Time. II* (New York, 1975), pp. 451–484. I am very much obliged to Silvio Bedini for his suggestions and his friendly encouragement. For other Jesuits who built clocks, see the items *horlogers, ingénieurs.* in Joseph Dehergne, S.J., *Répertoire des Jesuites de Chine de 1552 à 1800* (Paris/Rome, 1973).

31

mission built mechanical clocks; yet most of these were sent to China from Europe. Ricci then describes his entry into Peking and the presents delivered to the Emperor in 1601: two clocks with striking mechanisms, four oil paintings, mirrors and prisms, an atlas, breviaries and clavichords, European textiles and coins. For his admittance into the palace, which had till then been forbidden him, he had to thank the clocks alone, for when the presents were being brought before the Emperor the clocks had run down, and on that account the Jesuits were now urgently summoned. "The King then seeing that the large clock, for not having been wound, did not strike the hours, he commanded the fathers be called in much haste, that they should make them sound: racing always like a blast furnace on horseback to do this."[9]

To be sure, the success of the Jesuit missionaries' strategy did not depend upon two clocks that struck automatically, but his tactics did: Ricci had to explain the mechanics of the clocks and their care to four eunuchs within the palace. These functionaries, entrusted with keeping the clocks going and consequently worrying about being punished by the Emperor in case the machinery should fail, did what they could to keep the missionary in the palace. Thus Ricci was finally able to speak of Christianity, too, within the imperial palace. The message of the *annua* of 1603 was that the regard in which the fathers were held was increasing and that the clocks presented had moved the Emperor to view the missionaries with favor.[10]

Whereas Ricci reported the delivery of presents to Chinese dignitaries, Trigault, the most remarkable Jesuit missionary of the generation that followed Ricci, recounted the collecting of these gifts in Europe. Trigault had come to China in 1607 but was soon sent back to Europe by his superior Nicolo Langobardi, S.J., in order to publicize the mission. This trip to the European courts in the years 1615–1617 is probably the best known of the "missionary propaganda tours." After the honors paid by the court at Ricci's burial and after the Chinese request that Jesuits work with them in establishing a calendar, Langobardi took advantage of the favor of the Emperor and expanded missionary activity. At the time of Trigault's departure the mission comprised five stations—Nanking, Nan-ch'ang, Hang-chou, Ch'aochou, and Peking—and fewer than twenty fathers.

The first stop on Trigault's itinerary was Rome. In March 1615 he sought the understanding of the Holy Office regarding the goals of his fellow Jesuits: first of all permission to say mass in the Chinese language. Next he petitioned for a bishop of their own in China. Thus far the mission had been under the bishop of Macao, which belonged to the crown of Portugal, creating suspicion among the mandarins that the missionaries were Portuguese agents. From Rome Trigault traveled through Italy,

9 D'Elia, *Christianesimo in Cina.* Lib. II, p. 126, n. 594.
10 Streit, *Bibliotheca missionum.* Vol. V, No. 2039.

9. Jesuit missionary presenting scientific instruments to the Emperor K'ang-hsi. Copperplate engraving by Johann Wilhelm Meil, 1782, from Johann Matthias Schröckh, *Allgemeine Weltgeschichte für Kinder* (Leipzig, 1779–1784).

Spain, Germany, Belgium, and back again. In all the secular and ecclesiastical courts he pleaded for funds for the mission, funds that should ensure the mission's financial independence from the crowns of Portugal and Spain. Both monarchies supported the missionary activity, and provided money from customs receipts, primarily in India and Malacca, but the revenue was often delayed for several years.

Trigault reported to his fellow missionaries in China regarding his itinerary, his solicitations on behalf of the mission, and the funds, gifts, and books he had secured. Among his letters only the report of the period January 1616–January 1617, which was circulated in copies, has survived. Description of the gifts, always with an indication of value, makes up the major part of his report, and these gifts were primarily clocks. In fact, Lamalle, the editor of Trigault's *Relations,* has written, "elle l'encombre même au point de fatiguer le lecteur moderne par la description des horloges de taut formes que Trigault collectionna."[11]

The fact that the presentation clocks almost all originated at German courts was determined by Trigault's route, which coursed chiefly through Germany. Even more mechanisms than those specifically mentioned may have been given to Trigault, for he readily abbreviated lists (e.g., *munera, quae longum esset singula enumerare; aliqua reliquo ne fastidum sit*—"items to list which individually would take a long time; I leave the rest lest I tire the reader"). In Florence Trigault received from the Duke an automaton clock, almost two feet high, in the form of a dragon (the symbol of the Chinese emperor), which moved its jaws and wings and rolled its eyes when the hours struck, and also two gold clocks set with topazes and rubies. In Augsburg he acquired fifteen large tower clockworks, which he had shipped to Lisbon via Hamburg (Lisbon was the only port from which ships could sail for China—a longstanding privilege of the Portuguese crown).

On August 8, 1616, Trigault arrived at Munich and was supplied by the ducal family with precious gifts and a very considerable endowment. Let us enumerate these gifts, above all the clocks. Wilhelm V (since 1597 retired from the regency in favor of his son Maximilian) gave Trigault in addition to reliquaries a number of small and large clocks, including one in the form of a cross, another in silver with Mary, the Christchild, and the boy John. Duke Maximilian presented an elaborate cabinet containing silver mathematical instruments, including an odometer, clocks, an astrolabe, and other valuables. The Braunschweig artist Tobias Volckhamer, who had been appointed to the Munich court as mathematician and goldsmith in 1594, received payment for the instruments.[12] The duchess next added to Maximilian's clocks (which are not described in further detail) two pendant watches, one of gold, and an automaton clock representing

11 Edmond Lamalle, S.J., "La propagande du P. Nicolas Trigault," *Archivum Historicum Societas Jesu.* 1940, 9:49–120, esp. p. 75.
12 Ernst Zinner, *Deutsche und niederländische astronomische Instrumente des 11.–18. Jahrhunderts* (Munich, 1967), p. 574.

10, 11. Crèche automaton by Hans Schlottheim, Augsburg, before 1588. Back with dial and front with dials and opened heavens. Dresden, Staatlicher Mathematisch-Physikalischer Salon (destroyed in 1945).

a centaur. This automaton rolled by a drive of its own on the table, shot an arrow which could even penetrate a slab of wood held before it, struck the hours, and vigorously shook its savage head (cf. Catalog No. 105). Prince Albert, the youngest born of the Wittelsbachs, and Princess Magdalene parted with two pendant watches and one silver-gilt clock in the form of a cross. In addition, in order to promote the effectiveness of the Jesuit mission in China Wilhelm V promised 800 gulden annually, a sum which the Bavarian court did indeed pay until as late as 1687, with the exception of the war years 1632–1652.[13]

After this Trigault mentions four other clocks and mathematical instruments which he received from Ferdinand of Austria, a fifth being so beautiful there would be no end to describing it. After his departure from Munich he visited Dillingen, Neuburg, Ingolstadt, Würzburg, Mainz, Frankfurt, Trier, Bonn, Cologne, Liège, and Brussels, where he composed his letter of January 2, 1617. In all these cities he reported to the monasteries or bishops about the mission of his order and distributed his history of China, which had appeared in Augsburg in 1615. In Würzburg he received mathematical instruments from the bishop, at the Frankfurt fair he acquired books and mathematical instruments, and in Cologne a very large clockwork in addition to quadrants, astrolabes, and other instruments. The archbishop of Cologne, the Wittelsbach prince Ferdinand who was a brother of Duke Maximilian, gave Trigault an automaton which enacted the story of the nativity. The detailed description of this automaton with its scenery, the motions of the figures, and the accompanying music leads to the conclusion that this device was a copy of an automated nativity play built by Hans Schlottheim of Augsburg for the Saxon court at some time prior to 1588 (Figs. 10, 11).

Trigault returned to China in 1619, and the gifts arrived a year later. However, because of the recent Tartar war and the persecution of Christians, the presents remained for the time being in Macao, a situation that Trigault reported to Duke Maximilian once more in 1624.[14] To be sure, part of the valuables had earlier been sold in order to financially sustain the mission, which had been increased in 1619 by twenty-two missionaries in Trigault's company. Among them were Giacomo Rho, Wenzel Kirwitzer, Johannes Schreck (Terrentius), and Johann Adam Schall von Bell (see Fig. 8), all of whom soon won great esteem at the Chinese court by reason of their expertise in astronomy.

13 Fr. X. Kropf, S.J., *Historiae provinciae Soc. Jesu Germaniae Superioris,* Pars IV (Munich, 1746), p. 25 ff. Georg Leidinger, "Wilhelm V. von Bayern und die Jesuitenmission in China," in *Forschungen zur Geschichte Bayerns,* ed. M. Doeberl and K. von Reinhardstottner, Vol. XII (Munich, 1904), pp. 171–175. Herman Schneller, "Bayerische Legate für die Jesuitenmission in China," *Zeitschrift für Missionswissenschaft,* 1914, 4: 176–189. Heinz Dollinger, *Studien zur Finanzreform Maximilian I. von Bayern* (Göttingen, 1968), p. 443.

14 Streit, *Bibliotheca missionum,* Vol. V, No. 2114. Lamalle, "Propagande," p. 76. Among all the European gifts of clocks only work of the late 18th century has survived. All earlier automata and clocks have been lost; "no traces can be found of Matteo Ricci's clocks . . . ." See Simon Harcourt-Smith, *A Catalogue of Various Clocks, Watches, Automata and other Miscellaneous Objects of European Workmanship dating from the XVIIIth and the Early XIXth Centuries in the Palace Museum and the Wu Ying Tien, Peiping* (Peking, 1933).

Earlier, on December 15, 1616, Sabbathin de Ursis, S.J., had foretold an eclipse of the sun far more precisely than had the Chinese astronomers. In 1629 the official imperial prediction of an eclipse was in error compared with the prediction of the Jesuits by one hour as to the beginning of the eclipse, and by two hours as to its duration. "The prediction of the eclipse was one of the most important reasons for the credit the Jesuits were able to obtain at the imperial court."[15]

In the person of Father Schreck there came to China a missionary who in 1611 became the seventh member of the Accademia dei Lincei, immediately after Galileo. As early as 1626 Schreck published in Chinese a treatise on the telescope, with which Galileo had made his epochal astronomical observations a few years earlier, publishing the results in his *Sidereus nuncius* of 1610. The Jesuit missionaries in China were now learned men, true representatives of *propagatio fidei per scientias,* yet for them science was always a means to an end. In 1678 Ferdinand Verbiest, S.J. (Fig. 8), successor to Schreck in the office of president of the *Tribunale mathematicum* at Peking, was still writing to his fellows in Europe—as Ricci had done eighty years earlier—that only science and small gifts would overcome the resistance of the Chinese toward the Christian faith: "saltem hoc effecit ut rei christianae propagandae non obstent." He continues to mention among the presents particularly clocks: "praeterea omnis generis horologia."[16]

Better scientific achievement struck the Chinese as resulting from better religion. "Les Chinois sont portés croire que ceux qui excellent le plus dans les sciences ont la meilleure Religion, et ce préjugé est pour moins aussi bon qu'aucun de ceux on s'est servi contre les Protestans," as Leibniz had concluded.[17] Yet around the middle of the 17th century they realized in China that this science was not only "Christian," it was also perceived as *nuova scienza* in the West as well.[18]

Missionary activity, like history itself, is subject to ever-changing criteria of judgment. Joseph Needham and Wang Ling have deemed that "the coming of the Jesuits was by no means an unmixed blessing for Chinese science."[19] In this they were primarily balancing the advances achieved through the Jesuit reform of the calendar against the geocentric view of the universe which the missionaries advocated (or were obliged to advocate),[20] a concept which had by that time been rejected by all scholars with minds open to the facts. The ethical question of whether science ought to

15 Joseph Needham and Wang Ling, *Science and Civilisation in China.* Vol. III (Cambridge, 1959), p. 422.
16 Henri Josson, S.J., and Leopold Willaert, S.J., *Correspondance de Ferdinand Verbiest de la Companie de Jésus (1623–1688). Directeur de l'Observatoire de Pékin* (Brussels, 1938), p. 237.
17 Gottfried Wilhelm Leibniz, *Allgemeiner politischer und historischer Briefwechsel.* Ser. I. Vol. VIII (Berlin, 1970), p. 187; letter to Landgrave Ernst von Hessen-Rheinfels dated mid-November 1692.
18 Needham and Wang, *Science and Civilisation.* Vol. III, pp. 451 ff.
19 *Ibid.,* p. 437.
20 Pasquale M. D'Elia, S.J., *Galileo in China: Relations through the Roman College between Galileo and the Jesuit Scientist-Missionaries (1610–1640)* (Cambridge, Mass., 1960).

be taught solely in order to spread the faith had been posed to Leibniz as early as 1690 by Landgrave Ernst von Hessen-Rheinfels: "cela est non plus blamable que quant St. Paul dict en un endroit, sive per invidiam etc. dummodo praedicetur Christus."[21] Leibniz, on the other hand, did not see in the missionary activity in China merely a one-sided Western invasion of another culture, but instead the foundation for reciprocal cultural interchange between Asia and Europe. Even if science was for the Jesuits merely a bridge for the Christian faith, Leibniz recognized that in the long run science would not remain bound to faith but would instead be disseminated through faith. Accordingly he replied to the Landgrave: "Je ne blasme pas mais je loue les Jesuites de la Chine de ce qu'ils se servent de Mathematiques pour s'insinuer, mais je crois que l'Eglise n'en profitera gueres, et que les Chinois apprendront par là nos sciences."[22]

21 Leibniz, *Allgemeiner . . . Briefwechsel,* Ser. I, Vol. V (Berlin, 1954), p. 594; letter from Landgrave Ernst von Hessen-Rheinfels dated June 20/30, 1690.
22 *Ibid.,* p. 617; letter from Leibniz dated July 14, 1690.

5 The Role of Clocks in the Imperial Honoraria for the Turks

12. Augsburg clock in a portrait of an unknown personage wearing Persian costume (detail). Isfahan, c. 1650–1680. Private ownership.

In 1453 Constantinople, the ancient Roman imperial city on the Bosporus, after a long siege and an exceedingly valiant defense, fell into the hands of the Ottoman conquerors. The shocking news did not suffice to unite the states of the Christian West in a common defense. The bellicose doge of Venice, Francesco Foscari, was of a minority in the Signoria, and the first maritime power of the Mediterranean thought it would be better able to preserve its commercial interests in the Levant through a treaty relationship with Sultan Mohammed II than through military intervention. Under the arrangement arrived at in 1454 between the Serenissima and the Supreme Porte, Venice pledged to make regular payments to the Sultan's treasury.[1] his instrument of peace set the direction for the armistices regularly concluded between the Emperor and the Turkish Sultan down to the peace of Zsitvatorok on November 11, 1606.[2]

In the 16th century these armistices between the Hapsburgs and the overlords on the Bosporus were no more than short-term peaces which had to be negotiated repeatedly. Ferdinand, who had succeeded his brother-in-law Louis II as King of Hungary, never doubted that a pact with the Porte could be achieved only through the delivery of appreciable payments of money. As early as May 27, 1530, Ferdinand instructed his negotiators to offer the Sultan, Suleiman the Magnificent, between 20,000 and 100,000 ducats if in return he would renounce Hungary.[3] It was only in 1547 that a five-year armistice on the basis of the present holdings of each party was concluded. For this Ferdinand had to agree to a pension amounting to 30,000 ducats, to be paid annually.[4]

A permanent representative of the Emperor resided at the Ottoman Porte from the middle of the 16th century onward. The annual Turkish tribute was brought to the Golden Horn by emissaries from abroad and was there delivered to the Sultan by the imperial resident. Perhaps the most outstanding imperial diplomat in Constantinople was the Flemish humanist Ogier Ghiselin de Busbecq, who represented Hapsburg inter-

1 Johann Wilhelm Zinkeisen, *Geschichte des osmanischen Reiches in Europa,* Vol. II (Gotha, 1854), pp. 33–42.

2 Ludwig Bittner, *Chronologisches Verzeichnis der österreichischen Staatsverträge,* Vol. I (Vienna, 1903), pp. 33 f.

3 Text in Anton von Gévay, *Urkunden und Actenstücke zur Geschichte der Verhältnisse zwischen Österreich, Ungarn und der Pforte im XVI. und XVII. Jahrhunderte,* Vol. I: *Gesandtschaften König Ferdinands I. an Sultan Suleiman I. 1527–1532. Gesandtschaft König Ferdinands I. an Sultan Suleiman I. 1530* (Vienna, 1838), pp. 3–23.

4 Bittner, *Chronologisches,* p. 16.

ests in Turkey from 1554 to 1562. His travel reports, written for publication, provide a graphic picture of life and society in the Ottoman metropolis.[5]

From the second half of the 16th century there remain a number of travel descriptions and diary-like notes, for the most part from the pens of members of the retinues of imperial ambassadors. Together with the official dispatches, which fill many boxes among the holdings of the household, court, and state archives and the court treasury archives in Vienna, they constitute an indispensable source of information about relations between the imperial court and the Divan.

During the reigns of Maximilian II and Rudolf II, down to the outbreak of the great Turkish war in 1593, there were many Protestants among the imperial envoys. Among the large retinue which they took with them on their travels to Constantinople were preachers of their faith. Thus in 1573 Stephan Gerlach of Tübingen travelled to the Golden Horn with David Freiherr von Ungnad, designated by the Emperor Maximilian II as envoy to the Ottoman Porte to deliver the honorarium and instructed to renew the armistice with Selim II. From Gerlach's pen we have a compendious journal that runs from 1573 to 1578 and reports on the entire period of Ungnad's activity.[6] To cite a second example, Joachim von Sinzendorff was also an adherent of the Lutheran faith and was accompanied on his mission to the Bosporus by Salomon Schweigger, also from Tübingen, who published a well-received travel description in 1608.[7] Sinzendorff was the successor to Ungnad, and he represented the Emperor before the Sultan from 1573 to 1580.

The honoraria delivered to the Sultan by those European powers whom the Porte regarded as tributary were composed primarily of money and various prime products of the countries represented. These products were partly natural, such as spices and precious stones, valuable horses and hounds, and partly artifacts of high quality, such as exceptionally fine textiles. As a rule they were articles that were not found or produced with such high standards in the Ottoman empire. Thus the gifts not only had the character of duty-bound tribute; they also conveyed to the Porte an idea of the artistic and technical capabilities and the living standards of the representative countries.

In his journal Stephan Gerlach points out that many of the presents brought from abroad were not to be had in Constantinople. About those

<hr>

5 Cf. Ogier Ghiselin de Busbecq, *Omnes quae extant opera,* with an introduction by Rudolf Neck, reprint of the Basel 1740 ed. (Graz, 1968). A good survey of the more important travel descriptions is offered by Karl Teply, ed., *Kaiserliche Gesandtschaften ans Goldene Horn* (Stuttgart, 1968), pp. 451–467.
6 *Stephan Gerlachs dess Aelteren Tage-Buch der von zween Glorwürdigsten Römischen Kaysern, Maximiliano und Rudolpho. . . . An die Ottomannische Pforte zu Constantinopel Abgefertigten und durch den . . . Herrn David Ungnad . . . Glüklichst-vollbrachter Gesandtschafft* (Frankfurt a.M., 1674).
7 Salomon Schweigger, *Ein newe Reyssbeschreibung auss Teutschland nach Constantinopel und Jerusalem. Mit einer Einleitung von Rudolf Neck,* reprint of the Nuremberg 1608 ed. (Graz, 1964).

delivered by the Venetian envoy on November 19, 1577, he writes: "His gifts were of the most precious velvet, gold work, and beautiful textiles, such as one may always have in Venice but the like of which there is not in Turkey."[8] The entrance of the Persian mission with its presents in 1576 must have offered a particularly colorful spectacle. There were twenty manuscripts of the Koran bound in gold and precious stones, various historical manuscripts, gold- and silver-worked tents of silk, the poles set with precious stones, forty silken carpets, and six boxes containing diamonds, emeralds, rubies, turquoises, balases, pearls, and bezoars. On top of this there were sabers, bows, and black aigrettes decked with precious stones.[9]

The Emperor was also required to present gifts to high dignitaries in the Divan: first and foremost the Grand Vizier; the Reis Effendi, who carried out the duties of a minister for foreign affairs; the Mufti, who exercised the supreme religious function; the other viziers, the interpreter of the Porte, and other important officials; also particularly influential eunuchs; and sometimes the mother of the Sultan too. The Pasha at Ofen (now Budapest), on whom the ambassadors had to pay a call during the course of their trip, was also due to receive valuable honoraria.

The Emperor's annual deliveries of tribute commenced in 1548 after the conclusion of the five-year armistice already mentioned. An extract from the accounts of the court disbursing officer for the years 1548 to 1550 notes expenditures on behalf of the Sultan of 72,709 gulden 13 kreuzer in 1548, and that applies only to cash and textiles. The present for the Pasha at Ofen is recorded separately, and it came to 1,000 ducats and three silver-gilt drinking vessels, the total value being 2,100 gulden 6 kreuzer. In 1549 and 1550 deliveries to the Sultan totalled 72,779 gulden 35 kreuzer, and 68,508 gulden 16 kreuzer, comprising cash and silverware.[10]

It cannot be demonstrated conclusively that there were any clocks among the silver, and yet this seems to be suggested by an undated memorandum to the court treasury from the imperial envoy Justus de Argento,[11] who had brought the tribute for 1548 to Constantinople. He writes that in accordance with his instructions he would have to transfer 40,440 ducats in gold, and that over and above this he would need "simul horologia quatuor et Magistrum horologiorum" (at the same time four clocks and a master clockmaker).[12] This is the first concrete indication that

---

8 Gerlach, *Tage-Buch,* p. 412.

9 *Ibid.,* pp. 191–193.

10 Hofkammerarchiv, Vienna, Reichsakten 174, fols. 25–28. This source will hereafter be abbreviated HKA, RA.

11 The spelling of his name is given in various ways. Bertold Spuler, "Die europäische Diplomatie in Konstantinopel bis zum Frieden von Belgrad (1739)," in *Jahrbücher für Kultur und Geschichte der Slaven, Zeitschrift des Osteuropa-Institutes Breslau,* 1935, N. S. *11:* 332, cites him as Justi de Argento.

12 Cf. Memoriale ad Dominos de Camera pro expeditione Justi de Argento in Turciam, HKA, RA 190, fols. 26 f.

clock mechanisms were sought in Turkey and that envoys took along clockmakers of their own to care for them. This sort of thing is frequently documented from then on. Gerlach reports in his Turkish journal that a clockmaker had died during Ungnad's journey to Constantinople.[13]

The Vienna burgher and clockmaker Wolfgang Teissenrieder was sent to Ofen by the Emperor at the Pasha's request to repair the clocks there. He stayed from September 9 to December 9, 1562, and then again from March 26 to August 15, 1563. The military paymaster's office at the Vienna court had to put up his salary and travel expenses[14] (all business with the Turkish empire was conducted via the court military council, founded in 1556, down to the days of Maria Theresa). Teissenrieder also accompanied Caspar Minkwitz, who brought the honorarium to Constantinople in 1571, because according to his own account the clocks were seriously defective. On the trip back Teissenrieder fell gravely ill and petitioned the Emperor for prompt reimbursement of his expenses.[15] Minkwitz supported the clockmaker's petition, and it is apparent from his letter of recommendation that the clocks from Augsburg had arrived damaged to such an extent that Gerhard de Mase, the Dutch clockmaker to the imperial court, had not been able to put them in order in the short time before the departure of the mission. Minkwitz also indicates that every time presents were forwarded, it was the custom to send a clockmaker along to Turkey.[16]

From Minkwitz's letter it is clear that the clockwork mechanisms were made in Augsburg. Indeed, during the reigns of Maximilian II and Rudolf II and down to the outbreak of the Turkish war in 1593 the imperial governors in Swabia, all named Ilsung—Georg, Maximilian, and Johann Achilles Ilsung—were entrusted by the Emperor with the overall preparation of the Turkish honoraria. Their residence was in the free imperial city of Augsburg, where they had both to raise and to exchange appropriate sums in coined gold and to commission silversmiths' work and the clocks from the Augsburg artisans.[17] It is relatively seldom that one finds references to orders placed at Nuremberg, such as with Wenzel Jamnitzer.[18]

13 Gerlach, *Tage-Buch,* p. 21.
14 Extracts from the accounts of the military paymaster Mathes Fuchs, HKA, RA 190, fol. 708.
15 Undated petition of Wolfgang Teissenrieder to the Emperor with the official comment: "Ex Consilio Camerae Aulicae 22 Octobris anno 71." *Ibid.,* fols. 710 f.
16 Undated report from Minkwitz, earliest registry notation being of Oct. 28, 1571. *Ibid.,* fols. 712, 714.
17 It has been possible to exploit for this study only an infinitesimally small fraction of the rich documentary record available on this subject in the Vienna Central Archives. Systematic analysis of these sources could not fail to provide new facets for the history of south German artistic manufacturing, and it might above all clearly establish the contribution of the estates of the Empire toward the financing of the Turkish honoraria. It should be noted that a small but not insignificant part of this money remained in the Empire and found its way into the hands of south German art manufacturers and clockmakers.
18 Wenzel Jamnitzer, goldsmith at Nuremberg, received from Maximilian II on Feb. 20, 1574, an order to the amount of 2,000 gulden for silver work intended for the Turkish honorarium. HKA, RA 190, fols. 987 f.

13. Hans Ludwig, Count of Kuefstein, delivering presents to the Pasha of Ofen, 1628 (detail). Schloss Greillenstein, Lower Austria.

The imperial agents at Augsburg frequently received from the court detailed sketches and designs for the execution of the silversmiths' work and the clocks (aside from a few exceptions, these have unfortunately been lost). Once ready, the objects were brought by trusted representatives of the Augsburg middlemen to Vienna, usually by water, and there they were delivered to the military paymaster, who drew up precise lists, handed the presents over to the imperial envoy, and received an acknowledgment of receipt in return. The ambassador took the gifts from Vienna to Belgrade via the Danube, and thence by the difficult overland route to the Bosporus.[19] In Constantinople he delivered them to the permanent imperial resident at the Porte, who then first handed the honoraria to the Turkish dignitaries and finally laid the tribute before the Sultan in solemn audience.[20]

A halt was made at Ofen in Hungary on the way, and here the Emperor's presents were delivered in audience to the Pasha. Figure 13 shows the audience of the imperial ambassador extraordinary Hans Ludwig Count of Kuefstein before the Pasha on September 29, 1628. Embassies extraordinary were exchanged only after the conclusion of the peace of Zsitvatorok (1606), on the basis of reciprocity between the Emperor and the Sultan.[21] This peace settlement introduced a certain parity into the relationships between the imperial court and the Divan, and the yearly tribute was replaced by a one-time payment by the Emperor to the Grand Turk amounting to 200,000 talers.

In the mid-1560s we begin to find exact specifications and accounts for the Turkish honoraria; these provide information both on the total costs and on the distribution of the honorific gifts. On July 23, 1563, Paulus de Palina (also called Paul Diäckh) and Achaz Csabi (or Tschagi) took delivery in Vienna of the honoraria that were to be transported to Constantinople.[22] In the statement one finds along with the cash "seventeen silver-gilt double drinking vessels, likewise six large and small clocks, cost all together 2,437 gulden, 9 kreuzer, 3 pence."[23] It was expressly remarked that the silversmiths' work and the clocks were from Augsburg. Albert de Wyss, who represented the interests of the Emperor from 1562 until his death on October 21, 1569, was to deliver 30,000 ducats to the Sultan. For the Grand Vizier—at that time Ali Pasha, called the Fat—there were intended two of the most impressive double beakers in which there were

19 Precise data on the routes travelled are to be found in the various contemporary journals and travel descriptions.

20 See, e.g., Gerlach's report concerning the delivery of the honorarium for the year 1573. *Tage-Buch,* pp. 22 f.

21 Gottfried Mraz, "Von der Tributleistung zur Repräsentation. Die habsburgisch-türkischen Ehrengeschenke vom Frieden von Zsitvatorok bis zum Frieden von Karlowitz mit einem Rückblick auf das 16. Jahrhundert." Unpublished report, Institut für österreichische Geschichtsforschung, Vienna 1976, pp. 16–23.

22 Spuler, op cit., p. 324 does name Achaz Csabi as Nuntius, but not Paulus de Palina. In the HKA, RA 172, fols. 37–40 we find cost statements for the honoraria in the years 1559 and 1563, and here Paul Diackh and Achaz Tschagi are named as presentation envoys for 1563.

23 *Ibid.,* fols. 37v–38r.

41

1,500 ducats, then a further 2,500 ducats, two stallions, and a shirt of mail which the recipient had earlier personally ordered through the former envoy Busbecq. The clocks were received by dignitaries lower in rank, two of them by Mohammed Pasha, next in line below the Grand Vizier, along with two goblets and 1,200 ducats.[24]

The desire of the Sultan and his ministers for outstanding pieces from the Augsburg workshops mounted in the following years. The imperial envoy de Wyss reported to the court on May 1, 1567, that the Sultan was demanding, over and above the ducats, at least twenty double beakers filled with some thousands of gold pieces, and two or three clocks to boot. A clock, along with 4,000 ducats and three or four goblets, was also to be presented to the Grand Vizier Mohammed Sokolly.[25] The former envoy Busbecq, to whom the proposal was submitted for his opinion, found these demands by Sokolly too high; on the other hand, he thought, they should send the Grand Turk twenty-five beakers, "a little more or less, along with some clocks, that need not be so very intricate provided only that they be handsome and make a good appearance."[26]

Georg Ilsung, imperial governor in Augsburg, on June 6, 1567, received for the Turkish honorarium seven parcels which valued 787 gulden 34 kreuzer from the goldsmith Hans Flickher, nineteen pieces of silversmiths' work at 2,372 gulden 9 kreuzer from Martin Marquart, and fifteen objects that cost 2,045 gulden 14 kreuzer from Ulrich Meringar.[27] Ilsung additionally acquired a number of clocks from various masters.[28] The most expensive piece originated with Daniel Lotter and cost 250 gulden; similarly a quarter-striking mechanism from Caspar Buschmann came to 160 gulden, and another from Hans Connat, which is supposed to have been a masterpiece, was worth 145 gulden. Further specified were a work by Mathes Stosswender at 40 gulden, a large rectangular bracket clock by the same master at 50 gulden, a large bracket clock with a large bell by Simon Friderich at 50 gulden, a tall piece with a Hercules at 22 gulden, a small round clock by Asmus Pirenbrunner at 21 gulden, a monstrance clock by Hans Zeller at 166 gulden, and a desk clock by Georg Zorn at 45 gulden.

In his dispatch to the Emperor, Ilsung singled out Lotter's clock in particular: "Among these, however, a quarter-striking mechanism on which the case lid does not rise but is simply clapped down upon it and it has two panes that are slid back and forth; this was made by a particularly famous master and it is supposed to be very accurate. As he has hitherto

24 *Ibid.*, RA 190, fols. 111–114, 115–118.
25 The envoy wrote: "Further as to the moneys to be presented at that Porte, thirty thousand ducats that are asked for the one year elapsed, there may be sent for the Prince of the Turks at a minimum twenty beakers with several thousand ducats and two or three clocks. For Mohammed Pasha four thousand ducats or four beakers and one clock." *Ibid.*, fols. 368, 385, 369, 380–383.
26 *Ibid.*, fol. 372.
27 *Ibid.*, fols. 347 f., 349 f., 351 f.
28 "List of the clocks that I, Georg Ilsung, have purchased at Augsburg on June 6, 1567, but solely for the ratification and approval of his Imperial Roman Majesty, and which I have shipped off to Vienna" (*ibid*, fols. 353 f.).

never valued one otherwise than at four hundred gulden, it was only with great trouble that he let me have this one at 250 gulden."[29] Another clock that had been ordered by the court and was to cost 700 to 1,000 gulden had not yet been finished. Ilsung thought that it would take more than another twelve days for completion; at the same time he assured the Emperor that he had never seen such a piece of work before. On June 28, 1567, the honorarium was handed over from Ilsung to the Bishop of Erlau, Anton Verantius, and Freiherr Christoph von Teuffenbach. The reckoning of costs and the protocol of delivery show that the clock mechanism of the master Daniel Lotter and the monstrance clock of Hans Zeller were destined for the Sultan himself.[30]

In 1569 Albert de Wyss sent a memorandum to the court regarding the preparation of clocks for the Turks.[31] The designs which were once presumably attached have disappeared. Here there is mention only of a prohibition against representation of human beings, although one could apply ornamental etchings and representations of wild beasts or birds on clocks destined for the Sultan. The envoy cared little for the clocks having "spheres, motion of the planets, or striking of the quarters," but he emphasized that they should in each case "tell time, strike, and ring correctly and be very strongly and finely made." He also prescribed these conditions for certain pendant watches that were intended for distinguished officers.

The prescriptions regarding the appearance of the clocks become more and more exact over the years. The imperial envoy Ungnad reported to the court, on October 12, 1573, that at the most recent bestowal the clocks desired by Ahmed Pasha had been delivered in error to the Grand Vizier Mohammed Sokolly. Ahmed Pasha then contacted Ungnad to insist that the Emperor have two clocks made, based upon the plans which he enclosed.[32] The two sketches are still preserved. What is involved is an almost square sheet of paper, 30 × 29.5 centimeters with a Turkish text and its German translation, "Ein gevierte vhr in dises Papiers grösse, die gerecht sey. Dise vhr solle nuer schlechtlich geez (geätzt), ohne Monschein. Pildtwerch oder getribne arbeyt gemacht werden" (A quarter-striking clock of the size of this paper, which shall keep good time. This clock need be only simply etched, without phases of the moon, pictorial work, or modelled metalwork). The second sheet presents the plan for a round clock with a diameter of 41 centimeters and the same directives.

29 *Ibid.*, fols. 358–360.
30 "Hereinafter designated, that which his Imperial Roman Majesty directs shall be presented in the present Turkish honorarium to the Turkish Emperor himself and also to his Pashas and others . . ." (*ibid.*, fols. 290–293).
31 *Ibid.*, fols. 528 f.
32 *Ibid.*, fols. 854 and 857. Ungnad writes: "Ahmed Pasha, who last year vainly awaited clocks made in the form according to the model he had sent, . . . now asks that in his name we petition your Imperial Majesty that you order made for him and sent two clocks of the form designated by the pattern enclosed." The sketch of the circular clock follows after the heading indicated, fol. 855, and that of the rectangular clock at fol. 856.

14, 15. Pattern drawing for two clocks which were to be built in Augsburg. Constantinople, 1576. Vienna, Hofkammerarchiv.

The Islamic prohibition against the representation of beasts or men comes up again in many later reports.

In this same narrative by Ungnad, the Grand Vizier's desire for some smaller clocks is communicated to the Emperor: "clocks—or watches—more convenient for carrying about." Maximilian II reacted promptly to this wish and directed Ilsung on December 2, 1573, to commission for Ahmed Pasha two clocks after an enclosed pattern, likewise three pendant watches such as one wears about the neck, and three larger bracket clocks, that can be placed on a table, for Mohammed Pasha.[33]

Two other drawings of clocks have been preserved from 1576 (see Figs. 14 and 15). They are the patterns for two orders of spherical clocks for the Turkish Grand Vizier, which Georg Ilsung was to commission from Augsburg masters at the suggestion of Ungnad. On May 23, 1576, Ilsung sent the sketches to the Emperor Maximilian II at Vienna with the remark that of the total of seven clocks sought, only one could be made ready in time. One of the two designs (Fig. 14) is signed by the master clockmaker Oswald Khayser, resident in Constantinople and active for the Grand Vizier Mohammed Sokolly.[34] Four examples were to be made from this design, following Khayser's detailed instructions. The clocks were to indicate 12 or 24 hours, and the dial was to be provided with Turkish numerals. Here, too, great value was placed upon constant and regular functioning. The bells for the striking mechanism must have a clear tone. The sphere was to be equipped with handsome pierced foliage work having no figural representations. The upper hemisphere was to be transparent, so that one could see the bells, while the lower hemisphere was to be silver plated. The wheels were to be made of gilded brass. The size of the clocks was to correspond to the scale of the design; the height from the suspension ring on top to the ball-shaped feet was 14 centimeters; the diameter of the sphere itself was 12.3 centimeters.

The second sketch that has been preserved (Fig. 15) is very similar to the first, but instead of foliage work there is banded ornament and the dial is equipped with Roman numerals.[35] The Grand Vizier demanded three specimens of this model, two to be equipped with an alarm apparatus. In the introduction it was specified that "the bell should be made as clear and strong as possible." In addition, "He, the Lord Mohammed Pasha, would like very much for this work to come in with the presents; or if they should not be ready, they should be sent at the earliest date thereafter to the Pasha at Ofen." Ilsung indicated that a clock of this sort furnished with an alarm, and of the quality sought, might come to 55 gulden.

We are particularly well informed regarding the Turkish honorarium of the year 1578, for detailed delivery inventory of the silverwork and the clocks has been preserved. Through this instrument the court military paymaster Andreas Schnatterl, on April 18, 1578, confirmed the receipt

33 *Ibid.*, fol. 961.
34 *Ibid.*, RA 16, fols. 272, 277, illustration on fol. 276.
35 *Ibid.*, fol. 274.

44

of the objects to a value of 10,678 gulden 21 kreuzer 2 pennies, which Ilsung had sent to Vienna via his servant Daniel Manlich.[36] This information is particularly important, since there are extant makers' receipts for almost all the objects delivered. Hence we can trace quite completely the origin of the objects to specific Augsburg ateliers.

Thus we read that Melchior Schaller made a striking clock for 40 gulden.[37] From the atelier of Georg Walther came a gilded ostrich with built-in clockwork which set the bird's eyes in motion, and a striking mechanism which did the same with the bird's wings and beak; this at 100 gulden.[38] Hans Fronmiller made for 65 gulden a clock for the desk that had been commissioned from Conrad Stierlin for 2,171 gulden 20 kreuzer.[39] There also came from his shop a smaller clock by Gregor Beyrer, at 60 gulden, for the desk which cost 493 gulden 13 kreuzer 1 penny; a hexagonal gilded striking train for 77 gulden 35 kreuzer; and a round standing clock with five silver columns for 64 gulden 20 kreuzer.[40] Jeremias Metzger delivered a small, flat, round striking train for 33 gulden, another small clock for 8 gulden, a round striking clock in a gold case to be worn at the neck, for 54 gulden 20 kreuzer, a long rectangular clock in the form of a small chest with a case of ebony and glass, furnished with an enamelled silver dial at the top. In Metzger's itemized bill 55 gulden were paid for the mechanism, 16 gulden 30 kreuzer for the dial, 8 gulden for the wood, 1 gulden 40 kreuzer for the glass, and for the inside lining 2 gulden 35 kreuzer.[41] The principal clock mechanisms came from the atelier of Hans Schlottheim: a large gilded ten-day striking mechanism with four hour hands, side panes of glass, and on top an ingenious sundial, for a total of 208 gulden; further a striking mechanism in a hexagonal ornamental case of ebony with gilded ornamental ribbing for 180 gulden 30 kreuzer; and a third striking mechanism in a rectangular case of ebony and glass with a globe to show phases of the moon, the mechanism costing 125 gulden, the case 35 gulden, the ribbing 14 gulden, and the inside lining 6 gulden 30 kreuzer.[42]

At the beginning of 1584 a particularly rich clockwork mechanism for the Sultan was commissioned from the master Schlottheim. On January 28, 1584, Johann Achilles Ilsung reported from Augsburg to the Emperor who was at Prague that the master had already started with the work, but that he would need at least five months for completion.[43] Schlottheim was now demanding 1,200 gulden for the mechanism, but he would presumably let it go for 1,000. In July the Turkish honorarium was sent off to Vienna, but the great clock was not yet ready. On July 13, 1584, Ilsung

36 *Ibid.*, RA 192/I, fols. 20–33.
37 *Ibid.*, fol. 41.
38 *Ibid.*, fols. 46, 65.
39 *Ibid.*, fols. 42, 69.
40 *Ibid.*, fols. 43, 68, 50, 61.
41 *Ibid.*, fols. 52, 59, 53, 58.
42 *Ibid.*, fols. 54, 57.
43 *Ibid.*, RA 191, fols. 334 f.

expressed the hope that in eight days it would at last be ready, and he added that although Schlottheim was certainly an industrious worker, he was in the old-fashioned way a slow one.[44] On July 18 Archduke Ernest, at the time governor in the Austrian lands (his imperial brother Rudolf II resided in Prague), received from Augsburg the information that Ilsung wanted to send the clock, well packed, to Vienna with the next barge, for he could trust the vessel's crew.[45] But on July 31 the clock was still at Augsburg, because Ilsung would not agree to Schlottheim's already greatly reduced demand for 600 gulden.[46] By August 8 Ilsung received an imperial command to the effect that he should at once send the great Schlottheim clock to Vienna: he was exercising a false economy, for the mission was waiting to depart from Vienna and was obliged to wait for this clock—the major piece for the Sultan—and in the meanwhile maintenance costs were costing the Emperor dearly.[47] On August 13 Ilsung reported that he had sent the clock off to Freising with one of his own servants, and that from there onward it would go to Vienna by water.[48] But the shipment was made too late. On January 10, 1585, the court chamberlains at Vienna informed their colleague at the imperial camp at Prague that the Schlottheim clock was still in Vienna;[49] it had arrived late and in bad condition, requiring repairs costing 35 gulden. At the same time the councillors debated whether this clock should not be used for the next honorarium. On this point advice was sought of two former envoys in Turkey, David Ungnad and Joachim von Sinzendorff. Both were of the opinion that the piece was well suited for the Sultan, but one must consider that "for such a sum as this piece costs one could arrange for several others that would be more showy and win greater regard."

Even when he had been envoy in Constantinople, Sinzendorff had stressed the greatest possible thrift. With regard to the preparation of the honorarium for 1579 he gave advice on August 17, 1578, to the effect that clocks and silverwork should of course be commissioned, but one need devote special attention only to the wishes of the Sultan, then Murad III, and of the Grand Vizier. For, as he wrote, "with the other viziers and officers the usual thing is that they accept whatever one gives them."[50] And yet it is precisely with the officials of lower rank that this was not always the case. Paul von Eytzing reported to Rudolf II in November 1584 that his predecessor, the imperial orator Johann Freiherr von Preyner, who had died that same year in Constantinople, had promised

44 *Ibid.*, fols. 545, 552. The missive is accompanied by the specification of the silver utensils and the clocks that Ilsung had drawn up on June 4, 1584, in Augsburg; *fols.* 546–551.
45 *Ibid.*, fols. 444 f.
46 *Ibid.*, fols. 606 f.
47 *Ibid.*, fols. 604 f.
48 *Ibid.*, fols. 609 f.
49 *Ibid.*, fols. 725 f. Rudolf II had resided permanently at Prague since 1582. Yet the centers of administration at Vienna had not been completely disbanded. On this account one makes a distinction, in the administration of finances, between the Treasury present at the Emperor's camp and the Treasury left behind and remaining in Vienna.
50 *Ibid.*, fols. 51 f.

fine clocks to several aides to the Grand Vizier Ibrahim and to two eunuchs of the Sultan's seraglio.[51] These men were not satisfied with the silks or the money offered, "but instead they continue to insist vehemently upon the clocks that were promised them, desiring nothing else in their place." The ambassador therefore recommended that clocks of various types be sent at the earliest opportunity, each of them worth about 40 to 50 gulden.

As is apparent from what has been said, clocks destined for Turkey were by Western standards rather on the modest side, excepting the pieces reserved for the Sultan and the Grand Vizier. Ilsung reported to the Emperor on March 10, 1584, that he had seen in Augsburg the masterpiece of a young clockmaker; he sent the Emperor a wash drawing and a description of this "splendid and beautiful piece of work," along with the comment that it cost 200 gulden and would perhaps be appropriate for the next Turkish honorarium.[52] This was the masterpiece of Hans Fronmiller, who had perhaps himself drawn up the precise description, which still exists, though the drawing has been lost.[53] The particulars were submitted to the court paymaster for his response, which was to the effect that the clock

> . . . might be quite suitable for a great lord somewhere in Christendom. But for the Turkish honorarium, since the Turks understand very little indeed in connection with ingenious apparatus of this sort, like astrolabes, planetaria, diurnal circles, our whole calendar, etc., I hold it to be unsuited, save in the event that someone has information to the effect that the Turkish Emperor or the Grand Vizier Pasha takes delight in such objects, in which case it might be added to the honorarium.[54]

But the work of German artisans continued to be held in high regard in Turkey, as is underscored by one more episode. This was the occasion of a magnificent fete planned by Murad III for the spring of 1582 to celebrate the circumcision of his oldest son.[55] This celebration was intended to eclipse all others held before in Constantinople. The crowned heads of Asia, Europe, and Africa were invited to attend, and the preparations required more than a whole year. Naturally not only were the

---

51 *Ibid.,* RA 172, fols. 183–190.

52 *Ibid.,* RA 191, fols. 408, 411.

53 *Ibid.,* fol. 409. Also spelled Fromüller, the master used the form Fronmiller. The description reads: "The clock is arranged as the two Pieces show. Firstly, it will strike the quarters, and right alongside the quarter-indicator one adjusts the whole mechanism. Secondly, it will strike and show from one to 12. Thirdly, it will strike and show from one to 24. Fourthly, it will strike the whole astrolabe with sun and moon and will run the course through the twelve signs of the zodiac and the planets. Fifthly, it will show the length of the day and the night and the hours of sunrise and sunset on the polar height. Sixthly, it will show the calender with its most outstanding feast days. Seventhly, it will sound an alarm. On the other two sides that are not inscribed it will show how much has been struck, namely the quarters, the hours to 12 and to 24; and it is also of the same form as that one, and the next time such a one is bought it will cost 200 gulden."

54 *Ibid.,* fol. 410.

55 An exhaustive description of the festival of the circumcision is to be found in Joseph von Hammer, *Geschichte des osmanischen Reiches,* Vol. IV (3rd ed., Pest, 1828), pp. 118–134.

attending rulers obliged to send rich gifts through their special ambassadors, but also the Turkish dignitaries had to present valuable presents. The Pasha of Ofen believed he could best discharge this duty by having the Roman Emperor purchase these presents for him. So on October 29, 1581, Rudolf II charged Maximilian Ilsung with the ordering of these objects from Augsburg masters in accordance with the instructions from the Pasha.[56] Among other things, he expressed a desire for rings set with clockwork and for curious mechanical devices in the form of men and beasts that "can move and walk of themselves, also play with each other" (a most unconventional request, considering Turkish custom). Ilsung was rather in a quandary, and he wrote the Emperor that he had spoken about the rings with an Augsburg master (whose name is unfortunately not mentioned), but that the delivery date would have to be left entirely open, since this work could be done only on bright and fine summer days. Mechanical automata could not be hurried either, and he saw no possibility of supplying these within a year's time.[57]

The continued renewals of the armistice between the Emperor and the Porte and the regular delivery of tribute and honoraria could not stop the chronic skirmishes at the borders of the Ottoman empire. At the beginning of the 1590s military activity became more violent, until finally, after the painful defeat that the Turkish viceroy in Bosnia suffered in the summer of 1593 at the hands of the imperial forces at Sissek on the Kulpa, an open war broke out.[58] After more than ten years of dubious struggle, the peace of Zsitvatorok in 1606 placed relations between the Emperor and the Sultan upon a new footing.

In sum we may say that along with the work of the Augsburg goldsmiths and the Nuremburg armorers, the clocks from Augsburg were an important part of the Emperor's honoraria to the Ottoman Porte from 1550 to 1590. In addition, the clocks afforded eloquent testimony to a superior mechanical understanding and technical ability.

56 HKA, RA 191, fols. 42 f. The Turkish-language memorandum of the Pasha of Ofen is found in HKA under the heading RA 174, fol. 170.
57 *Ibid.,* RA 174, fols. 160, 187.
58 See Zinkeisen, *Geschichte des osmanischen Reiches,* Vol. III, pp. 588 f.

6 Astrolabe Clock Faces

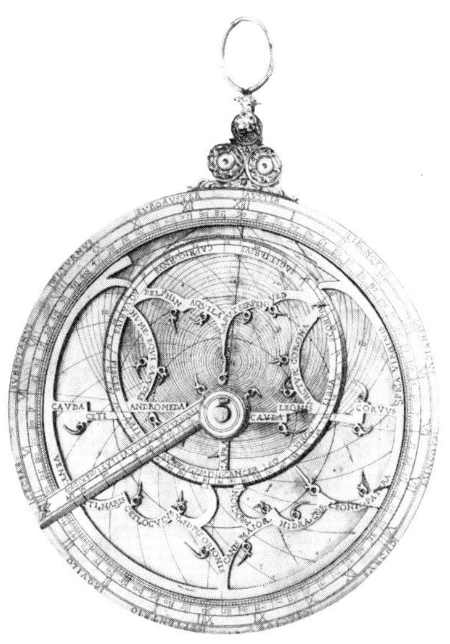

16. Astrolabe by Georg Hartmann, Nuremberg, 1537. Washington, D.C., National Museum of History and Technology.

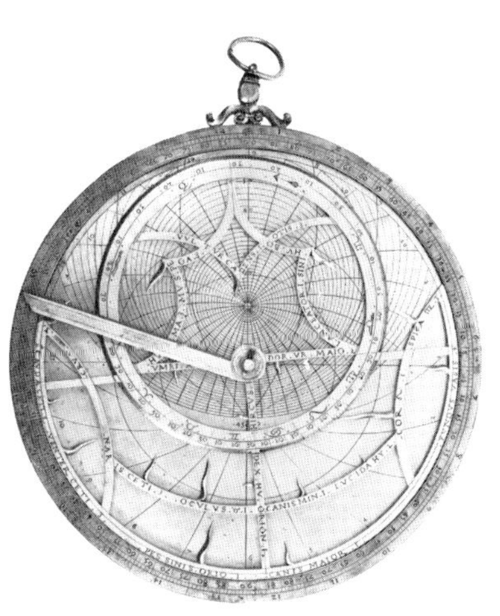

17. Astrolabe signed MP. German, 1542. Washington, D.C., National Museum of History and Technology.

There is much evidence to suggest that the astrolabe enjoyed particular success as a scientific instrument in Europe during the 16th and 17th centuries. Almost three quarters of the extant European examples of the astrolabe date from the period between 1500 and 1700. Figures 16 and 17 show two examples of astrolabes made in Germany in the 16th century: one by Georg Hartmann of Nuremberg, dated 1537, and one by M.P., possibly Markus Purmann of Munich, dated 1542. An impressive number of early published books discuss the construction and use of the astrolabe; some works, like the *Elucidatio fabricae ususque astrolabii* of Johannes Stöffler, first published in 1513, appeared in dozens of editions. Renaissance artists were careful to include an astrolabe among the paraphernalia associated with learned men. For example, several astrolabes furnish the studio of Saint Augustine depicted by Vittore Carpaccio about 1500. Given the apparent popularity of the astrolabe, it is easy to understand why many European clockmakers might have wished to incorporate elements of its design into their own works.[1] The astrolabe faces on a large number of Renaissance table clocks well illustrate the effect of combining familiar science with a developing instrument-making industry.

Scholars have concluded that the basic principle upon which the design of the astrolabe is based, the principle of stereographic projection, was developed by Hipparchus about 150 B.C. In the course of his astronomical investigations Hipparchus discovered that he could reduce problems of spherical trigonometry to problems of plane trigonometry by projecting figures on a sphere from one of its poles onto a plane parallel to its equator (Fig. 18). While the interval between principle and practice may not have been as great as remaining evidence indicates, no one has yet uncovered an earlier discussion of the construction and use of an astrolabe than that which appears among the works attributed to Theon of Alexandria (*fl.* 350).[2] The instrument Theon described consists essentially of two disks of equal diameter pierced in the center; an alidade arm of length equal to the diameter of the disks, also pierced in the center; a pin joining all three pieces by means of their central holes; and a method for suspending the assembled apparatus from the edge of one of the disks.

1   In an article entitled "The Plane Astrolabe and the Anaphoric Clock," *Centaurus.* 1959, *3:* 183–189, A. G. Drachmann contends that a relationship between astrolabes and mechanical (i.e., water-driven) clocks has existed since antiquity. Drachmann suggests that, in fact, the astrolabe was patterned after the anaphoric clock. Renaissance clockmakers appear to have reversed this process, patterning a face of their instrument after an instrument with an already well-established separate identity

2   Theon's work on the astrolabe is preserved only indirectly, in the works of later authors. Otto Neugebauer has traced the transmission process and reconstructed the original work of Theon in an article on "The Early History of the Astrolabe," *Isis.* 1949, *40:* 240–256.

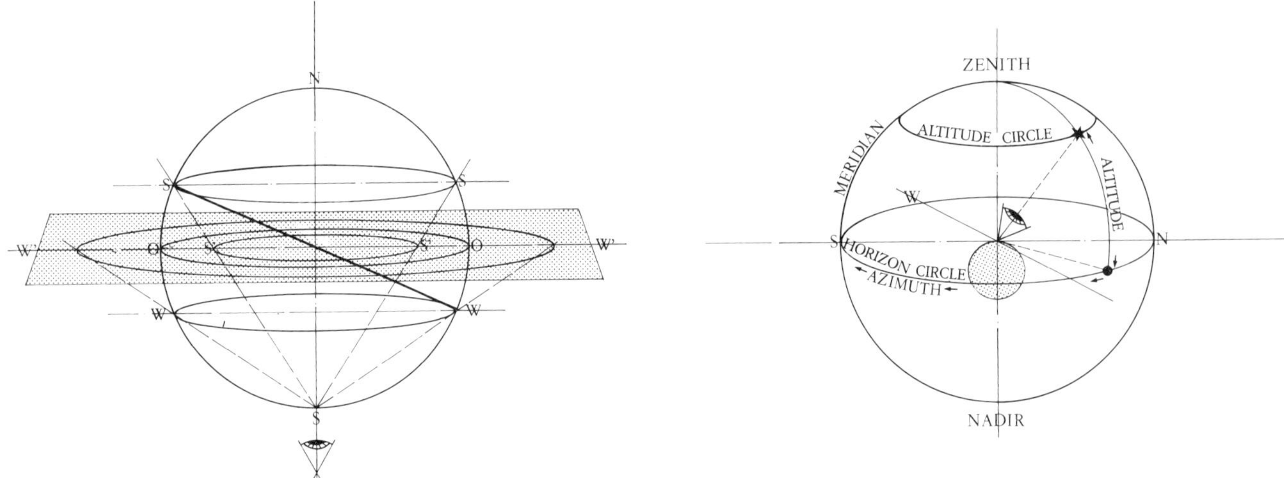

18. Examples of stereographically projected circles. The stereographic projection of the circle *SS* is the circle *S'S'*.

19. An observer at the center of an apparently spherical universe can use circles to designate the position of a star.

Theon's astrolabe was designed for use by astronomers and astrologers who envisioned themselves at the center of a vast celestial sphere whose axis was coincident with the axis of the earth. By further envisioning circles on the inner surface of this sphere, these centrally located observers created frames of reference which could be used to specify celestial positions (Fig. 19). The lines and circles engraved on one of the disks of Theon's astrolabe represent the stereographic projection of a miniature model of one of these reference frames, that based on an observer's horizon circle. Since the orientation of this horizon circle relative to the poles of the celestial sphere varies with latitude, the projection of it and of the reference frame based upon it also vary with latitude.

The second disk of Theon's astrolabe, like the first, is engraved with the stereographically projected elements of a celestial frame of reference. This second reference frame incorporates points and circles whose orientations relative to the poles of the celestial sphere are independent of any changes in the observer's location. The projected points represent stars; the projected circles include at least the ecliptic and the Tropic of Capricorn. In order for the combined disks to function according to Theon's design, the unengraved portions of the second, geographically independent plate must be cut away. The circular skeleton, or rete, thus formed, when placed over the solid backplate, enables a user to relate two reference frames in a miniaturized universe (Fig. 16). By rotating the rete about its center, which is the projection of the north celestial pole, the user of an astrolabe can replicate the apparent motion of the stars or sun (at a point on the ecliptic) across his meridian. Why would he wish to do so? As Theon recognized, the astrolabe's ability to function as a model of the heavens made it a particularly effective timekeeper. His and every subsequent discussion of the use of the astrolabe includes instructions on how to find the time. To paraphrase these sources:

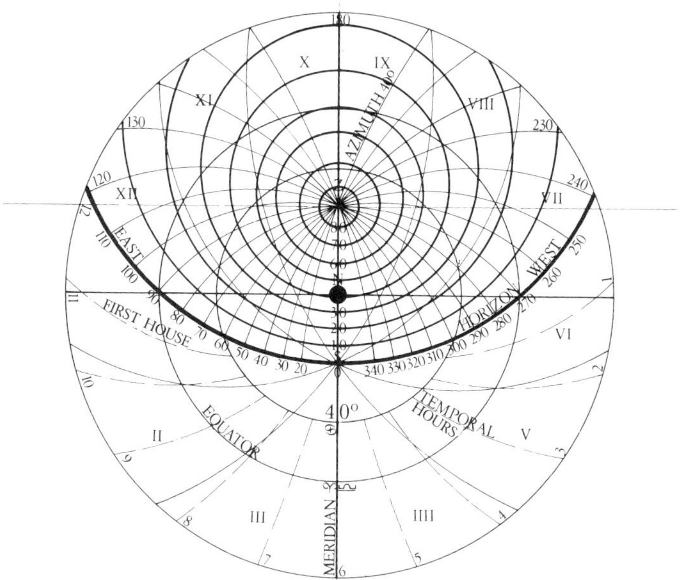

20. Horizon plate of an astrolabe face for latitude 40°.

When the hour of the night is required, the assembled astrolabe (disks and alidade) is suspended so that it hangs plumb. The alidade is adjusted so that one of the stars whose projected image is incorporated into the rete is centered in the pin holes of its sighting plates. The elevation of this star is read from a scale engraved along the margin of the suspended disk. The rete is rotated about the pin over the face of the disk engraved with the horizon-based reference frame, so that the projected stellar image is above the projected horizon at the elevation measured by the alidade. Once the rete has thus been properly oriented, the time is indicated by the position of the point on the ecliptic then occupied by the sun.

The process is basically one of using the astrolabe to convert a measure of stellar position into a measure of time.

The reputation of the astrolabe as both a model of the heavens and as a timekeeper was long established by the time the first geared clocks were constructed in Europe. In the mid-15th century an astrolabe might be found hanging near a mechanical clock.[3] Remaining evidence suggests that by the early 16th century the astrolabe and the mechanical clock were just as likely to be combined into one.[4] The act of rotating the rete over the fixed solid plate could easily be accomplished by mechanical means.[5]

3   In a miniature illustrating the *Horloge de sapience* of Henri Suso (now in the Bibliothèque Royale Albert I in Brussels) an astrolabe can be seen hanging just below the face of a gothic mechanical clock. Henri Michel has recently published this illustration in *Images des sciences* (Brussels, 1977).

4   Among the hundreds of clocks illustrated by Klaus Maurice in *Die deutsche Räderuhr* (Munich, 1976) are about 70 examples with astrolabe faces. Half of these were made in Augsburg.

5   Derek Price has examined the origins of the application of the geared clock to the problem of duplicating celestial motion in Bull. 218, Paper 6 of *Contributions from the Museum of History and Technology* (Washington, D.C., 1959). Price points to geared astrolabes made in the 13th century as important antecedents of astrolabe clocks. In his view, a second antecedent and a link to the anaphoric clock can be found in an astrolabe-faced clock attached to a mercury drum and described in the *Libros del Saber.*

21. Astrolabe face of a table clock by Jeremias Metzger, Catalog No. 26.

22. Rete from the astrolabe face of an Augsburg table clock, end of the 16th century. Berlin, Kunstgewerbemuseum.

When it was attached to a properly adjusted geared mechanism, the rete —a projected model of the starry sphere—could be rotated in apparent harmony with the stellar universe, one rotation every 23 hours 56 minutes.[6]

The background horizon plate on the astrolabe face of a mechanical clock cannot be distinguished on the basis of its content from the plate of an astrolabe. (Fig. 20 and cf. Figs. 16 and 21.) The plate is fixed in place on the clocks so that the ends of its two engraved diameters mark the equal hours of 6 A.M. and 6 P.M., noon and midnight. The latitude of geographic position at which the engraved network would have meaning is always clearly marked, often with the phrase *Elev*[*atio*] *pol*[*i*] *grad*[*us*] . . . . Lines representing celestial altitudes are generally engraved at intervals of 5°; lines representing celestial azimuths are engraved at 10° intervals. The area on the plate below the horizon is inevitably divided by a series of arcs which mark the ancient seasonal (or unequal) hours. These hours are of equal length on any given night, but they vary in length during the year, being longest at the winter solstice and shortest at the summer solstice.[7] Finally, every plate face is segmented by the stereographic projections of the boundaries of the twelve astrological houses.

Recognition of the astrological significance of dividing the heavens into twelve houses can be traced to antiquity.[8] The particular method of establishing the boundaries of the houses found on the astrolabe faces of 16th-century clocks was first described by Regiomontanus in an essay completed in 1467. Rejecting as inaccurate the methods proposed by the "Chaldeans," by Alchabitus, and by Campanus, Regiomontanus recommended that the heavens be divided by great circles running through the north and south points on the horizon, which, beginning with the horizon, cut the celestial equator into twelve equal parts. In Regiomontanus' system, numbering of the houses begins with the segment bounded above by the projection of the eastern horizon (i.e., on the left) and continues counterclockwise. Each house is identified with a facet of human affairs susceptible to the influence of the "stars": I physique and appearance, II money, III relations with kin, IV lineage, V progeny, VI health, VII spouse, VIII death, IX religion, X employer, XI relations with friends, XII enemies.

There is an obvious family resemblance between the rete of an astrolabe and the rete of an astrolabe clock face. In some cases the resemblance is manifest in the patterns of the strapwork left to tie star pointers to the

6 The rate of rotation of the rete of an astrolabe clock effectively ensures that a different area of the projected universe culminates at midnight every night during the year. Such a condition agrees with observation.

7 Aesthetics and convenience are most often suggested as motivation for a decision to project only the nocturnal seasonal hours. It is also true, however, that the relative sizes of the seasonal hours are most effectively demonstrated by the stereographic projection of the hours of the night.

8 Ptolemy alludes to the existence of a house system in Bks. III and IV of the *Tetrabiblos* (written about A.D. 125), one of the earliest well-developed discussions of astrology, available in Latin translation in 16th-century Europe. See F. E. Robbins' translation into English in the Loeb edition (Cambridge, Mass., 1964).

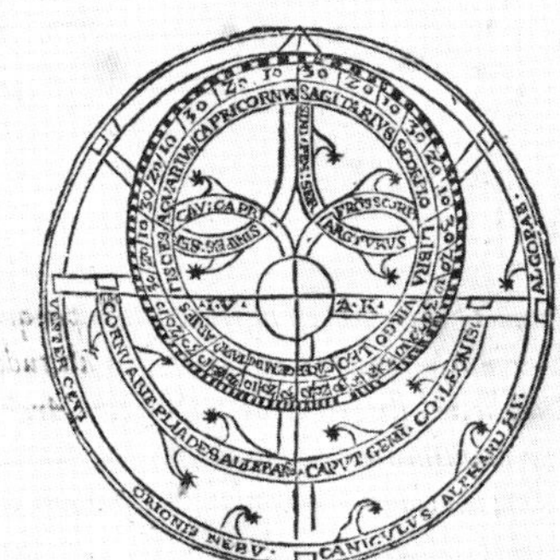

23. A page from the Paris edition (1552) of Jacob Koebel's *Astrolabii declaratio* (Nuremberg, 1517).

central projection of the celestial pole. The anonymous Augsburg maker responsible for the clock rete shown in Fig. 22 appears to have incorporated designs also available to the astrolabe makers whose works are depicted in Fig. 16. There is some evidence that both types of instrument makers may have been influenced by "construction and use" literature. The page from Jacob Koebel's *Astrolabii declaratio* (Paris, 1552) shown in Fig. 23 depicts a rete pattern with elements of design adopted by many clockmakers at work in southern Germany. (A German edition of Koebel's work was published in Nuremberg in 1517.) Jeremias Metzger's rete, shown in Fig. 21, is a typical adaptation of Koebel's published design.

In general, the resemblance between the retes is evident in the selection of circles represented (ecliptic, Tropic of Capricorn) and in stars projected. Table 1 compares and identifies the stars projected onto the astrolabe rete in Fig. 17 and the clock rete in Fig. 21.[9] In the midst of the

9 The surviving literature which had been available to 16th-century rete makers does not present hard and fast rules for the selection of projected stars. Most extant treatises include a table of about 40 well-distributed stars, all located north of the Tropic of Capricorn. Celestial position and relative magnitude are given for each entry. Makers could select from

24. Rete signed AS from the astrolabe face of a table clock, Catalog No. 33.

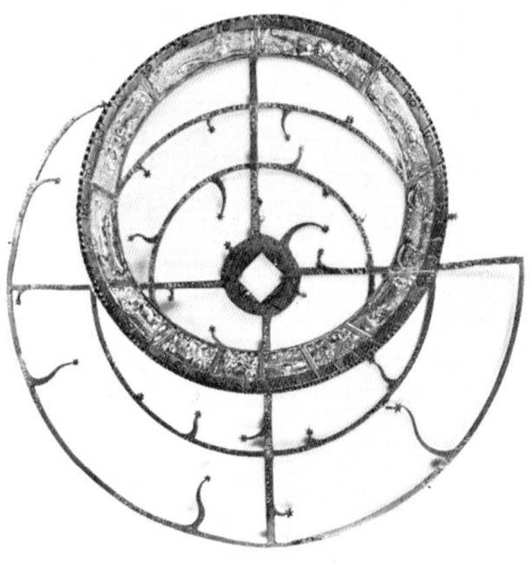

25. Rete from the astrolabe face of a table clock made by Steffen Brenner, Catalog No. 49.

similarities, however, there are distinctive differences between astrolabe retes and clock retes. Whereas the ecliptic divisions on an astrolabe rete are identified with symbols or names, these same divisions on a clock rete are identified with graphic images of the zodiacal signs. A phrase deemed unnecessary by makers of astrolabe retes, *Stellarum fixarum nomina et magnitudinis numeros notant,* is often used by makers of clock retes to decorate the Capricorn band. When the Capricorn band of a clock rete does not carry a title, it may sometimes be divided into a scale. In the latter cases subdivisions of the ecliptic band have been omitted. On no astrolabe rete do divisions on the Capricorn band substitute for subdivisions of the ecliptic band.

The mixture of clues supplied by the rete of an astrolabe clock face suggests that it was not just a transplanted astrolabe rete. The clock rete maker may have had access to an astrolabe model or, more likely, he may have consulted one of the available treatises on the astrolabe's construction and use. It is apparent from many clock retes, but especially those constructed in southern Germany, that their makers interpolated an emphasis on decorative effect not likely communicated to them by model or text. The strikingly symmetric selection and arrangement of star pointers is the best evidence of this characteristic emphasis. In some cases attention to decorative details seems to have distracted a clock rete maker from his attention to astronomical or geometrical details. The maker AS, for example, conflates stereographic and radial projection in his clock rete (Fig. 24). The zodiacal divisions of Steffen Brenner's ecliptic (Fig. 25) coincide with twelve equally spaced radii, a circumstance inconsistent with the requirements of stereographic projection.

The elaboration of the ecliptic band in clock retes and the consistent inclusion of house divisions on the horizon plate suggests that clockmakers' clientele had definite astrological interests. Indeed, by mechanically orienting the rete, a clock became a timesaver as well as a timekeeper for those interested in astrology. Someone attending the birth of a child need only glance at a clock equipped with an astrolabe face in order to determine the ascendent, the mid-heaven, and the zodiacal intersections of the house boundaries.[10] The ascertation of propitious moments need not be thwarted by inclement weather.

Even as they illustrate celestial motions, volvelle attachments to the astrolabe clock face stress astrological themes. It is astrological tradition

this list. It is sometimes possible to match a maker with his published source or to determine that he has used a nonstandard source. E.g., the rete of some clocks made in Augsburg include the star *Caput Antinoi* (see Fig. 21), a star not mentioned in any of the familiar treatises on the astrolabe's construction and use. Richard Hinckley Allen (in *Star Names Their Lore and Meaning,* New York, 1963) says the revival of European interest in the constellation is due to the fact that Mercator included it on the celestial globes he made in 1551.

10 The ascendent is the degree of the ecliptic (or zodiac) which is on the horizon at the moment of birth. It is the starting point of any horoscopic prediction. The mid-heaven is the degree of the zodiac which is culminating (intersecting the meridian) at the moment of birth. Some astrologers, including Ptolemy (*Tetrabiblos,* Bk. III, Ch. 10), have considered the mid-heaven to be the most important (or dominant) astrological position.

Table 1. Comparison of star pointers on an astrolabe rete and a clock rete

Star named	Astrolabe by Georg Hartmann (Fig. 16)	Clock by Jeremias Metzger (Fig. 21)
β Ceti	CAVDA CETI	3[a] CAVDA CETI
β Andromeda	ANDROMEDA	2 VMB[ILICUS] PEG[ASI]
ζ Ceti	VENTER CETI	3 VENTER CETI
α Ceti	NARIS CETI	3 NARIS
γ Eridani		3 ERIDANVS
β Persei	GORGO	
α Tauri	OCVLVS[TAURI]	1 OCVLVS[TAURI]
α Aurigae	CAPRA	1 CA[PRA]
β Orionis	SI: PES ORIONIS	1 SINIST[ER] PES ORIONIS
α Canis Majoris	CANIS MAIOR	1 CANIS MAIOR
α Canis Minoris	PROCION	1 CANIS MINOR
α Hydrae	HIDRA	2 LVCIDA HYDRAE
α Leonis	COR LEONIS	1 COR LEONIS
α Crateris	PATERA	4 FVN[DAMENTUM] VAS[IS]
β Leonis	CAVDA LEONIS	
α Corvi	CORVVS	3 ROSTR[UM] CORVI
α Virginis	SPICA VIRGINIS	1 SPICA VIRGINIS
α Scorpionis		1 COR SCORPIONIS
η Ursae Majoris	VRSA	(2) CAVDA VRSA
α Bootis	ARCTVR	(1) ARCTVRVS
α Corona Borealis	CORONA	(2) CORONA
δ Ophiuchi	YED	3 PALMA SINIST[RA]
α Ophiuchi	CAPVT SERPEN[TARII]	3 CAPVT SERPENT[ARII]
α Lyrae	LIRA	1 LI[RA]
α Aquila	AQVILA	2 AQVILA VVLTVR VOLANS
η Aquila		3 CAPVT ANTINOI
ε Delphini	DELPHIN	
α Capricorni		5 COR[NU] CAPRICORNI
α Cygni	HOLOR	2 CAVDA CY[GNI]
δ Aquarii		3 CRVS AQVARII
ε Pegasi	PEGASVS	
α Pegasi	HVME[RUS] EQVI	2 SCA[PULA] PE[GASI]

[a] Numbers refer to stellar magnitude.

which identifies the two intersection points of the orbit of the moon and
the apparent orbit of the sun (the ecliptic) with the head and tail of a
dragon. The regression of these intersection points, or lunar nodes, along
the ecliptic is traced on an astrolabe clock face by a dragon-shaped arm

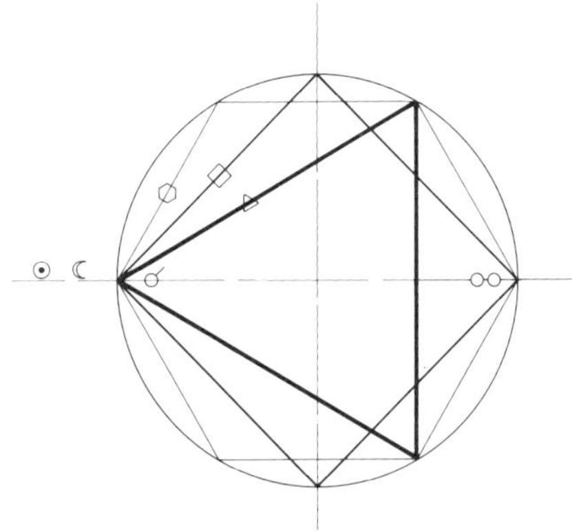

26. The aspect diagram often engraved on the central disk of the volvelle on the astrolabe face of a clock. One point of each geometric figure coincides with the arm which indicates the position of the sun or moon.

  ♂ conjunction (favorable aspect)
  ∞ opposition (unfavorable aspect)
  △ trine (favorable aspect)
  □ quartile (unfavorable aspect)
  ○ sextile (favorable aspect)
  ☉ sun
  ☽ moon

which revolves clockwise, taking 23 hours 55 minutes and 48 seconds to complete one revolution.[11] The index arm which traces the apparent motion of the sun revolves in a clockwise direction once every 24 hours. By moving just 4 seconds slower than the ecliptic band, the sun's image effectively progresses through the zodiac in the course of a calendar year.

On many clocks the solar arm is attached to a central disk whose circumference is divided into 29½ units, just the number of days in a synodic month. The significance of this scale and the decorated area it surrounds is revealed by an index arm which traces that motion of the moon. The moon arm revolves in a clockwise direction once every 24 hours 48 minutes. The central disk to which it has been attached is pierced, off center, by a small circular hole. The circular hole is placed so that while the fiducial edge of the moon arm indicates a day of the synodic month, the hole reveals an illustration of the corresponding lunar phase. By moving more slowly than the ecliptic band by 52 minutes per day, the lunar arm duplicates the relatively rapid movement of the moon through the zodiacal signs (13° per day).

The importance of the relative positions of the two luminaries, sun and moon, is stressed by most astrologers, including Claudius Ptolemy in Book IV of his *Tetrabiblos*. According to Ptolemy, these two bodies are especially influencial in determining questions of dignity and happiness, in establishing the quality of an individual's actions, in settling uncertainties about marriage. The obvious influence of sun or moon, however, may be mitigated by stars, zodiacal signs, or planets in aspect with the body (at geometric angles from it). Thus many of the central components of volvelles are engraved with an aspect diagram (see Fig. 26) which originates at the point where the sun or moon arm joins its central disk.

The assembled astrolabe face of a geared mechanical clock can be viewed as a kind of monument to classical astronomy and astrology. It is a face that would have been familiar to Ptolemy, who, in fact, considered the horoscopic astrolabe to be the only reliable instrument available to astrologers. In Ptolemy's view, no other mechanism could replace the observations made with an astrolabe's alidade arm.[12] By perfecting the geared clockwork mechanism, Renaissance clockmakers undoubtedly believed that they had eliminated the need to make celestial observations with the astrolabe's alidade. In their clocks the gearworks literally displace the alidade. Perhaps the astrolabe face is thus most appropriately viewed as a reminder of Renaissance achievement.

11 The relatively fast rate of rotation of the dragon arm ensures that its head and tail (representing the lunar nodes) regress along the ecliptic band (i.e., move in the sense opposite to the zodiacal sequence) at a rate of about 20° in 29½ days. The moon's nodal points are astrologically significant: e.g., in Bk. III, Ch. 12 of the *Tetrabiblos*, Ptolemy points out that "when the moon is at the nodes . . . there come about deformations of the body."

12 In Bk. III, Ch. 2 of the *Tetrabiblos*, Ptolemy asserts that "in general only observation by means of horoscopic astrolabes at the time of birth can for scientific observers give the minute of the hour, while practically all other horoscopic instruments on which the majority of the more careful practitioners rely are frequently capable of error."

7 The Augsburg Clockmakers' Craft

27. The Augsburg trades render homage to the Emperor Leopold I. Note Augsburg's trademark, the pine cone. Copperplate engraving (detail) by Joseph Werner. Augsburg, c. 1680.

Lewis Mumford has written, "The clock, not the steam-engine, is the key-machine of the modern industrial world."[1] In view of the ornate, almost fantastic creations of early horological art, Mumford's pronouncement seems to lose plausibility, especially upon realizing that historically timetelling was merely a secondary function of the clock. The apparent contradiction is reconciled by Moscovici's thesis that "The most important elements of the scientific and technical revolution were at first items of luxury, which only gradually became the vital necessities of modern society as new cravings assumed definite form."[2]

Insofar as the expansion, exploration, and supplying of a first, primarily elite market for the clock are concerned, it is unquestionably to Augsburg that we must look. Starting about the middle of the 16th century the quality and beauty of Augsburg clocks won a rapidly increasing number of admirers at home and abroad, indeed far beyond the confines of Europe.

The imperial free city offered ideal circumstances for the transmission, application, and effective presentation of technical knowledge as a result of the high state of development of mercantile and guild life, artistic and scientific life, which had arisen in the same context. A rich treasure of contemporary archival materials makes it possible to cast light upon this milieu of the Augsburg clockmaker's art, and most of all upon the guild as the social foundation of its development and organization, with consequences, whether good or bad, for the individual master and for production as a whole.

An historian of Augsburg clockmaking can begin to tread firm ground and build a reliable historical basis starting with the year 1392. This key date, hitherto unpublished, is found in an "old book of accounts of the master builders": "1392 . . . about mending the clock on the Perlach Tower. . . ." So a clock already existed on the Perlach Tower, and its being repaired is noted here. Further on, emoluments (plainly already on a regular basis) for the custodian (*Halder*) of the clocks are mentioned. A broken clock is put in order by one Wessisprunner. 1393: Master Hans (perhaps Hans von Ketz frequently named thereafter) is working on clock bells. Payment is made for a rope for the clock at the Court; payment is

This contribution is based on archival research in Augsburg and Vienna which was made possible by a generous grant from the Volkswagen Plant Foundation.

1 Lewis Mumford, *Technics and Civilization* (New York, 1934), p. 14.
2 S. Moscovici, *Essai sur l'histoire humanine de la nature* (Paris, 1968). Paraphrased in W. Krohn and Edgar Zilsel, eds., *Die sozialen Ursprünge der neuzeitlichen Wissenschaft* (Frankfurt, 1976), p. 25.

made for a rope plus oil for another clock. An unnamed master fits the wheels of a clock. 1398: Frans Amman is to provide a small bell for the alarm movement of the clock in the council chamber, and so forth.[3] It is notable that in these sparse accounting memoranda at least three clocks (unquestionably ones with clockwork mechanisms) and several clockmakers are mentioned within a decade. It can hardly be supposed that we are simply dealing with itinerant artisans. Therefore the history of Augsburg clockmaking commences while the 14th century is still in progress.

From these beginnings it was to be more than two hundred years before the guild organization was fully developed. Among the earliest Augsburg guilds sanctioned in 1368 and recognized by the Emperor as well in 1374, we find the general guild of the smiths, with which painters, saddlers, goldsmiths, and others had associated at the start. The craft of the clockmaker was incorporated into this guild, as it was in many other comparable cities. The first acceptance of a clockmaker is entered in the earliest guild register for the smiths, that of 1442, but for a date as early as 1441.[4]

By the 16th century the clockmakers were still so far integrated with the smiths—at least so far as large clocks were involved—that there is frequently the mention of a double qualification, as locksmiths and as clockmakers. When around the middle of the century the clockmakers began to constitute an increasingly more powerful and more self-confident group within the craft community,[5] the smiths complained repeatedly, insisting that the locksmiths' craft remained the foundation of clockmaking; in this way they strove to retain their position of leadership, which sought to exercise power partly by representation, partly by far-reaching arguments about rights. The other crafts incorporated into the smiths (four after the middle of the century: locksmiths, gunsmiths, ringmakers, and winch or windlass makers) formed a common front against the clockmakers. Thus in 1564 the clockmakers were accused of an elitist attitude in that they kept only to the articles of the common guild code that suited them but "cut loose from those they did not care for."[6] Specifically, the clockmakers interpreted the new law in the interests of dependents of masters by giving preference only to the sons and widows of clockmakers, not to the dependents of their guild colleagues. Similarly they resisted the custom of having their products stand proof by nonclockmakers, something that had been in effect up to 1564.

---

3 P. von Stetten, *Excerpta aus ainem alten Rechnungsbuche der bauMeister, welches von mir . . . 1778 . . . gefunden worden . . .*, Stadtbibliothek Augsburg, 4° Cod., p. 84. The *Geschichte der Heil. Röm. Reichs Freyen Stadt Augsburg* by von Stetten had gone to press 35 years before this discovery; it is to that history that the "earliest" date, within the 15th century, quoted in more recent literature goes back. Von Stetten's son published this discovery in a supplementary volume of his *Kunst-, Gewerb- und Handwerks-Geschichte* of 1788, p. 63.
4 Stadtarchiv Augsburg (hereafter STAA), Zunftbuch der Schmiede (SZB) for 1442, fol. 20v. These records were transcribed in 1569 from older ones.
5 ". . . now that in our day there have come to be as many clockmakers as there used to be locksmiths and gunsmiths," STAA, Handwerksakten Uhrmacher (hereafter U), I, fols. 77–82 (1564). See also Table 1 below.
6 STAA U I, fols. 61, 62.

Here, as in most of the other major legislative codes, it was primarily the masterpiece that was at issue. This was the central event in a clockmaker's career: except in rare cases, nonadmission deprived him of a livelihood. For an already established master this naturally meant one competitor the less and one cheap working hand more. Almost every decision or article directly or indirectly involved the masterpiece. Almost all the guild's wrangles—about apprentice time, journeyman time, residence time, marriage—were kindled by it. This makes it all the more surprising to find how late the masterpiece took legislative root in guild organization. Whereas this was usual as early as 1541 in Munich,[7] as late as 1569 in Augsburg there were masters occupying responsible positions who had to admit to never having produced a masterpiece at all. Thus ran the arguments of Jakob Marquart, a member of the established Augsburg clockmaking family, when he returned from far-ranging sojourns abroad and was confronted by the newly created masterpiece articles of the guild organization.[8]

It was not the clockmakers who promoted the introduction of masterpieces, as one might expect; instead, the locksmiths were "the first who cried out for the masterpieces."[9] But in the subsequent drafting of the articles it was the Augsburg clockmakers who took a leading part (not the least for their own interest) and thus provided a model for many cities. Outstanding among others was Nuremberg in modelling itself on the Augsburg regulations. The evolution of the clockmakers' craft within the guild of the smiths over the course of a century is clearly apparent in two hitherto-unpublished craft regulations of 1552–1554 and 1560.[10] The first is a compendium of general articles with few pronouncements affecting clockmakers in particular; the second shows a subdivision into makers of small clocks and makers of large clocks. Some of the provisions of these regulations will now be discussed, especially as they bear on the career of a typical clockmaker in Augsburg.

The fundamental prerequisite for learning the craft was proof of birth both legitimate and free, that is, unencumbered by obligations of serfdom. An apprentice must be at least twelve years old,[11] must have stood up during the fortnight's trial period, and must finally be proposed and registered by a master of the guild.[12] The duration of basic instruction, the apprentice period, is constantly prescribed as three years.[13] After that

7 Klaus Maurice, *Die deutsche Räderuhr* (Munich, 1976), Vol. I, p. 134.
8 STAA, U I, fols. 137 ff. Those assailed by Marquart were the so-called *Vorgeher* or foremen, who held the highest offices in the guild of the smiths taken as a whole. "Elder Masters" who had never produced a masterpiece are mentioned as late as 1582. *Ibid.,* fol. 328.
9 *Ibid.,* fols. 83–86 (Aug. 12, 1564).
10 Stadtbibliothek Augsburg, 2° Cod. Aug. 442, fols. 333v–344v (1552–1554). STAA, U II, fols. 385–392 (1650). Both regulations are reproduced in Appendix I and cited hereafter as Regulation of 1552/1554 and Regulation of 1650.
11 STAA, U II, fol. 455 (no year).
12 Regulation of 1552/1554; article of June 19, 1554.
13 E.g., July 29, 1564 (STAA, U I, fol. 65) and Mar. 19, 1650 (U II, fols. 385 ff.).

the apprentice was "absolved"[14] and could go out as an itinerant journeyman. The period between absolution and admission for the masterpiece
was the time when guild regulations were the most effective. Above all,
the number and the qualifications of journeymen and future masters could
be controlled. This was especially true after 1558, the year that the masterpiece was first declared mandatory.[15]

As early as 1552 three masters were selected from among the craft of
the smiths, sworn by the town council, and "instituted" as responsible
administrators and trustees of the guild regulations and as arbitrators and
treasurers.[16] "Then the inspection and evaluation of the piece at issue shall
take place prior to its being certified as a masterpiece."[17] The first point
of the regulation of the smiths that applies expressly to the clockmakers
has to do with limitation of the authorization to gild copper and brass.[18]
At first the inspection was entrusted to all the "prescribed masters" jointly.
In subsequent years the structuring and manning of the positions according to the laws became differentiated.

The whole guild of the smiths was headed by four "appointed foremen"
(*verordnete Vorgeher*) who were selected from among the ranks of all five
crafts. In order to judge the quality of work in contested cases, starting
in 1562 four "sworn inspection masters" (*geschworene Geschaumeister*) were
elected from each individual craft for approval of masterpieces; annually
on Saint James' day the one in service the longest was replaced by another.[19] It was only in 1564 that the clockmakers managed to establish sole
authority over decisions regarding their masterpieces in questions of admission, awarding, and inspection.[20] Thereafter the journeyman admitted
was assigned to the special care of one of the four inspection masters, in
whose shop he was to produce his masterpiece.[21]

Before the pieces were assigned to a journeyman, he had to satisfy
requirements which, in accordance with a flood of articles which accumulated particularly in the second half of the 16th century, were deci-
sively governed by consideration of the social origin of the petitioner.
Naturally the most propitious case was that of the son of an Augsburg
master clockmaker. After the three apprentice years he was prescribed at
first three journeyman years,[22] four later on, which he could pass where
he wished, "whether here, or outside the city."[23] Thus, quite in contrast

---

14 Up to the prohibition of 1650 it was customary to honor the new journeyman with a mock
    ceremony in the course of which "no one . . . could imitate the priest more wickedly than
    M. Niclas Rugedass" (*ibid.,* U II, fol. 422).
15 Dec. 1, 1558 (*ibid.,* fol. 2).
16 Regulation of 1552/1554: article of June 25, 1552. A "guild master" who was replaced
    annually is in evidence from as early as 1441 (STAA, SZB, fol. 20v).
17 STAA, U II, fol. 77.
18 Regulation of 1552/1554: article of Jan. 8, 1554.
19 STAA, Handwerksakten Schmiede (hereafter S), I, fol. 229.
20 *Ibid.,* U I, fol. 69.
21 *Ibid.,* S I, July 3, 1568.
22 *Ibid.,* U I, fols. 83–86 (1564).
23 Regulation of 1650: article 4.

to others, he was bound neither to itinerant journeymanship nor to many years of work in Augsburg. After six or seven years of training he could register for masterpieces. The law, in force from 1564 at the latest, to the effect that annually only one journeyman should be "admitted to the pieces,"[24] made special provisions for the sons of masters. As late at 1565 it was stated bluntly that sons of masters were to be admitted before newcomers.[25] Numerous written protests to the city council brought about a decision by lot which was more equitable, at least at the start,[26] although its execution was finally regulated as follows: whereas the alien journeyman had to draw lots at Christmas for the good fortune of admission, the sons of masters had a special term that lasted till Pentecost.[27] Given the large number of sons of masters (see Table 2), there were usually at least two "piece masters" each year, that is, journeymen who were working on their masterpieces.

The prescribed period for clockmakers was six months—as one might expect, a far more extended period than for the other crafts among the smiths. For comparison: ring and winch or windlass makers, four months, locksmiths and gunsmiths, three months.[28] During this time the piece master was to be engaged in no other work[29] and was not to be married.[30] After an inspection of the masterpieces that was satisfactory to the sworn delegates,[31] the young master had to fulfill two further conditions in order to qualify to have his own shop:

1. The prohibition against marriage now turned into to an unavoidable obligation. Even before the introduction of the masterpieces, marriage had become a precondition for working independently.[32] Above all, this rule had the purpose, in addition to imposing a certain selectivity,[33] of providing for a fair apportionment of citizenship duties such as the payment of taxes.[34]

2. He must possess the smiths' eligibility (*Schmiedegerechtigkeit*). With masters' sons this was no problem, as it was automatically inherited from

24 STAA, U I, fols. 69–72.
25 *Ibid.,* fols. 49 ff.
26 The development and course of these proceedings are described and documented in detail in Maximilian Bobinger, *Kunstuhrmacher in Alt-Augsburg* (Augsburg, 1969), pp. 14 ff.
27 Regulation of 1650: article 7.
28 STAA, U I, fols. 69–72.
29 E.g., Michael Chaspar, who along with his masterpieces had made a "striking" clock, was punished with a fine and confiscation of this clock. Hans Wörner, who during the masterpiece work had helped Hans Schlottheim with commissions for Duke Wilhelm V, had his masterpieces stopped. *Ibid.,* fols. 369 ff. (July 5, 1582).
30 Older ordinance (undated) quoted Oct. 31, 1564. *Ibid.,* fol. 69.
31 In 1602 an article was promulgated that had the purpose of preventing corruption in this field. *Ibid.,* U II, fol. 394.
32 In 1553 the bachelor Ulrich Klieber the younger was not permitted his own shop. STAA, 1530–1569, according to Maximilian Bobinger, *Alt-Augsburger Kompassmacher* (Augsburg, 1966), pp. 78–80.
33 Jeremias Schmidt, who did not marry allegedly because of a "bodily defect," secured only a special temporary license. STAA, U II, fols. 537 ff.
34 A master "must . . . do everything that pertains thereto, and help to bear all burdens as a burgher." *Ibid.,* U I, fol. 96, no date (1564–1566).

the father (this applied also to daughters) and normally became fully effective upon assuming the status of master (through payment of 1 gulden 8 kreuzer[35]) and being registered in the guild book of the smiths. In the 1550–1650 period there were only three sons of masters upon whom the smiths' eligibility was conferred as an exception while they were still in the status of journeymen.[36] The father had by no means any obligation—contrary to Maximilian Bobinger's understanding[37]—to give up his own workshop in connection with "inheritance" of the smiths' eligibility.

Given observance of all the laws and with luck in the drawing of lots, the son of an Augsburg master had a chance of becoming a master at the age of twenty, qualified to head a workshop. Among the unfortunately small number of biographies of masters' sons from which we know both the exact year of birth and the date of the smiths' eligibility, the youngest was twenty-three years old upon becoming a master, the oldest thirty-eight.[38] Thus the average age lies somewhere above twenty-eight (see Table 2). This result, at first glance rather surprising, surely springs—although partly from voluntarily extended itinerant years[39]—primarily from the fact that despite all preferential treatment, even a master's son had a long period to wait before being licensed to make a masterpiece, since the number of his competitors at his own level was considerable (see Table 2).

The corresponding average age for newcomers to the guild or even to the locality lay between three and four years higher. After the three apprentice years they had to be active as journeymen for seven more years, of these at least three with one or two masters at Augsburg. If a newcomer appeared with ten years training behind him—that is, a completed term —he nevertheless had to work out the further three years of residence with one or at most two Augsburg masters.[40] This last limitation was in existence by 1564[41] and was supposed primarily to counteract the luring away of journeymen. Over the whole period under study journeymen were in great demand, and apparently there were always far too few available.[42] Surely this situation resulted in a great part of the articles

---

35 Regulation of 1552/1554: article of Dec. 12, 1553.

36 ". . . and though he would wish to receive the craft eligibility, neither he nor the others may do so in that men of the craft may not work alone save the craft eligibility shall have been conferred through lease." STAA, U II, fol. 139 (1569).

37 Bobinger, *Kunstuhrmacher,* p. 80. See also STAA, Musterregister, 1610 ff.

38 The enumeration is chronologically as follows (distribution is relatively regular) from 1550 to 1650, from which one can see that there is no obvious connection between age at eligibility and the overall course of time: 26, 24, 27, 29, 24, 36, 27, 31, 38, 26, 30, 29, 27, 27, 33, 23, and 26 years.

39 Concrete prescriptions for the itinerant years were not formulated. But from statements like the following, one sees that they were attributed great importance: "they may . . . not be abbreviated . . . as regards itinerancy, in the interest of greater experience in the art." STAA, U I, fols. 78 ff. (1564). ". . . if one is not practiced in the craft then one will have had no experience or will have learned but poorly." *Ibid.,* U II, fol. 312 (1644).

40 Regulation of 1650: article 2.

41 STAA, U I, fol. 84 (1564).

42 E.g., *ibid.,* fols. 319, 321, 336, 339.

prolonging journeyman time. To ensure a fair apportionment, it was decided in 1582 to limit each workshop to one apprentice and two journeymen (masters' sons not included), something that greatly hampered the acceptance of large commissions and frequently gave rise to long-drawn-out negotiations.[43] Subsequently there was an "assignment master" (*Umschick-Meister*) who had to assign the newly arriving journeymen.[44]

Thus after at least ten but up to thirteen journeyman years,[45] the newcomer could register in the Christmas term for the drawing of lots for the "piece-master." If he was lucky, he could at the age of twenty-two to twenty-five commence production of a masterpiece. If the lot fell to another, he must wait at least a year, but mostly quite a bit longer, and he was kept on as a journeyman in Augsburg, for it was forbidden for him —again in contrast to the masters' sons[46]—to work outside Augsburg.[47] There was, however, an escape route which circumvented some of these handicaps: engagement to the daughter or widow of an Augsburg master was rewarded with the waiver of three years of residence.[48] Even if the masterpiece was unsuccessful and was turned down, the waiting period of half a year before a new attempt, otherwise prescribed, was dispensed with under these circumstances.[49] The purpose of this special rule is clear: "there has been a consideration for widows, so that wife and children might be best provided for"[50]—a proto-social-security for widows and orphans. It is true that widows might in many instances continue to run the shop for some time, but upon expiration of the permitted term, remarriage became necessary for their survival. For out-of-town or out-of-guild journeymen such marriages provided several opportunities: to achieve master status without waiting as long as others, to take over an already established workshop, and, not the least important, to acquire smiths' eligibility through the marriage. The smiths' eligibility of the *Schweher* (father-in-law, but not brother-in-law; *Schwiegervater, nicht Schwager!*)

---

43 *Ibid.*, fols. 355 ff. (Jan. 4, 1582). The first dispute came as early as 1582: the "old masters" had put the new article through and had constructed for themsleves the escape clause that they could count their workshops as double, one for large clocks and one for small clocks, and could thus justify a double number of journeymen. The younger masters among the small-clock makers, already trained in more specialized fashion, protested in vain. One of their petitions, in which they name as their grounds the demanding and urgent work for "potentates" and especially for the Turkish honorarium, is signed by Ulrich Klieber, Hans Connat, Hans Schlottheim, Andreas Rain, Georg Walter, and Georg Degen. These six makers of small clocks were outvoted by the remaining clockmakers (apparently 40). *Ibid.*, fols. 327 ff.

44 *Ibid.*, U II, fol. 122. See also Regulation of 1650: article 16.

45 Following are some figures from the cabinetmakers' craft, for payment and working hours for journeymen: "for a journeyman per week, 5 to 6 batzen wages [= 20 to 24 kreuzer], plus board, bed, tools, and other subsistence"; "on a summer day they are to work 12 hours and also part of the time over that, since on a winter day they do not work over 8 hours." STAA, Handwerksakten Kistler (hereafter K), I, fols. 92r and 94r (1557).

46 Regulation of 1650: article 9.

47 STAA, U II, fols. 121–125 (1626).

48 *Ibid.*, S I, fol. 356 (Nov. 29, 1565), U I, fol. 35 (1566), and fols. 65–68 (1564).

49 *Ibid.*, U I, fols. 69–72 (1564).

50 *Ibid.*, fol. 79 (1564).

or of the *Vorfahrn* (the deceased husband of the wife) could be acquired through marriage (*Erheurat.*)[51] The fee at marriage was 8 gulden 4 kreutzer and 16 pennies to begin with. Journeymen related neither by blood nor marriage to the guild had, with all the other disadvantages, to "purchase" the smiths' eligibility at a cost of 16 gulden 8 kreuzer and 16 pennies.[52] Over a hundred years this outlay remained almost constant, while the cost of acquisition through marriage was reduced by 1639 at the latest and was made the same as "inheritance."[53] The "grain-gulden" (*Kornguldin*) introduced in 1624 was also deducted from the basic fee in the cases of marriage and inheritance.[54]

Tables 1–4 and Fig. 28 present some statistics on clockmakers practicing in Augsburg, which in turn reveal some of the effects of the various regulations discussed above. Tables 2–3 deal with the period 1550–1650, an interval which also includes clockmakers who acquired smiths' eligibility before 1550 and were thus active during this period. The upper terminus is oriented toward the dates 1650 for smiths' eligibility, 1630 for birth. This explains the differences that arise in comparison with Table 1 and Fig. 28. Tables 2–3 also include smiths' eligibility evidenced in numerous cases other than entry in the Zunftbuch der Schmiede.

Table 1. Smiths' eligibility acquired by clockmakers in Augsburg, 1500–1700 (from Zunftbuch der Schmiede).

	Number of clockmakers	Percentage of total
Time period		
1500–1550	16	7.84
1550–1600	81	39.70
1600–1650	70	34.31
1650–1700	37	18.13
	204	99.98
Type		
Inherited	72	47.68
Through marriage	43	28.47
Purchased	36	23.84
	151	99.99

51 These conclusions have been drawn from a comparison of numerous biographical data with entries in the SZB. Smiths' eligibility can also be acquired through marriage and passed on by a person not pursuing any one of the crafts incorporated into the smiths' guild. But in such cases it is merely "loaned" to him (see n. 36).

52 Regulation of 1552/1554: article of Dec. 12, 1553.

53 STAA, SZB, fol. 197b (Nov. 13, 1639). Dating by means of this *terminus ante quem* is too imprecise to allow more extensive conclusions, e.g., as to the number of widows.

54 *Ibid.*, fol. 181b (Feb. 25, 1624).

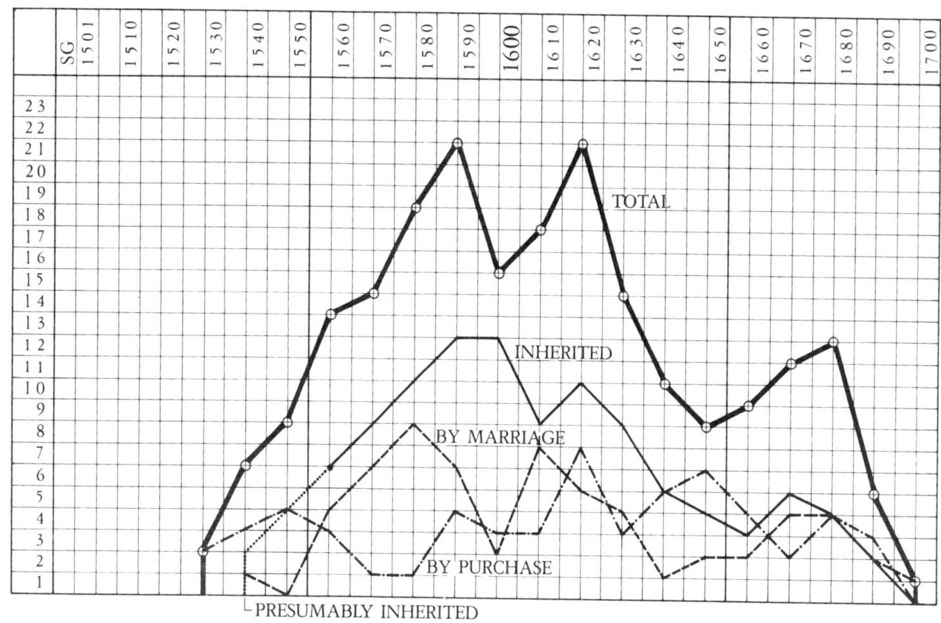

28. Graph showing the acquisition of smiths' eligibility by Augsburg clockmakers.

Table 2. Clockmakers in Augsburg, 1550–1650

Total number	284
Without smiths' eligibility	102 (35.91%)
With smiths' eligibility	182 (64.04%)
Inherited, at 28.41 years (av.)	88 (48.35%)
Clockmakers' sons . . . 51[a]	
Locksmiths' sons . . . 32	
Others . . . 25	
Through marriage, at 29.76 years	46 (25.27%)
Purchased, at 32.33 years	45 (24.72%)
No indication	3 (1.64%)

Religion
Protestant . . . . 165 (87.30%)[b]
Catholic . . . . 24 (12.69%)[b]
Unknown 95

Average longevity
Computed from exact birth and
death dates for 26 clockmakers . . . . 68.96 years
Clockmakers with recorded smiths'
eligibility . . . . 65.71 years

Average working life
Computed from exact records of
smiths' eligibility and death
dates for 47 clockmakers . . . . 35.55 years

[a]Based on reliable tradition and limited to sons of Augsburg clockmakers who were themselves active in Augsburg. If one includes sons of clockmakers who received smiths' eligibility while still journeymen and clockmakers' sons whose fathers were not active at Augsburg, the number is 66 (75%).

[b]This is the percentage of those whose religion was known.

Table 3. Origin of clockmakers in Augsburg, 1550–1650

	Augsburg	150 km radius	Beyond 150 km	No indication
Overall (284)	138 (48.59%)	46 (16.19%)	37 (13.02%)	63 (22.18%)
With smiths' eligibility (182)	115 (63.18%)	26 (14.28%)	16 (8.79%)	25 (13.73%)
Through marriage (46)	8 (17.39%)	18 (39.13%)	11 (23.19%)	9 (19.56%)
Purchased (45)	16 (35.55%)	8 (17.17%)	5 (11.11%)	16 (35.55%)

Table 4. Some comparative numbers of craftsmen in Augsburg, 1555–1669

Year	Number	Type	Source of data: Stadtarchiv Augsburg
1555	150	locksmiths, gunsmiths, clockmakers	Handwerksackten Schmiede I, fol. 109
1560	130	cabinetmakers	Handwerksakten Kistler (K) I, fol. 135r
1582	46 and 15	clockmakers and journeymen	Handwerksakten Uhrmacher (U), I, fol. 319
	45	small- and large-clock makers	U I, fol. 317
	40	clockmakers	U I, fol. 336
1589	130	cabinetmakers	K III, fol. 72
1594	130	cabinetmakers	K III, fol. 275
1610	40	clockmakers	Musterregister
	115	cabinetmakers	
	187	goldsmiths	
	7	turners	
1615	30	clockmakers	
	39	cabinetmakers	
	9	goldsmiths	
	57	turners	
1619	42	clockmakers	
	106	cabinetmakers	
	186	goldsmiths	
	20	turners	
1631	15	clockmakers' shops	U II, Jan. 30, 1631
1644	6	large-clock makers	U II, fol. 321
1645	7	clockmakers	Musterregister
	23	cabinetmakers	
	45	goldsmiths	
	4	turners	
1652	3 and 25	master clockmakers and journeymen	U II, fol. 489
1669	70	cabinetmakers	K VII, fol. 61v

After this treatment of biographic and statistical factors bearing on Augsburg guild life, we will turn to the clocks themselves, beginning with the first work that a clockmaker is fully responsible for—the masterpiece. The specifications for this are known from 1558, the year of the statutory introduction of masterpieces.

Masterpieces of 1558:
Small-clock makers: A clock a span high, without weights, to strike each quarter-hour. The astrolabe runs as part of the clock. A small flat clock or a spherical clock with the phases of the moon; the latter to move forward in time with the hand.

Large-clock makers: A tinned clockwork mechanism striking each quarter-hour, with sun and moon running through the twelve signs of the zodiac.[55]

In December 1577 the inspection masters for the clockmakers, supported by the foremen of the guild, called for the introduction of new masterpieces, because the old ones had been "so sketched off and copied from that, what with tracing and copying [the pattern], everyone knows the why and how of them."[56] The construction of masterpieces had thus been so facilitated that anyone who had completed his apprentice term could manage them easily, and "it was no longer any art at all."[57] In order to obviate these abuses, five new masterpiece tasks were drawn up, to be revealed to the journeyman only shortly before the start of his work. In addition to the "great piece" selected from the five choices, a pendant clock (*Halsuhr*) was also to be produced, as before.

Masterpieces of 1577:
1. A clock of the dimensions as hitherto, about a span high, which strikes the hours and the quarters. It shall also have an alarm and shall likewise show the astrolabe, the length of the days, the calendar, and the planets with their signs. When the quarter-hand is moved, all hands shall move in time with it, and in addition the clock shall strike the hours both to 12 and to 24, as one may select.

2. A clock of the sort called mirror (*Spiegel*), which shall show all things and strike as with the above-mentioned clock.

3. A squared clock, also about a span high, that likewise indicates and strikes as in the case of clock No. 1.

4. A hexagonal clock, also about a span high, strikes hours and quarters, also alarms, and shows the astrolabe, the length of the day, and the planets.

5. A squared standing clock (*Stotzen*) that strikes the hours and the quarters, and alarms. On all its four vertical faces it shall have separate indicators, and on top it shall show the astrolabe or the length of the day. And when the quarter-hand is turned, all the others shall move with it.

In addition, along with each clock that is referred to above there shall be made a clock showing the phases of the moon or a little sphere, as has been usual hitherto.[58]

55 STAA, Ratsprotokolle, Dec 10, 1558, fol. 93v, cited in Bobinger, *Kunstuhrmacher,* p. 16.
56 STAA, U I, fol. 377.
57 *Ibid.,* S II (Dec. 5, 1577). Cited in Bobinger, *Kunstuhrmacher,* pp. 66, 67.
58 STAA, U I (dec. 7, 1577). Appendix to the Schlottheim-Wörner suit cited by Bobinger in

The list of masterpieces of 1577 was destined to have a long life—it was in effect for 155 years. The enumeration of the obligatory masterpieces for the year 1732 is identical.[59] The archival sources have thus fully confirmed the range of clocks of standardized types compiled by Maurice for 1616–1703 and interpreted as craft masterpieces.[60]

The journeymen repeatedly petitioned to have the masterpiece specifications updated, mainly for financial reasons. Thus, for example, they complained in 1702 that they have antiquated (*altfränckische*) and hence unsalable masterpieces.[61] High manufacturing costs combined with limited sales prospects amounted to "uncertain gambling."[62] A petition of 1725 mentions a total cost of 217 gulden for completing a masterpiece, which comes pretty close to the value of 200 gulden specified by Hans Fronmiller the younger for his masterpiece of 1584.[63] The 217 gulden are broken down as follows:[64]

One needs brass around	fl. 20.
Silver according to how one wishes to ornament the work and how big one wants the dial to be; but one can get by with	15.
engraving costs around	36.
chains and springs	9.
the two bells	5.
steel, around	5.
gilding, including mercury and cream of tartar	30.
for dividing the astrolabe	3.
for an extra seat and the masterpiece assignment involved	7.
three inspections, total	3.
inspection of springs	3.
one's promulgation as a master	9.
added . . . for board	64.
rent to one's master for room, 8 months	8.

Kunstuhrmacher, p. 17. The regret Bobinger expresses (pp. 16–17) over the loss of the regulations of 1667 and 1677 "as a whole," only individual articles still being at hand, is to my mind needless, as this "whole" was presumably not created anew in these years, but the old regulation (see Regulation of 1552/1554 in Appendix I) was constantly augmented through new articles, the enumeration of which, it is true, had become very confusing after 1650.

59   STAA, U III, fols. 390 ff., July 29, 1732. See the complete text of 1732 in Appendix II. From another passage it transpires that 8 months are now allowed, without cash fine, for the same masterpieces that before were allowed only 6 months working time. *Ibid.*, fol. 391 (1732).

60   Maurice, *Räderuhr*, Vol. I, p. 98; all with illustrations in Vol. II.

61   The response is to the effect that "it is in fact currently permissible to add perpendiculars and other ornaments calculated to improve sales, in accordance with present fashion." STAA, U III, fols. 31 ff. Permission to install masterpieces in wooden cases was until 1703 a privilege accorded sons of masters. The petition of a "stranger to the guild" to enjoy this permission as well describes products already completed by cabinetmakers and others. *Ibid.*, fols. 64 ff.

62   *Ibid.*, fol. 391.

63   See Appendix II.

64   STAA, U III, fol. 301r,v.

to the master's wife at the celebration; although
voluntary for journeymen, she ought to be satisfied
with one thaler, but usually   8.
Case can indeed be made expensive and fine, but also
for 8 or   10.

Total	fl. 217.

A calculation of costs by Fronmiller might have looked very much the same, all in all. Two items would definitely be different: first, in 1584 he would certainly not yet have been able to buy the clock springs ready-made.[65] Second, in Fronmiller's day it would have been out of the question to house his masterpiece in a case for 8–10 gulden. The latter subject, the production of clock cases, and the adjunct crafts, will next be investigated.

For the production of "large clocks"[66] the number of auxiliary crafts was limited: in a large-clock maker's account for 1616, out of a total of almost 3,000 gulden there is mention only of paying a blacksmith and the eight journeymen working with him;[67] in another case paying a sculptor for "a little human figure to show the hour."[68] Presumably the "hand-makers"[69] worked on tower clocks as well. For "small clocks," however, because they required cases, a number of other crafts were involved.[70] Two groups of participating craftsmen will be considered: those working in wood and those working in metal.

The most important representatives of the wood workers were turners and cabinetmakers. The latter offered serious competition to the turners, because they were constantly exceeding the concession granted them in 1589[71] which permitted them to turn ebony but only to ornament their own output; that is, they performed work not appropriate according to the laws of the guild. The cabinetmakers, however, were principally responsible for the making of wooden cases even when accessory parts were bought from turners,[72] and it was to them that the clockmakers addressed

---

65 The first man to specialize as a maker of clock springs was Abraham Scheurlin. He had earlier been active as a maker of large and small clocks, and he received special permission to take up this production on account of age. His idea soon found imitators, e.g., Elias Pfaff in 1682 (*ibid.*, U II, fols. 811 ff.) and Jakob Blessing in 1694 (fols. 991 ff.), who speaks of commissions from England and Italy and of merchants who "send springs out in quantity."

66 This means (mechanical) tower clocks, a classification that did not, however, exhaust the field of activity of the "large-clock maker." That field comprised all clocks with weights, irrespective of size.

67 STAA, Bauamts Berichte, 1616–1625: June 21, 1616—restoration of the great clock on the Perlach Tower by the large-clock maker Hans Schlem.

68 Sebastain Loscher, 1516. STAA, Buff-Regesten, "Bildhauer," fols. 1, 2, 6.

69 STAA, "Beschreibung der Stadt Augsburg," Musterregister 1610, etc.

70 See the enumeration of the craftsmen participating in the masterpiece of Caspar Langenbucher prior to 1650, Catalog No. 45

71 STAA, Handwerksakten Kistler (K) XXIV, fol. 30r.

72 At times also from merchants who imported work made from woods unavailable in Augsburg, from Esslingen and Göppingen, as in 1567: STAA, Kramer-Akten, I, May 13, 1567.

their orders.[73] There are individual instances where the cabinetmakers, faced with major commissions from the clockmakers, sought permission to expand the notal number of their journeymen.[74]

Within the metalworking crafts that produced most of the housings for Augsburg clocks of the period studied,[75] matters were more highly differentiated. Along with the silver-turners, who slowly parted company with the wood-turners and who increased in number rapidly in the first half of the 17th century,[76] the coppersmiths, and the brass-founders, it was the goldsmiths who played the most prominent role. The goldsmith journeymen, whose work was less expensive because they were not yet independent, were preferred by the clockmakers, yet toward the end of the 16th century the journeymen were repeatedly forbidden to do any such work.[77] On the other hand, up to 1560 the clockmakers had been allowed two goldsmith journeymen through express permission.[78] The frequently attempted specialization (five specialists are named for 1567)[79] in many cases ended with the "clock-case makers"—who took over the main etching, engraving, and cutting—having to settle outside Augsburg.[80] In any event collaboration with the clockmakers was made very difficult. So the clockmakers had frequently to turn to local master goldsmiths despite the

73 STAA, U III, fols. 64 ff.
74 Thus Jakob Schaur on July 14, 1584: "but now I have been commissioned by the Honorable Georg Rollen [*sic*], local citizen, to make a large and intricate clock case." STAA, K III, fols. 1 ff.
75 *Ibid.*, U III, fols. 64 ff.
76 *Ibid.*, Handwerksakten Drechsler (hereafter D), I, Feb. 26, 1643, and July 1, 1651.
77 A. Weiss, *Das Handwerk der Goldschmiede in Augsburg bis zum Jahre 1681* (Leipzig, 1897), pp. 145 ff. STAA, Handwerksakten Goldschmiede (hereafter G) II, March 23, 1568; Jan. 23, 1560 (Hans Baiss of the Netherlands); Jan. 23, 1564 (Jörg Kraus from Frisia); Sept. 24 and Oct. 4, 1569 (Hans Bossierer, who had already worked 30 years for Augsburg clockmakers, and also for the compass maker Christoph Schissler).
78 STAA, G II, Jan. 23, 1560.
79 Some specialists are listed below:
"Rot von Niernberg" and his journeymen (G II, June 17, 1567).
Mattheus Mair: Before 1571, collaboration with Christoph Schissler (Dresden Hauptstaatsarchiv, Loc. 4418, Bk. 1, 1st half, sheet 226; see Maximilian Bobinger, *Christoph Schissler,* Augsburg, 1954, p. 30). About 1602, collaboration with Hans Schlottheim (STAA, Ratsprotokoll, March 28, 1602).
Georg Schöttlin: Clock-case maker of Lechhausen who, according to a communication of Nikolaus Rugendas and Hans Buschmann, had been of great service to the Augsburg clockmakers and was allowed to live temporarily at Rugendas' house. May 3, 1632. (STAA, U II, fols. 215–220). Evidently there were three case makers in that family: "clock-case makers for a hundred years since, namely Gëorg Schötlen the elder, Kleebühler, Gëorg Schlötlen the younger, and the still living son of the last were all established here." This is the reasoning for the recommendation of a working permit for the following man, Kayser.
Ludwig Kayser (1630–1706): The best worker on cases "delicately pierced and engraved." Against the objection of the goldsmiths it was rejoined that "he makes them hold up as well as anyone could desire," so that a question of stability was involved. According to the supporting signatures, Kayser worked for Jakob Widenmann, Elias Weckerlin, Hans Sommer, Johann Buschmann the younger, and Wilhelm Pepfenhauser. Aug. 31, 1662 (STAA, U II, fols. 617–624). By this time it was easier to put specialization across; in the Augsburg register of deaths Kayser's trade is specified as "clock-case maker." (A. Haemmerle, *Ev. Totenregister zur Kunst- und Handwerksgeschichte Augsburgs,* Augsburg, 1928).
80 E.g., at Lechhausen (STAA, G II, Apr. 3, 1578) and Oberhausen (Apr. 16, 1562).

bemoaned high prices and long waiting time,[81] something which was probably always true about intricate embossed figurework. That good collaboration between these two crafts[82] was usual and more frequent than one might suppose from the accumulated records of lawsuits is also demonstrated by the fact that as early as 1578 six master goldsmiths were already specializing on clockmaker commissions.[83] Of course, in the 16th century and also in the 17th individual clockmakers are named in repeated instances as producing their own cases.[84]

A special position is occupied by the trade of the brass-founders, who were more and more hemmed in by the goldsmiths. Finally in 1588 only a group limited to seven masters was allowed to continue working in open shops, with the further condition that they might cast only for clockmakers.[85] The series production demonstrated by Maurice for the cast figural components of cases was presumably supplied by the shops of these brass-founders, insofar as the articles were not imported into Augsburg.[86] A hint at mass production also lies in the astonishingly low price, which within a total bill of more than 300 gulden for a masterpiece constitutes an item barely worth mentioning: "on top an image of brass," for 8 gulden.[87]

Coppersmiths who were incorporated into the hardware (*Kramer*) guild at Augsburg[88] played a surprisingly minor part. On the one hand they entered complaints against the smiths' guild that the latter were buying up brass and copper for casting,[89] and on the other hand they themselves

81 *Ibid.,* Sept. 24, 1569; March 30, 1578.

82 Examples:

Hans Helmbrecht, goldsmith and embosser, is supported by six clockmakers for whom he has just commenced work: Caspar Buschmann, Jeremias Metzger, Hans Zeller, Simon Friderich, Matthäus Stosswender, and Philipp Müller. Feb. 7, 156 (STAA, U I, fols. 131 ff.)

The goldsmith *Hans Wergler* and the clockmaker Thomas Geiger received in joint account a payment for two clocks they made for the Emperor Rudolf II. July 30, 1594. *Jahrbuch der Kunsthistorischen Sammlung,* Vol. VII/2 (Vienna, 1888). Reg. No. 5540, Hofzahlamtsrechnung 1594, fol. 33v. Cited in Friedrich Streng, "Augsburger Meister der Schmiedegasse um 1600," *Blätter des Bayerischen Landesvereins für Familienkunde,* 1963, 26· 276.

The goldsmith *Jakob Knoll* received from Hans Schlottheim a certificate, with reference to his collaboration with clockmakers, to the effect that, his high prices aside, "he was entirely qualified to perform such work." Sept. 26, 1577 (STAA, U I, fol. 204). One should assume that Schlottheim speaks from his own experience, even though in other cases he employed goldsmith journeymen or made clock cases himself. Presumably Jakob Knoll also worked with Georg Roll; after Roll's death he assumed guardianship over the bereaved family (Bobinger, *Kunstuhrmacher,* pp. 40, 46).

83 STAA, G II, April 3, 1578.

84 *Ibid.,* Jan. 27, 1560. May 20, 1568: Christoph Kitzmögel. STAA, papers of Maximilian Bobinger: Hans Jacob Schönauer (1575, clockmaker and clock-case engraver) and Johann Sommer the younger (small-clock maker and clock-case maker, 1662). STAA U I, fol. 203: Hans Schlottheim: "for which he himself made the case with his own hands" (Sept. 26, 1577).

85 Weiss, *Handwerk der Goldschmiede,* pp. 114 f. Jakob Knoller is named. They were also allowed to process their product by turning. STAA, D I, Dec. 22, 1609.

86 Maurice, *Räderuhr,* Vol. I, pp. 99 ff.

87 STAA U III, fol. 456v. No indication of year, yet still dealing with the masterpiece as applicable to our period.

88 STAA, G I, Sept. 6, 1567.

89 *Ibid.,* G II, Nov. 4, 1561.

traded in ready-made articles from competing Nuremberg.[90] Painters were also brought in to decorate the clock cases; they had permission, alongside the goldsmiths, to etch on silver and other metals.[91] In 1602 a decision of the Senate took effect which obliged the Augsburg nail-makers to provide for the clockmakers, among other things, fastening pins since importation from Carinthia and Styria had been prohibited.[92]

The extensiveness of trade connections is also highlighted by archive material that deals with the production of clock bells. It was reported in 1596 that it had earlier been customary to bring clock bells from France, predominantly from Lyon. The Augsburg turners sent patterns there and got back rough-cast bells,[93] which they finished themselves—a division of labor bridging a great geographical expanse. By reason of disturbances to this trading connection through war, and also because of the alleged inability of the turners to turn—that is, to tune—the bells as desired, a dispute arose between turners and clockmakers. The clockmakers then had a specialist trained[94] in order to cut out the turners (it was even said in some quarters that they had taught the turners how to tune bells[95]).

The packing-case makers, part of the bookbinding craft, were engaged in the safe and attrative packaging of clocks. The importance of these craftsmen—particularly for shipments, such as the Turkish honoraria—was emphasized in the letters of recommendation by clockmakers in connection with proposals for the granting of privileges to them.[96]

While describing the exportation of Augsburg clocks we must by no means neglect the trade within Augsburg itself, which should not be underestimated. Here the clock was not only considered an exclusive luxury item, but was also on the brink of becoming an article of extensive everyday use. Purchasers were already recruited from among many social strata. This tendency is strikingly apparent in the statement of Georg Roll, who worked mainly for the imperial and princely courts, to the effect that he also made deliveries to farmers.[97] Evidence of the relatively widespread use of household clocks is, aside from a remark of a clockmaker on this point,[98] the important part played by "domestic and chamber clocks"[99] in the negotiations of large-clock and small-clock makers about

---

90 H. R. Weihrauch, "Studien zur Süddeutschen Bronzeplastik," *Münchner Jahrbücher der bildenden Kunst,* 1952/1953, 3/4: 199–219, esp. pp. 214 f.

91 STAA, G II, Nov. 13, 1561.

92 *Ibid.,* Handwerksakten Kramer VIII, May 14, 1602.

93 Rough castings came from Nuremberg as well. STAA, D I, Nov. 19, 1596. As early as 1559/1560 quarter-hour bells were imported from Nuremberg and Innsbruck, because "when Maisperger cast four bells none was any good." Stadtbibliothek Augsburg, 4° Cod., p. 84.

94 Elias Mair. STAA, D I, Dec. 7, 1596–Jan. 16, 1597.

95 *Ibid.,* Dec. 7, 1596: "Old Mr. Guetermann and S. Boschmann" were named. This is probably an insinuation against the turner journeyman Hans Artzendorfer, who was turning bells for clockmakers around 1575 (STAA U I, fol. 153).

96 STAA, Handwerksakten Kramer, IX, Oct. 1, 1668.

97 Bobinger, *Kunstuhrmacher,* p. 30.

98 STAA, Bauamts-Berichte 1570–1601, Dec. 17, 1595.

99 STAA, U II, fol. 527 (Nov. 5, 1564).

their respective competency.[100] Large clocks were not limited to a few prominent buildings but were also to be seen at a hospital, for instance, and in private houses as early as 1577.[101]

There is also evidence of the use of a sort of "stopwatch" in Augsburg at an extremely early date. On the occasion of a shooting contest on October 6, 1567, it was announced on the invitation that "a small clock will be set up at the designated target point and if one or more persons fire a shot after the timing signal of this clock, and even if they hit the target, it will not be counted."[102] As early as 1558 small round striking clocks worn as pendants (*Halsuhren*) were in favor as personal accessories among elegant young men.[103]

If one wanted to purchase a clock, he could go to the clockmaker's shop and pick one from the stock available, for as early as 1562 it was quite usual to produce clocks for stock. In most cases the workshop and the retail store of the clockmaker were in the same building.[104] Aside from selling over the counter, clockmakers could offer their wares at weekly markets and annual fairs, but the door-to-door sale of clocks was forbidden, both to them and to others.[105]

Strong competition for the clockmakers arose from among the ranks of those working in ancillary crafts, first the goldsmiths and then in the second half of the 17th century also the engravers and others. Until 1578 the right to sell was expressly limited to the clockmakers.[106] The continuing efforts of the goldsmiths to gain access to the clock trade were finally successful in 1603,[107] so that now a controlled commercial interchange on a basis of placed orders ensued. However, the authorization included the proviso that the goldsmiths were to work on and sell clocks exclusively of Augsburg manufacture.[108] In this way Augsburg clocks came to the Frankfurt fair and the markets of Leipzig and Linz, which were probably attended by clockmakers but demonstrably by Augsburg goldsmiths.[109]

---

100 From 1558 onward. *Ibid.*, fols. 1–4 (Apr. 22, 1608).

101 Clock by Benedikt Marquart the younger at the "Saint Sebastian sickhouse." STAA, Wochenbuch 1593, July 11. Striking clock at Hans Schlottheim's house. STAA, Bauamtsprotokolle, Oct. 10, 1577.

102 STAA, single sheet dated July 29, 1567, but otherwise unkeyed, at the front of the weekly report for 1595–1596 but not belonging to it.

103 P. von Stetten, *Kunst-, Gewerb-, und Handwercks-Geschichte der Reichs-Stadt Augsburg,* 2 vols. (Augsburg, 1788), cited by Bobinger, *Kompassmacher,* p. 82.

104 "Nor shall anyone allow any journeyman to work outside the house; but if a man has a *permanent sales shop,* he shall have work done not in the house and the shop, but in the shop alone or else in the house, and only in the one place, whichever he wishes." Senate resolution of Jan. 8, 1562. STAA, S I, fol. 234. See also the passage in a letter of Johann Buschmann the elder to Schloss Ambras in 1637: "I have nothing in the way of indicator-clocks and alarm-pieces ready-made at the moment, as I constantly have to make works commissioned by others." Cited in Bobinger, *Kompassmacher,* p. 142.

105 STAA, Handwerksakten Kramer III. Senate resolution of Aug. 6, 1580, as appendix to a suit dated Dec. 4, 1599.

106 STAA, U II, fol. 75 (Aug. 19, 1578).

107 Weiss, *Handwerk der Goldschmiede,* pp. 145 ff.

108 STAA, K IV, fols. 110r,v.

109 *Ibid.*, D I, Feb. 28, 1651.

The reverse case of foreign clocks being sold at Augsburg was permitted only at the two fairs of Saint Michael and the "Translatio Vdalrici." The prohibitions against bringing alien clocks into Augsburg, formulated in various articles,[110] found their first and ever-recurring impulse in the infringements by colleagues from Friedberg, a neighboring town to be sure, but foreign territory both politically and in relation to guild matters. The situation reached its most acute phase when the whole Friedberg clockmaking colony had to be taken in by Augsburg in 1648, "in exile until more peaceful times." There was even black-marketing and counterfeiting, in which both Augsburgers and Friedbergers shared. The cheap Friedberg wares were given Augsburg names guaranteeing quality, thus providing a considerable margin of profit.[111]

The greatest problem which Friedberg presented to the Augsburg clockmakers was Georg Roll and his idiosyncratic ideas and claims. Born in Silesia, Roll had come to Friedberg in 1560 as a journeyman clockmaker and had settled in 1578 at Augsburg. Difficulties with guild officials had begun even earlier, and they were not to cease until his death in 1592. His combination of outstanding artistic talent and mercantile ability contributed in good measure toward establishing the renown of the Augsburg clockmakers' art. Despite this, his activity was hampered—in no small part because of competitive fears—by rigid adherence to tradition and guild regulations. Roll was neither the first nor the last who had to fight against such restrictions in order to put his ideas into practice. Even Augsburgers by birth, like the artist-engineer Johann Philipp Treffler about a century later, were restrained. Here we find displayed the negative side of guild organization, which indeed guaranteed a standard of quality but which also had a levelling effect. Officially Roll was entered among the hardware merchants and built up an extensive factory and international trade enterprise, while his talents as a clockmaker have made him famous.[112]

In additional to Roll there were other clockmakers who struck out as entrepreneurs, as the unpublished court records of Hans Ferdinand Mehrer show.[113] As a journeyman Mehrer carried on export trade to Italy with the participation of his master, Johann Ott Hallaicher. In 1639–1640 Mehrer acquired finished and semi-finished works of other masters at Augsburg, shipped them to Venice with his and Hallaicher's clocks, and there completed and sold them. This undertaking came to an early end through interventions on the part of guild officials, as Mehrer had neg-

---

110 From 1565 onward. STAA S I, fol. 361. See also Regulation of 1650: article 18.
111 To the accusation of falsification, the Friedberg men replied that Augsburg clockmakers had bought their finished and semi-manufactured output and had their own names or those of other masters engraved on them. (Oct. 15, 1648, STAA, U II, fol. 363.) As early as Oct. 29, 1641, a Senate resolution had been made against the Friedberg clockmakers with a prohibition against "engraving the names of our local masters on their work and through such forgery, selling their clocks as Ausburg work." (*Ibid.*, fols. 295–297.)
112 See Bobinger, *Kunstuhrmacher*, pp. 29 ff., and the recent treatment, with bibliography, by Maurice, *Räderuhr*, Vol. I, pp. 168 ff.
113 STAA, U II, fols. 266 ff.

lected to secure authorization in proper form. This journeyman, in whose interest no imperial letter of recommendation interceded, as in the cases of Roll and Treffler, had to consider himself lucky to be allowed to make his masterpiece and not to be ejected completely from his Augsburg career.

Large-scale export transactions were handled through the mediation of merchant princes, governmental administrators, and art agents. The role of these middlemen was often not confined to commercial matters but also included far-reaching influence upon formal and indeed even technical considerations.[114] The following were the most significant of these intermediaries: the Fuggers,[115] Zacharias Geizkofler, Georg Ilsung, and Philipp Hainhofer.

Georg Ilsung (1510–1580), imperial counselor and governor in Swabia, was the middleman in delivering numerous Augsburg clocks to the Emperor, whose need was enormous by virtue of the Turkish honorarium[116]: "in this way a great number of clocks of each and every sort have been produced and supplied here for the lord deputy Ilsung in the name of His Imperial Roman Majesty our most gracious Lord."[117] Zacharias Geizkofler (1560–1617), appointed imperial paymaster in 1589, took over this function as middleman after Ilsung's death.[118]

The highly cultivated and widely travelled Augsburg art collector and agent Philipp Hainhofer (1578–1647) was a much-sought adviser at royal and princely courts, given his supreme knowledge of the whole European art world. For our subject his relationship with Duke August the younger of Braunschweig is of particular interest. As a great amateur and connoisseur of clocks the Duke acquired more than forty pendant clocks and more than twenty table clocks. The regular and intense exchange of correspondence between Duke August and Hainhofer is a rich source of information of the most diverse kind.[119] On the basis of many expressions of expert opinion we see here a strong influence of the patron upon the specifications and configuration of the clocks.

Some of Hainhofer's letters convey an impression of the personalities of Augsburg clockmakers, for example of Johann Buschmann the elder.

114 Merchants acting as entrepreneurs now and then took matters of production so far in hand that they themselves placed the orders for the clock cases. *Ibid.,* fol. 617.

115 E.g., in 1533 King Ferdinand promised the Fuggers the sum of 2,000 gulden for a silver- and gold-mounted clock mechanism. Maurice, *Räderuhr,* Vol. I, p. 172. In 1574 they conducted business on behalf of Duke Wilhelm V, with delivery of 400 gulden to Georg Roll. B. C. Baader, *Der bayrische Renaissancehof Wilhelm V.* (Leipzig, 1943). See also N. Lieb, *Die Fugger und die Kunst im Zeitalter der hohen Renaissance* (Munich, 1958).

116 Discussed in detail in Ch. 5.

117 STAA, U I, fol. 203 (1577).

118 F. Blendinger, "Zacharias Geizkofler," *Neue Deutsche Biographie,* Vol. VI (Berlin, 1964), p. 167. A. Sitte, *Kunsthistorische Regesten aus den Haushaltungsbüchern der Gütergemeinschaft der Geizkofler und des Reichspfennigmeisters Zacharias Geizkofler 1576–1610* (Strasbourg, 1908).

119 See A. Fink, "Die Uhren Herzog August d.J. von Braunschweig," *Kunsthefte des Herzog-Anton-Ulrich-Museums,* Vol. 8 (Braunschweig, 1953), and for more detailed and extensive evaluation, Maurice, *Räderuhr,* Vol. I, pp. 172 ff.

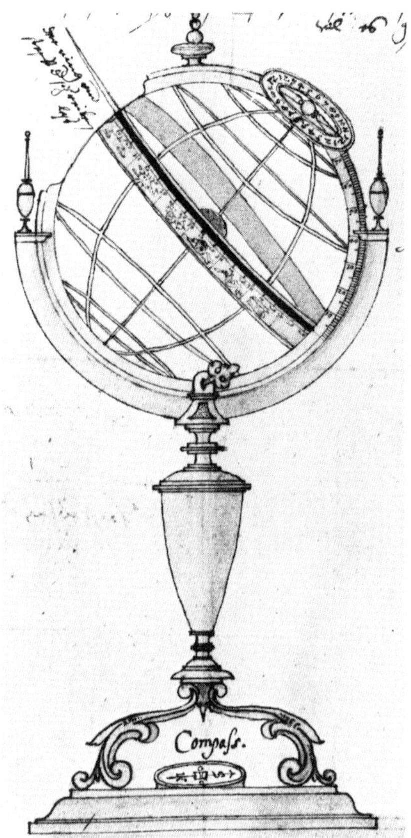

29. Drawing of a sphere by Johann Buschmann (presumably the younger, 1632–1676).
Augsburg, middle of the 17th century. Wolfenbüttel, Niedersächisches Staatsarchiv.

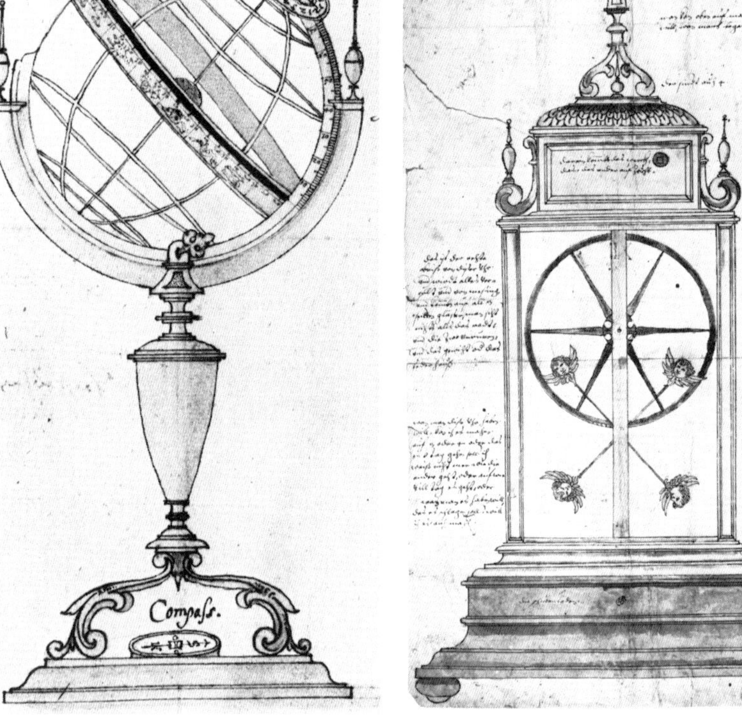

30. Drawing of a clock with cross-beat es-
capement: "Prague Clock." Hans (Johann)
Buschmann the elder. Augsburg, c. 1652.
Wolfenbüttel, Herzog-August Bibliothek.

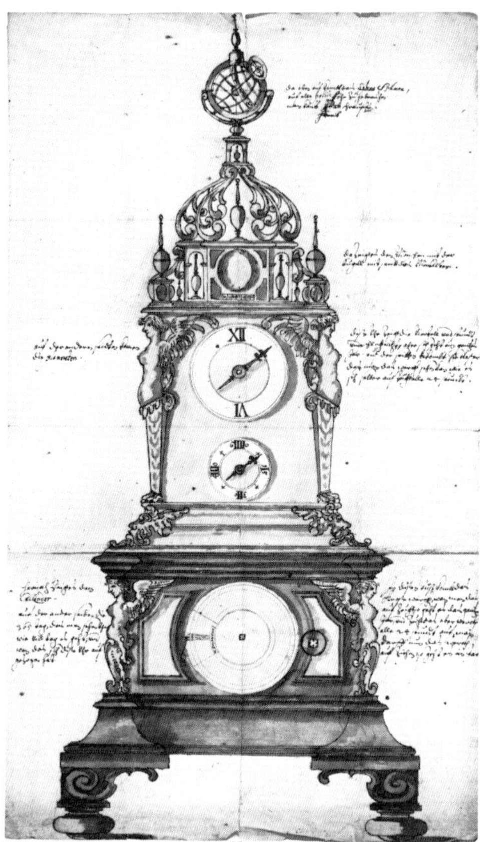

31. Drawing of the year clock, Catalog No.
57, by Hans (Johann) Buschmann the elder.
Wolfenbüttel, Herzog-August Bibliothek.

"Dealing with this worthy master is for the Court's agent 'a juggling with
a shell-less Egg.' He is obstinate, a man with a fearfully mean tongue and
a hopeless drunkard, does not hold to agreed delivery dates, and frequently drives his customer to despair."[120] Presumably Buschmann
showed a better side when he accepted the Emperor's invitations, for "the
Emperor Ferdinand III was so gracious as to have Johann Buschmann
brought before him frequently during his stay here in 1564, in order to
converse with him on matters of the art."[121] Here inevitably the question
arises: what scientific level could such a conversation have achieved? What
school education, what mathematical knowledge, would a clockmaker of
that time have had?

For Buschmann the question is answered partly through another description of the meetings with the Emperor:

120 Fink, "Die Uhren Herzog August," pp. 7 ff.
121 Stetten, *Kunst-, Gewerb-, und Handwercks-Geschichte*, Vol. I, p. 171. Stetten supposed Johann
Buschmann the elder was also the man commissioned for the large clock that Ferdinand III
had delivered, by Jesuit missionaries, to the Emperor of China as a present in 1655.

76

. . . old Johannes Buschmann was for many years a renowned master of the clockmakers' craft, who, however, had become so famous for his expertise in the production of mathematical instruments that His Imperial Roman Majesty *Ferdinandus* the Third, of most glorious memory, on a number of occasions during his sojourns here had Buschmann called before him in order most graciously to be informed in this science.[122]

Furthermore, his sound theoretical training can be inferred from his share in the estate of his father, "the frequently named Caspar Buschmann's books, in such part as they are of astronomy and geometry, and also his design drawings and his writings."[123] Some of Buschmann's own drawings have survived, as well as the drawing of an "invention" by his son Johann.[124] (See Figs. 29–31.)

It is surely not admissible to generalize from one of the stars of the clockmaking craft to the whole trade. Yet in general one can demonstrate indirectly that clockmakers in Augsburg attended the public school of Saint Anna, which besides its elementary curriculum also prepared pupils for university studies.[125] The director of the school managed the library and simultaneously also the Fugger library. Books not available at Saint Anna could be borrowed from the Fugger library. One may conclude from the old library catalogs that have survived that the necessary natural science literature was available in a public library in a public school.[126] In addition, from 1575/1576 a "Sphaera u. zwey Globos" and from 1592/1593 two further "Globos mathematicos" were at hand.[127] In the student rosters of the Jesuit Gymnasium there are a number of surnames that were borne by clockmakers.[128] In the absence of given names or of a specification of calling or trade, identification is not possible, and it is also very unlikely that clockmakers should have gone to a Catholic school since at the time under investigation more than 87 percent of those whose religion was known were Protestant (as shown in Table 2). The clockmaker Christoph Hayden[129] willed the respectable sum of 600 gulden to Saint Anna with the proviso that the school must remain Protestant. Beyond the interesting religious aspect of this act,[130] it was evidence of the

122 STAA, U III, fol. 27v (1702).
123 STAA, marriage certificate of Caspar Buschmann and Catharina Zieglerin, Aug. 20, 1629.
124 Niedersächsisches Staatsarchiv, 1 Alt 22 No. 170, fol. 129r,v. I am grateful to Mr. R. Gobiet of Salzburg for the reference to this drawing.
125 Julius Hans, "Beiträge zur Geschichte des Augsburger Schulwesens," *Zeitschrift des Historischen Vereins für Schwaben und Neuburg,* 1876, *4:* 17 ff.
126 O. Hartig, "Die Gründung der Münchner Hofbibliothek durch Albrecht V. und Johann Jakob Fugger," *Abhandlungen der königlich Bayerischen Akademie der Wissenschaften,* 1917, *18* (No. 3): 244 f.
127 Stadtbibliothek Augsburg 4° Cod., p. 84. 1575/1576 ("Tobias Klibern eine Sphaerum und Zwey Globos so man bey St. Anna braucht . . ."), 1592/1593.
128 STAA, Societas Jesu-Gymnasium, lists of pupils for 1592–1614. P. Braun, *Geschichte des Kollegiums der Jesuiten in Augsburg* (Munich, 1822).
129 STAA, Spreng'sche Notariatsakten, Aug. 14, 1573. Christoph Hayden is actually called only a "scales and balance master," but it transpires from later statements that scales and weight makers were always clockmakers from early times, so he can with high probability be numbered among these. STAA, Bauamts-Berichte, April 16, 1640, June 19, 1688.
130 See Robert K. Merton's attempt to explain the rise of the scientific way of thinking in the

high esteem in which this institution was held.[131] Among the teachers at
Saint Anna were luminaries such as Georg Henisch, the "Medicus and
Mathematicus" who was appointed teacher of mathematical sciences after
a trial period demonstrated "how splendidly his pupils had been instructed in arithmetic and geometry."[132] Henisch worked as theoretical
adviser with two clockmakers in the making of Augsburg Cathedral's
monumental clock.[133]

An indication that clockmakers normally could develop designs without
the help of others is afforded by the deprecatory comment about a certain
journeyman that "he could neither make a globe nor any compass *on his
own,* but had to make use of *models* carved in wood by others more
experienced in the art."[134] Yet within the workshop this sort of "model"
and drawings were the basis for the transmission of knowledge, mastery
over which was tested by the masterpieces. For this purpose there was
repeated used of *Risse,* construction drawings in which standardized work
was drawn in a large diagram. The *Risse* were "pricked through" repeatedly and remained the same for generations (see Fig. 33). Few have
survived, but their existence can be established in the manner of a textual
reconstruction, from the surviving masterpieces which Maurice has studied and evaluated from this point of view.[135] That series of clockworks,
which are identical in size and function, attests to a standardization
through the medium of the drawing, which is mentioned in the sources
and which accompanied the transmission of knowledge in the shop.

The period under consideration was in general a time of practical artisans. The following century was one of theory, of the academies and the
scientific societies. It is true that theoretical works in mathematics, astronomy, and geometry appeared both in new texts and in editions of ancient
authors edited notably in the free cities of the Empire; but a professional

<hr>

17th century on the basis of the forward surge of the Protestant ethical system, in *Science,
Technology, and Society in Seventeenth-Century England* (1938; reprint, New York, 1970). On
the original form of the Protestant thesis see M. Weber, *Gesammelte Aufsätze zur Religionssozi-
ologie* (Tubingen, 1968). I am grateful to Dr. Lenz Kriss-Rettenbeck for the suggestion of
considering the religious affiliations of the clockmakers.

131 A document regarding the exceptional schooling of a clockmaker is, e.g., the application of
1616 by Matthäus Degen, who in his petition for appointment to a Board of Works position
says that he surpasses his competitors in intellectual attainments. STAA, Bauamts-Berichte,
Jan. 7, 1616.

132 Hans, "Augsburger Schulwesens," p. 39.

133 With Tobias Klieber and Benedikt Marquart the younger. Bobinger, *Kompassmacher,* p. 92.
On Henisch see also M. Radlkofer, "Die humanistischen Bestrebungen der Augsburger
Ärzte im 16. Jahrhundert," *Zeitschrift des Historischen Vereins für Schwaben und Neuburg,* 1891,
*18:*25 ff.

134 STAA, U III, fol. 29. The following shows how great the mathematical knowledge of
Augsburg clockmakers was in comparison with those in other cities: "and one must not
forget that the great Bohemian clock of Prague could not have been devised anywhere else
except here in Augsburg, since there were also many mathematical movements to be found
here." *Ibid.,* fol. 384v (June 28, 1732). For the 18th century we have the reverse case: "since
hardly any of the local masters is practiced in mathematics, it follows that such mathematical
work as is required on the old assignments of the masterpiece must be borrowed . . . from
an outside artist." *Ibid.,* fol. 294v (no year named).

135 Maurice, *Räderuhr,* Vol. I, p. 98, and the appropriate figure in Vol. II.

literature of the specialized kind which became characteristic of the 18th century was missing. Despite the lack of horological treatises and of a theoretical exposition of craft knowledge, there emerges from the documents of the 16th century onward, surprisingly enough, a comprehensive functional terminology: in very systematic fashion craft skills are designated by technical terms which continue in use to this day. Although the complex working procedures which are precisely described by these terms are recognizable, the formulations never achieved an abstract, scientific quality. They became, however, the foundations for the treatises of the following century. The artisan was familiar with the phenomena but not with their basis in physical science. For this reason it was not yet possible to transcend a level of conceptualization based on experience alone—as is reflected by this statement for the era in Augsburg: "It is vain to suppose that one could put together clockwork that would not be subject to this mutation" (i.e., the effect on a clock caused by changes in temperature).[136]

Thus Augsburg in 1550–1650 was not in a position to introduce innovations into devices for the measurement of time. That development took place after 1650 and outside of Germany. This shift in the center of gravity of European clockmaking became manifest in Augsburg toward the end of the 17th century when counterfeiters began engraving on clocks (mostly imported from Geneva) "London and the names of the best English masters."[137] Earlier they had achieved their highest profit by falsifying Augsburg signatures.

Appendix I. Clockmakers' Regulations of 1552/1554 and 1650

Regulation of 1552/1554

"Regulations of the Locksmiths, Gunsmiths, Clockmakers, Ring-makers, Windlass-makers, File-makers, Planers, Nail-makers, and Spur-makers."[138]

The greater part of this twenty-three-page collection of laws affects the locksmiths in particular. Below are assembled all the articles that are also valid for the clockmakers; only a single one applies to clockmakers directly.

136 STAA, Bauamts-Berichte 1570–1610. Dec. 17, 1595. Expert opinion expressed about the Rathaus clock, by the foremen Hans Connar the elder and Hans Marquart the younger.
137 STAA, U II, fol. 1043 (1696).
138 From *Beschreibung aller Handwerksordnungen in Augsburg*, Stadtbibliothek Augsburg, 2° Cod., Aug. 442, fols. 333v–344v.

Folios 336r–337r (June 25, 1552):

Die drej Maister betrefent.

Vnd Zue mehrer Vnd Stattlicher handthabung diser obgemelten ordnung
Vnd satzung hatt ein Ersamer Rath drej maister verordnet dieselben beaidiget
vnd Jnen eingebunden ein fleisiges aufmerken darauf Zue haben, was sie
hörren sechen oder Jnnen werden das einem Ersamen Rath oder den hant-
werkeren Zuer verklainerung raichen Vnd komen möchte, dasselbig einem
Ersamen Rath oder den darzue Verordneten herren bej Jren ayden vnd
Pflichten anZueZaigen, auch sonst darob sein sollen damit obgemelte hant-
werker niemandts er seie burger oder frembdt in ainchicherlaj arbait oder
Lohn Vbernomen Werde Wo aber derohalben Clag Von burgeren oder
frembden für sich keme also das sich die Partheyen des Jbernemens oder
Lohns halber nit Vergleichen oder Vertagen khündten, alssdan dieselb arbait
oder Lohn Zue wurdigen, Zue schezen, Vnd also gueten friden, rueh Vnd
ainigkeit Zue erhalten. was aber in dem Jnen Zue schwer sein vnd fallen wolte
alssdan an einem Ersamen Rath gelangen Lassen.

Die drej obgemelten Maister sollen hiemit Jn sonders auch Verbunden vnd
schuldig sein alle die brief so einem hantwerkh Zuegeschriben es seien Vmb
wasserlaj sachen es Jmer, wölle die Jnen Zuekomen einem ersamen Rath
verschlossen vnd Vngeöfnet furZuebringen Der es alssdan Verantworten,
Vnd sonst die notturfft die ess erfordert darinnen furnemen wurdet . . . Actum
Jn Sennatu 25 Junÿo Anno 1552.

Folio 337v (December 12, 1553):

Jtem ein Jeeder der des hantwerks Gerechtigkait von schmiden kauffen will,
der solle darfur bar bezallen, Sechtzehen gulden Reynisch in mintz Vnd acht
kreitzer dem knecht sechtzehen Pfennig Vnd den Verordneten Vorgeern auch
sechtzehen Pfennig.

Jtem so einer des hantwerks Gerechtigkait erheurat der solle darfur mehr nit
dan acht Gulden ZuebeZahlen Schuldig sein, Vnd Vier kreitzer vnd einZue-
schreiben Sechtzehen Pfennig.

Wöllicher aber dise hantwerks Gerechtigkeit ererbt Vnd maister werden will
der solle ein Gulden Vnd acht kreitzer Vnd nit mehr Zue erlegen schuldig sein
. . . Actum Jn Sennatu 12 Decembris 1553.

Folios 338v, 339r (January 8, 1554):

Ain Ersamer Rath hatt Zue Furkhomong allerlej gefahr Vnd betrugs erkant
dass die Vhrmacher furohin, Weeder Kupfer messing noch ainich ander der-
gleichen mettal wölliche fur guet silber oder gold Geachtet, Vnd die Leut
dardurch betrogen werden möchten, ausserhalb dessen so Zue Jr der
Vhrmacher hantwerkh Vnd arbait gehört nit Vergulden sollen bej derro straf
Zechen kreitzer will sagen Gulden. Actum Jn Sennatu 8 January Anno 1554.

Folios 339r and v (June 19, 1554):

Jtem Nach dem Jn deren Von Schmiden ordnungen vnder anderem begrifen
ist, wan einer einen Leren Jungen annemen Wil das er in woll VierZechen tag
Lang Versuchen möge, das er auch Volgents denselben fur die Verordneten
Vorgeer stellen, Vnd sechtzig Pfennig in die Verordnete bix Leegen soll
. . . Actum Jn Sennatu 19 Junio Anno 1554.

Regulation of 1650

As a consequence of the separation of the craft of the large-clock makers from that of the small-clock makers, resolved upon May 15, 1649,[139] the large-clock makers needed a set of regulations of their own. They drew this up—according to their own statement[140]—after the regulations of the small-clock makers, which have not survived but can thus be deduced.

The Regulation of 1650 remained almost unaltered until 1673.[141] At that date a few points were added, but for the greater part these can be identified as binding decisions of the Senate from a time before 1650. For this reason, and also because of the more comprehensive formulation, we shall reproduce here the formulation of 1673. The concordance numbers or the date and the source indication for the Senate decision are given in parentheses after each article. Articles 8, 9, and 14 of the Regulation of 1650 were no longer at hand in 1673, and for that reason they are prefixed here.

> Articul, die Erbare von Grossvhrmacheren alhier betreffendt . . . Zum Achten, wenn die Zeit kombt, dass die frembde Gesellen vmb die Stuckh lösen werden, so soll ein ieder Maisters Sohn (so anderst ainer Verhanden ist) auch macht haben, mit denen frembden gesöllen einZustehen, vnd Zulössen, vnnd wass dann dass vngefahrliche loss gibt, dass es auffgeschriben werde, vnd die Stuckh Jhnen also nacheinander volgen. Doch wann der Maisters Söhn Zeit vff Pfingsten Zulösen herbey kombt, solle Jhnen der frembden Gesellen halben nit fürgriff beschehen, sonder Sie (Maisters Söhn) desselbigen loss allein für sich selber befrey sein. Zum Neündten, Wann dann den Maisters Söhn die Maister Stückh Zumachen vergondt, Vnnd sie in dass Loss kommen, mögen Sie wohl ausserhalb der Statt arbeiten vnd wandern, biss die Jenige ferttig, so vorJhnen in den Stuckhen sizen thuen.
>
> . . . Zum Vierzehenden, wann einem die Maisterstuckh geschaut worden, vnd Er darmit bestanden ist, soll Er alssbald widerumb ainen gulden Jnn Münz auflegen. . . .[142]

Regulation of 1673

"Craft regulations and articles for the crafts incorporated into the smiths' eligibility, which have been renovated and confirmed by the Noble and Most Wise Council on September 7, 1673."[143]

> 1. Jeder ist schuldig 3 Jahr lang zu lernen.
>
> Wer oder welcher sich fürhin des Gross-Uhrmachen gebrauchen wollen, die sollen zuvor 3 Jahr lang aneinander gelernt haben u. ehe nicht zugelassen werden, Sie haben dann solche 3 Jahr u. sonderlich bey einem Meister, es seye allhier oder ausserhalb der Stadt, der auch das Gross-Uhrmacher Handwerck redlich erlernt völlig erstanden. (= No. 1/1650)

---

139 U II, fol. 371 (taken over word for word in the Regulation of 1673 as No. 18).
140 *Ibid.*, fol. 378.
141 *Ibid.*, fols. 385–392, "Approved and confirmed in a session of the Honorable Council on March 19, Ao. 1650" (fol. 392).
142 *Ibid.*, fols. 388, 391.
143 Undated transcript, compared in 1807 with the original and authenticated before a notary as a true copy. STAA, U, 1812–1859 (folder entitled "Uhrmacher Ordnungen").

2. Vor Aufnehmung der Stück, soll jeder Gesell bey einand oder zwey Meister 3 Jahr arbeiten.

Da ein fremder Gesell des Gross-Uhrmacher Handwercks, die Meisterstück zumachen begehrt, dem sollen Sie nicht aufgegeben werden, Er habe dann auch zuvor ausserhalb seiner Lehrjahren, noch 3 Jahr allhier, bey einem oder zwey Meister aneinander vollkommentlich gearbeitet, auch soll ein jeder fremder Gesell sowohl als ein Meisters Sohn, die Meisterstück lediges Standes zumachen schuldig u. allein diejenigen, so sich zu Wittfrauen verheyrathen, desselbigen befreyt seyn u. welcher das über fährt dem sollen die Stück abgeschafft werden u. nichts gelten. (= No. 2/1650)

3. Ueber die 3 Lehrjahr noch 7 Jahr auf dem Handwerck zu seyn ehe Jhm die Stück aufgegeben werden.

Es solle hinfüro keinen fremden Uhrmacher Gesellen aufgegeben oder derselbe zum Meister zugelassen werden, Er habe dann über seinen 3 Lehrjahren, noch 7 Jahr lang, u. darunter 3 Jahr aneinander bey einem, oder zwey Meister u. also mit den 3 Lehrjahren in allem zehn Jahr erstanden es seyn gleich allhier oder ausserhalb der Stadt, doch welcher fremder Gesell zu einer Wittfrau der fünf Handwercker, sich verheyrathen würde, so Er gelernt an den zehn Jahren die Er sonsten zuerstatten schuldig, 3 Jahr nachgelassen werden seyn. (= No. 3/1650)

4. Meister söhn sollen über die Lehrzeit noch 4 Jahr erstehen ehe sie zum Meisterstuck gelassen werden.

Kein Meisters Sohn, der fünf Handwercker solle hinfüro, Er verheyrath sich zu einer Wittfrau oder Meisters-Tochter, der jetztgedachten fünf Handwercker oder nicht, auf seinem erlernten Handwerck zum Meister nicht zugelassen werden Er habe dann über seine Lehrjahr noch 4 Jahr u. also in allem Sieben ganze Jahr auf jedem Handwerck besonder so Er gelernt, erstanden, es seyn allhier, oder ausserhalb der Stadt. (= No. 4/1650)

5. die so die 10 Jahr ausserhalb erstanden u. nicht unter die 5 Handwercker heyrath. wurden, sollen die 3. Ersitz Jahr nichts desto weniger erstehen.

Da es sich aber begebe, dass fremde Gesellen oder Bürgers Söhne dieser Stadt allhero kämen u. zuvor allhier nicht gearbeitet, aber doch die 10 Jahr ausserhalb der Stadt schon erstanden hätten, u. sich ausser der 5 Handwercker verheyrathen würden, die sollen der obbestimmten Ordnung nicht fähig, sondern nichts desto weniger schuldig seyn, die 3 Jahr ohngeachtet dass Sie die 10 Jahr schon erstanden bey einem oder 2. Meister aneinander zuerstatten, wie Handwercks-Gebrauch u. Ordnung in sich hält. (= No. 5/1650)

6. Jeder soll in eigener Person um die Stuck losen.

Wann ein Meisters Sohn unter den fünf Handwerkern, auf dem grossen Uhr macher Handwerk die Meisterstück aufzunehmen begehrt, So sollen erstlich, wie Handwerksgebrauch, allhier in Arbeit seyn, u. eigner Person um die Stück losen. (= No. 6/1650)

7. des Jahrs nur einen zu den Meisterrecht Zulassen.

Es soll auch auf dem Gross-Uhrmacher Handwerck ein Jahr nicht mehr, dann ein Meister gemacht werden. Nehmlich wann etlich fremde Gesellen vorhanden wäre so der Meisterstuck begehren würde dass Sie untereinander zu lösen schuldig u. wie Sie im Loos herauskommen, also verzeichnet u. aufgeschrieben werden u. dem Loos nach die Stuck machen es solle aber denn fremden Jhre Zeit zu lösen allwegen auf Weynachten u. den Meisters Söhnen, welche mit den fremden nicht zulösen, sondern jedesmal den Vorzug haben sollen, auf Pfingsten angesetzt seyn. (= No. 7/1650)

8. der fremde ist zu warten schuldig bis des einen Stuck fertig seyn.

So einem fremden Gesellen die Stück zumachen vergönnt werden, die-
selbe sollen allhier darauf zu warten schuldig seyn. bis diejenige, so vor
Jhnen in den Stucken sitzen, fertig worden sind. (= No. 10/1650)

9. Bei Aufnehmung der Stuck 1. fl. zu bezahlen.

Wann einer die Meister Stuck zu machen aufnimbt soll Er alsbald ein
Gulden in Münz auflegen. (= No. 11/1650)

10. Wie viel ein Stuck Meister, dem wobey Er die Stuck macht wochentlich
schuldig.

Ein jeder, dem die Stuck aufgegeben werden soll dem Meister, bey dem
Er die Stuck macht alle Wochen 15 Kreuzer zu bezahlen schuldig seyn u.
dieselbige erlegen, ehe Jhm die Stück geschaut werden. (= No. 12/1650)

11. Vor der Geschau die Stuck nicht zuverkaufen anbieten.

Es soll auch Keiner, wer der auch seye, wan Er in den Stücken sitzt, Seine
Stück ehe Sie gar ausgemacht oder gar geschaut sind, nicht zu kaufen anbie-
then noch viel weniger besehen lassen es seye heimlich oder öffentlich, bey
Verliehrung der Stuck. (= No. 13/1650)

12. Neben den Stück keine andere Arbeit unter handen nehmen.

Welcher Meisters Sohn, oder ledige Gesell in den Stücken sitzt, u. zwischen dieser Zeit ehe Er die Stück ausgemacht, eine andere Arbeit neben der
unter die Hand nimmt u. daran macht, solle Jhm die Stück alsobald auf-
gehoben werden u. nichts mehr gelten. Doch wo einem Meisters Sohn Vater
oder Mutter kranck u. die Werckstatt darnieder lege, mag einer auf 14. Tag,
u. länger nicht, Jhnen Beystand thun, doch das Ers in allweg den geschwornen Meistern im Amt zuvor anzeigen u. vermelden sollen. (= No.
15/1650)

13. Ohn umgeschaut keinen Gesellen, oder Jungen befürdern.

Welcher Meister des grossen Uhrmacher handwercks ein Gesellen oder
Junge ohnumgeschaut, wieder des Handwercksgebrauch in seiner Werkstatt
hielte, der soll von dem ersten Verbrechen um 4 fl u. von dem andern um
8 fl. all in Eines Ers. Raths Büchsen gestraft werden u. so Er zum drittenmal
betreten würde, die gebührende Straf von Einem Ehrsamen Rath gegen Jhn
vorzunehmen, in allweg vorbehalten seyn. (= No. 16/1650)

14. Keinem kein Gesellen aus der Werckstatt kaufen.

So ein Meister einen Gesellen mit Geld einem anderen Meister aus der
Werckstatt kauft oder demselben Gesell, weil Er noch dem andern Meister
in der Arbeit ist Geld fürstreckt u. leiht oder sonst verbotten practiquen übt,
u. gebraucht derselbige Meister, solle eben so viel zur Straf in eines Ehrsamen Raths-Büchsen zubezahlen schuldig seyn. (= No. 17/1650)

15. Fremde Uhren Zu kaufen verbothen.

Keiner solle keine fremde Uhr hereinkaufen und allhier wieder ver-
kaufen, welcher aber das wieder handelt (crossed out and corrected illegi-
bly), der soll um den dritten Theil der Arbeit als so (crossed out) viel Sie
werth ist, in Eines Ehrsamen Raths-Büchsen gestraft werden. (= No.
18/1650)

16. Stählerne Federn in den holzernen Uhren sind ganz abgeschafft.

Die stählernen Federn in den hölzernen Uhren sind hiermit ganz abgeschafft. (not in effect in 1650)

17. Wieviel des Stückmachers Vermögen sein soll.

Welcher fremde Gesell die Meister-Stuck aufnehmen will, derselbige
eben messiger Gestalt den geschwornen Meistern, so im Amt seyn werden,
angeloben soll, das Er zu Verfertigung (corr.: Machung) derselben fünfzig
Gulden baar Geld ohne einige daraufhabende Schulden vermöge. Da Er
aber über oberwähnte 50 fl. (die Er zur zeit der Anstehung des Loos zweyen
angesessenen Mitbürgern allhier verbürgen solle). was weiters, dadurch Er
etwa an seiner Wohlfahrt, u. den Stücken verhindert werden möge abliesse,
soll Jhm alsdann unversagt (corr.: -währt) seyn solches bey guten freunden
aufzunehmen, u. kein Verleg (?)herrn zu haben, sich einiges gefährlichen
Betrug zugebrauchen. (Dec. 17, 1583. U II, fol. 393)

18. Die Gross-Uhrmacher haben ohne Einred der Klein Uhrmacher auf
    Jhrem Handwerck zu verfahren.

Deren von Klein u. Gross-Uhrmacher halber bleibts bey der Herrn, ob
Jhrer Ordnung Bericht u. Gutachten, u. solle denen von Gross-Uhrmachern
wie von Alters nicht allein Uhren so gut Sie immer könten u. mögen zumachen, sondern auch dessentwegen, eigene geschworne Jhres Handwercks
zuverordnen vergönnt u. zugelassen werden: Hingegen aber die Klein-
Uhrmacher bey Jhrem Handwerck verbleiben, auch denen von Gross-
Uhrmachern weder in Jhren Artikeln noch auch in Mach- u. Beschauung
Jhrer Meisterstück keinen Eintrag thun, sondern sich dessen alles gänzlich
verzeihen u. Jhnen von Gross-Uhrmachern zu ihrer Disposition lediglich
überlassen. (May 15, 1649. U II, fol. 371)

19. Wie lang einer an denen Stücken zuarbeiten?

Ein jeder Gesell so die Meister-stück machen will soll zu Machung u.
Vollendung derselben in allem nur 15 Wochen, nehmlich 14. Tag zum
Werckzeug u. ein viertel Jahr zu denen Stücken haben. da aber einer länger
daran arbeiten u. innerhalb den 15 Wochen damit nicht fertig werden
würde, solle Er als dan für jede folgende Woche 30 Kreuzer zur Straf
verfallen seyn. (no year, but before 1650. U II, fol. 381)

20. Wie man sich mit den Lehrjungen u. den Versprochenen zu verhalten?

Auf der Deputirten Herrn über die Gross-Uhrmach. Ordnung Bericht u.
Gutachten ist denen drey supplicirenden Meistern von Gross-Uhrmachern
allhier, aus denen von Jhnen fürgebrachten erheblichen Ursachen hiemit
bewilliget, das Sie ins künftig einen Lehrjungen drey- u. einen versprochenen (completed: Junge) zwey Jahr oder zween versprochene Jungen
neben ein ander, den einen drey u. den anderen zwey Jahr, doch mit diesen
Unterschied lernen sollen, welche sich nur zwey Jahr bedienen wollen,
allhier nicht, ausserhalb aber etwa die zwey Jahr gültig passirt werden sollen.
(not yet in effect in 1650)

Appendix II. Masterpieces

Masterpieces of 1577–1732

Small-clock makers:[144]

1.mo Eine Uhr in der Proportion ohngefehr einer Spannen hoch so die Stunden und viertl Stunden schlage, auch wecke, und das astrolabium, die Tag länge, den Calender und die Planeten zeüge, und wann der viertl Zeüger gerieben wird, dass alle Zeüger des ganzen Werks mitgehen, auch darneben 12. oder 24. Stund, welches man haben will, gehen,

2.do Eine Uhr welche man einen Spiegel nennet, der auch alle dienge Zeüge und schlage, wie obgemeldet.

3.^{tio} Eine gevierte Uhr die auch ohngefehr einer Spannen hoch, und alle ding Zeuge und schlage wie gemeldet.

4.to Eine Sexekigt Uhr, die auch ohngefehr einer Spannen hoch, soll schlagen die Stund und Viertel Stund auch wecken und das astrolabium die Tag Länge und Planeten Zeüge wie oben gesetzet,

5.to Einen gefuterten Stozen der die Stund und Viertel schlage oben das Astrolabium oder Tag Länge Zeügen, doch dass beede daran gemacht werden wie oben gemeldet, und wann der Viertl Zeiger gerieben wird, sollen alsdann alle Zeiger des ganzen Werks mitgehen, es solle auch zu jeder uhr wie gehört ein zeigend flach ührlein mit dem Mondschein, oder Äpfelein gemacht werden, wie zuvor auch gewesen.

Masterpiece of Hans Fronmiller the younger for 1584[145]

Item di vr ist in der fierung wie die Zway stuckh aus weyst (= wie die beiden Zeichnungen zeigen)

firs erst wirt sy schlagen die fiertel vnd mit dem fiertel zayger richt man das ganz werckh

firs ander wirt sy schlagen vnd zeigen von ain bis auf 24

firs fiert wirt sy zaygen das ganz astrolabio mit son vnd mon vnd drackhen durch die 12 himlischen zaychen vnd planeten

firs 5 wirt sy zaygen die tag vnd nacht lengen mitsamt den stunden von dem auf vnd nyder gang der sonen auf die polus heche

firs 6 wirt sy zaygen den calender mit seinen fürnemesten festen

firs 7 wirt sy auch weckhen

auf den zwuo andern seyten die nitt ver zaychnett sind wirtt sy zaygen wie vil es geschlagen hatt namlich die fiertel die 12 vnd die 24 vnd ist auch in der form wie die vnd wirt kosten 200 fl den necheste kauf

Masterpieces of 1650

Large-clock makers:[146]

Verzaichnuss der Maisterstuckh, wie Sie ainem Jeden sollen auffgeben werden. Erstlich, Eine flache Vierthel- oder gewicht Vhr auff Sechs Stundt vbersetzt, sambt einem Gewichtwöckher, welche inwendig von Eisen oder Mössing, aussenher aber von Mössing vnd im feuer verguldt sein, auch die

---

144 *Ibid.,* U III, fols. 390 ff. (July 29, 1732).
145 Appendix to a document of March 10, 1584, Hofkammerarchiv, Vienna, Reichsakten, 191, fol. 409, in which this clock is commended to the Emperor.
146 STAA, U II, fols. 389–390 (March 19, 1650).

Viertel vnd Stundt schlagen, vnd an allen vier seiter Zaigen solle; alss nemblich am vordern blat die Viertel ohn ein lufft vnnd die Vier vnd Zwanzig, wie auch Sonn vnd Mond durch die Zwölff himlische Zaichen: auff der rechten seiten soll SieZaigen, wievil es virtel, auf der linckhen seiten aber, was für ein Stundt es geschlagen hat, So dann auff dem hindern blat die Stundt vnd die Siben blaneten gestochen.

oder aber

Ein gewichtVhr, welche inwendig vnd ausswendig von Eisen, auch aussenher gemahlt, das Zifferblat von Mössing vnnd gestochen, vnd inn verguldt sein, auch die Stundt schlagen vnnd Zaigen solle, wie oben gemelt, doch soll dise GewichtVhr am hindern blat vnnd auff den Seiten keine Zaiger, beynebens aber ein Jedes Werckh seinen aigenen Gewichtwöckher haben.

8 Jost Bürgi, or On Innovation

32. Jost Bürgi. Detail from the title page of Benjamin Bramer's *Bericht,* 1648 (Marburg, 1684), by Anton Eisenhoit.

The free imperial cities made continuity in the crafts possible; indeed, they guaranteed it. The city-republics of south Germany stood constant guard over the uniformity of their guilds—over uniformity of training, number of journeymen, sizes of workshops, and type of production. The cities would even countermand an imperial majesty in Prague or in Vienna who attempted to create privileges for individual craftsmen based on their dexterity, diligence, or native intelligence. The city-republic distrusted the individual as soon as he overstepped conventional boundaries. Tools of production which were faster and more efficient were destroyed, for they threatened the artisans who continued to use time-honored methods. Extraordinary talents operating outside the guild-regulated crafts were censured, for this was disruptive of conventional training in the crafts. The law of tradition determined every action in the city-republic.

A prince, on the other hand, was autonomous and thus might support talent. Those who worked for and were salaried by the court were free from all city regulations, including the obligation to pay taxes and to serve as town watchman. It was the court which made possible, indeed demanded, the search for the new and unusual, the creation of the excellent.

Two residencies in Germany promoted the construction of scientific instruments and thereby increased astronomical knowledge: the princely residence at Kassel and the imperial one at Prague. The clockmaker Jost Bürgi (Fig. 32) served at both. Landgrave Wilhelm IV and Emperor Rudolf II did not just let their agents know they wanted mechanisms; instead, both of them inspired mechanical inventions, Wilhelm by his own scientific work as an astronomer, Rudolf by inviting Tycho Brahe, Johannes Kepler, and also Jost Bürgi to his court.

The articles Bürgi created were original and well thought out; not one of his works followed the stereotypical drawings that were handed down in the workshop. What is more, none of his works was modified from its original concept. Aside from his introduction of the cross-beat escapement and the remontoire, it would be difficult to describe his ingenuity as a craftsman: unprecedented precision in the cutting of teeth and in the paper-thin stacking of numerous gear trains, the perfect and simple articulation of one element into another. This technical skill is not easily visualized; its principles are revealed only as the device is taken apart. What is, however, immediately apparent is the multiplicity of functions his mechanisms could accomplish in a confined space and their precisely calculated form and dimensions, the clarity of design, symmetry of layout, and the superb engineering skill reflected especially in his machines which reproduced celestial movements. Bürgi must have had an outstanding spatial perception in order to fit his mechanisms with their gear trains into a very

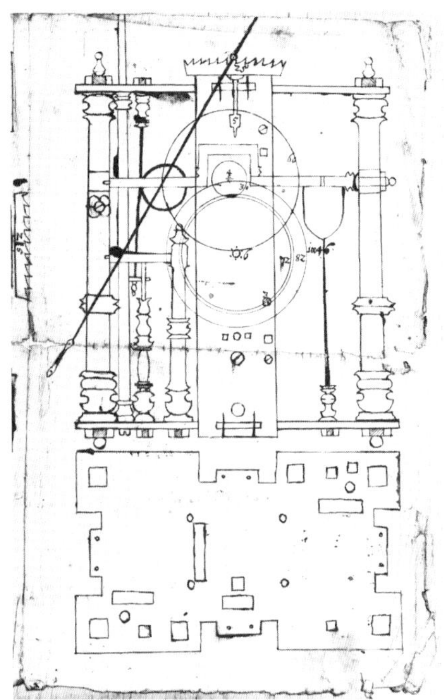

33. Working drawing by the Zug clockmaker Johann Michael Landtwing (1688–1776) for the masterpiece he made at Augsburg. Private ownership.

restricted working space; at no point does one element interfere with the working of another. His construction drawings must have been executed in perspective, not as the plane renderings of elevation, plan, and side view which were used to lay out the Augsburg guild masterpieces (see, e.g., Fig. 33).

The fact that Bürgi acted as supervisor over the actual production of cases—the woodworking, the engraving of metal, the grinding and polishing of rock crystal—does not set him apart from other Augsburg masters. The difference is only that he worked together with court artists, not with artisans.

Jost Bürgi was born on February 28, 1552, at Lichtensteig in the Toggenburg (Switzerland);[1] we know the date through the copper engraving for the title page of Benjamin Bramer's report (Fig. 32). It is only with his appointment at the court in Kassel that we know the next date: on July 25, 1579, he was employed by Wilhelm IV as clockmaker, which included using and maintaining astronomical instruments. Everything that lies between 1552 and 1579 is obscure. Presumably he left Lichtensteig because of religious differences. Where he got his training remains unknown: whether in Strasbourg, a city which actually became famous for its clockmaking craft only after completion of the cathedral clock (see Fig. 6), or at the Kassel court, where Eberhard Baldewein (see Catalog No. 115) and Hans Buch had preceded Bürgi in building complicated astronomical instruments for the Landgrave.

Wilhelm IV of Hesse invited well-known astronomers to attend him, and famous ones dropped in of their own accord: in 1575 Tycho Brahe visited, in 1577 Christoph Rothmann was installed as court mathematician, in 1584 Paul Wittich sojourned at Kassel, and in the same year Nicolas Reimarus Ursus arrived and remained four years. It was here that the Hessian catalog of the fixed stars was undertaken,[2] with the Landgrave himself taking part in the observations: the positions of the stars were determined with the help of an improved set of scientific instruments and a corrected version of the traditional *Alphonsine Tables.* In 1584 the Landgrave called Bürgi "a second Archimedes" because of his mathematical talents and his mechanical ingenuity, and in 1606 Kepler wrote that a later generation would rate Bürgi as high in his field as Dürer was in painting.

At Kassel Bürgi created a series of complex clocks, especially some automatically operating celestial spheres (Catalog No. 116) which Wil-

1 C. Alhard von Drach, "Jost Bürgi, Kammeruhrmacher Kaiser Rudolf II. Beiträge zu seiner Lebensgeschichte und Nachrichten über Arbeiten desselben," *Jahrbuch der Kunsthistorischen Sammlungen des A. H. Kaiserhauses,* 1895, *15:* 15–44. Armin Müller, "Jost Bürgis Herkunft und Verwandtschaft," *Toggenburgerblätter für Heimatkunde,* 1968, *27:* 53–68. Armin Müller, "Jost Bürgis Herkunft," in J. H. Leopold and K. Pechstein, *Der kleine Himmelsglobus 1594 von Jost Bürgi* (Lucerne, 1977), pp. 107–113. Ludolf von Mackensen, *Die erste Sternwarte Europas mit ihren Instrumenten und Uhren: 400 Jahre Jost Bürgi in Kassel* (Munich, 1979).
2 Rudolf Hallo, *Die Sternwarten Kassels in Hessischer Zeit* (Kassel, 1929). Paul A. Kirchvogel, "Wilhelm IV, Tycho Brahe, and Eberhard Baldewein—The Missing Instruments of the Kassel Observatory," pp. 109–121 in *New Aspects in the History and Philosophy of Astronomy,* ed. Arthur Beer (Oxford, 1967).

helm IV presented to other princes. At length his clockmaker became so famous that the Emperor called him to Prague, and in 1592 Bürgi delivered a celestial sphere driven by clockwork, which showed the motions of the planets and the fixed stars, and also a special pair of compasses which he had designed. He departed from the audience, richly rewarded, on July 4. On his homeward journey to Kassel Bürgi passed through the merchant cities of Nuremberg and Augsburg, procuring ready-made parts for clocks from goldsmiths and ordering, through German merchants, bells from Lyon for a musical machine. But before he reached Kassel, Wilhelm IV, "a prince among astronomers, too,"[3] had died. His successor, Landgrave Moritz, at once renewed the contract with Bürgi. A further trip to Prague followed in 1596, and in 1603 the Emperor Rudolf II asked the Landgrave to allow Bürgi to transfer to Prague for good. The Emperor had already invited Tycho Brahe to his court in 1599, Johannes Kepler in 1600. Bürgi now became clockmaker to the imperial court, with a workshop and lodgings at the imperial residence. In 1610 he was made a burgher of Prague, without, however, giving up his Kassel citizenship or his house there. He frequently visited Kassel, and in 1631 he returned permanently; he died there on January 31, 1632.

At Kassel Bürgi discovered the logarithms around 1588,[4] somewhat later a variant of the proportional compass (a kind of slide-rule).[5] In 1592 he commissioned the goldsmith Anton Eisenhoit to build his newly conceived triangulation instrument,[6] and in 1598 he completed his "Canon sinuum," or table of sines (which was never published).[7]

At Prague Bürgi was ennobled in 1611 and his coat of arms was bettered. In 1620 he published his table of progressions *(Arithmetische und geometrische Progress-Tabulen)* and also constructed Kepler's theoretical design for a toothed-wheel pump (a vane-type pump). While Bürgi was at this Catholic court in Prague, the Protestant Landgrave requested an imperial favor through Bürgi's mediation. Bürgi's skill as a teacher was evidently no less than his diplomatic accomplishments: at Kassel he was instructor in astronomy to Prince Hermann, and as early as 1586 Ursus called Bürgi his teacher.

3 Paul A. Kirchvogel, "Landgraf Wilhelm IV. von Hessen und sein astronomisches Automatenwerk," in *Index zur Geschichte der Medizin, Naturwissenschaft, und der Technik* (Munich/Berlin, 1953), p. 17, quoting Friedrich Wilhelm Bessel, 1848.

4 Luboš Nový, "Joost Bürgi," *Dictionary of Scientific Biography,* Vol. II (New York, 1970), p. 602. Friedrich Klemm, "Die roten Zahlen von Prag, Mathematik und Naturwissenschaft am Hofe Rudolf II.," *Die BASF,* 1969, *19:*137–145.

5 Ivo Schneider, *Der Proportionalzirkel, ein universelles Analogrecheninstrument der Vergangenheit* (Munich, 1970), p. 22.

6 Anna Maria Kesting, *Anton Eisenhoit, ein westfälischer Kupferstecher und Goldschmied* (Münster, 1964), pp. 16, 64. Rainer Rückert, "Anton Eisenhoit," *Neue Deutsche Biographie,* Vol. IV (Berlin, 1959), p. 415. Zdeněk Horský, "Bürgiho Sextant ve Sbírkách NTM," *Sborník Národního Technického Muzea v Praze,* 1968, *5:*280–300 (I am grateful to Prof. Dr. Walther Gerlach of Munich for this reference).

7 Martha List and Volker Bialas, *Die Coss von Jost Bürgi in der Redaktion von Johannes Kepler,* Bayerische Akademie der Wissenschaften, Mathematisch-Naturwissenschaftliche Klasse, Abhandlungen, 154 (Munich, 1973).

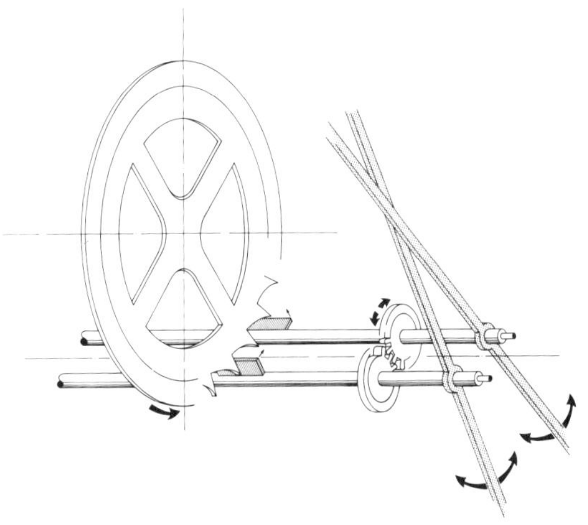

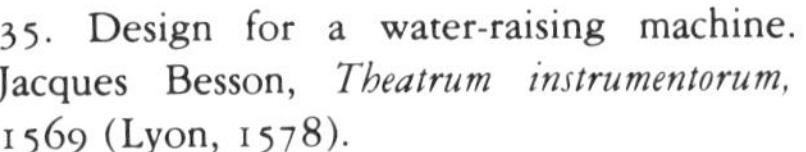

34. Diagram of a cross-beat escapement.

35. Design for a water-raising machine. Jacques Besson, *Theatrum instrumentorum,* 1569 (Lyon, 1578).

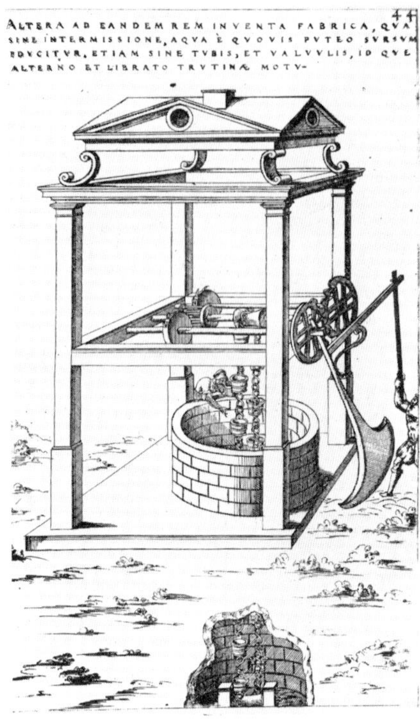

In the introduction to his unpublished "Canon sinuum" Bürgi described the encouragement he had received from the courts and the scholars in residence there. They had constantly been of assistance to him in making various instruments, in preparing works of instruction in astronomy, and also in translating ancient authors and current works in Latin. This in turn had inspired him to think out geometrical problems for himself, to ponder and probe and seek solutions of his own. In this process Jost Bürgi underscores the theme advanced here: innovation requires talent, but also encouragement and support.

Now in order to realize what Bürgi's contribution, his innovation, actually was we must first make it clear that it was only after his time that the clock began to develop into an instrument for the measurement of time. This was brought about through the application—by a scientist, not a clockmaker—of the pendulum as an isochronically swinging regulator for the clock. Up to the end of the 16th century the practical man alone had been the decisive factor in mechanics; but thereafter this field of knowledge—Galileo's *nuova scienza*—began to be approached theoretically. Theory led to the introduction of scientific method into the crafts, particularly into clockmaking, by scientists working at the courts or within the academies. The artisans did continue to hand down mechanisms in a traditional fashion, unchanged, until the middle of the 17th century; in our own exhibit the uninterrupted series of masterpieces demonstrates this immutability.

90

In the free imperial cities commerce received support, but not investigative research. The artisan had neither appreciation nor time for innovation or for technical development; indeed, he was usually not even aware of such development. Volker Himmelein has illustrated this lack of awareness in the following characteristic episode. In April 1660 Duke August in Braunschweig, an important source of business for the Augsburg clockmakers, desired clocks that "instead of a balance wheel have a perpendicular, as are made in Holland." This was shortly after the invention and publication of the pendulum clock by the Dutchman Christiaan Huygens in 1657, but the revolutionary technical innovation had remained totally unknown in Augsburg. Caspar Langenbucher, one of the outstanding artisans working for the Duke, replied that for his part he would not bother his head with inventions of this sort. The clockmaker Hans Buschmann, too, along with his sons David and Johann, was not impressed with this discovery; he was even convinced that if he were paid for such a job, he would be quite well able to make this notable invention himself.[8] At the court, on the other hand, Bürgi shared in the accomplishments of Copernicus, Kepler, and Tycho in astronomy, and in those which Simon Stevin, Isaac Beeckman, and Galileo discussed in the field of mechanics.

Bürgi had made clocks that showed the seconds and could therefore be used for astronomical observations. At this time clocks with seconds-hands were rare, yet not unknown,[9] but Bürgi's clocks functioned with certainty, reliability, and precision. He achieved reliability and regularity of operation through three improvements. First, through painstaking handwork he reduced all disturbing effects of the transmission elements (i.e., of the gear trains) upon the traditional regulator—the verge escapement. Second, he improved this regulator itself, in its oscillating motion, through the cross-beat escapement. Finally, he introduced the remontoire, an automatic device which holds the drive force for the clockwork mechanism, the going train, at a constant level. (Such painstaking techniques reveal a skillful hand, but it is also possible that he made tools for himself which provided greater precision—if only a stronger magnifying glass.)

By means of the cross-beat escapement he articulated the verge escapement with its regulator, the foliot or the balance wheel, into a system consisting of several distinct elements (Fig. 34). He replaced the vertical verge with its two pallets by two arbors meshing with each other, each bearing one pallet; the balance arm now sits on the ends of the arbors. (As the gear train is checked, a cross is formed with each beat through the intersecting motion of the two balance arms; the cross then breaks apart and forms again repeatedly, thus giving this escapement its name.) Like the verge escapement, the cross-beat is a recoil device and thus no real step

8　Volker Himmelein, "Die Uhren," in *Sammler, Fürst, Gelehrter—Herzog August zu Braunschweig und Lüneburg, 1579–1666* (Wolfenbüttel, 1979), pp. 164 f.

9　Klaus Maurice, *Die deutsche Räderuhr* (Munich, 1976), Vol. I, p. 146, n. 87, referring to a clock before 1557 which indicates the seconds.

36. Rock-crystal clock by Jost Bürgi, made between 1622 and 1627. Vienna, Kunsthistorisches Museum.

37. Rock-crystal clock with cap removed.

forward in technical respects. It is an advance in practice, however, since it was made of several units and thus permitted a more precise adjustment of the intervention of the escapement pallets into the escapement wheel, which now was larger in diameter and had a greater number of teeth.[10]

Bürgi introduced this cross-beat escapement into the construction of clocks, but he did not invent it. Jacques Besson had published, as early as 1569, a system which brings about an evenly timed interruption of a continuous rectilinear motion through a lever device (see Fig. 35), though actually his design produces an effect opposite to Bürgi's escapement: in Besson's system two meshing arbors alternately engage with a hooklike wheel in a gear train, each time converting the to-and-fro motion of a pump handle into a continuous rotation. A chain of buckets can be raised by means of this rotary motion. Thus Besson's pump mechanism and Bürgi's cross-beat escapement consist of the same mechanical elements; but in Besson the cross-beat motion produces continuous rotation, in Bürgi it continuously restrains a rotary motion. Besson's book of machines, which had been translated into several languages, appeared in an

---

10   The escapement wheel has the teeth cut in the plane of the wheel, in contrast to the crown wheel of the verge escapement, in which the teeth are at right angles to the plane of the wheel (hence the name "crown wheel").

92

expanded edition in 1578, and it is probably from one of these editions that Bürgi got the inspiration of applying this "escapement" in clock construction. Subsequently the cross-beat escapement was imitated by the Augsburg masters under the name "Prague clock" (see Fig. 30).

Bürgi's third improvement was the remontoire, or self-winder, in my consideration his real innovation in clockmaking. The problem Bürgi solved was the functional dependence of the output motion of a clock upon the driving torque: if the torque changes, the accuracy of indication will change also. The variance in motion is less in weight-driven clocks, yet even here—as Tycho had already lamented—the driving torque becomes greater in the course of operation, since there is added to the intended weight also that of the cord as it unwinds. The variation is still greater when a mainspring is used, even though the fusee does provide

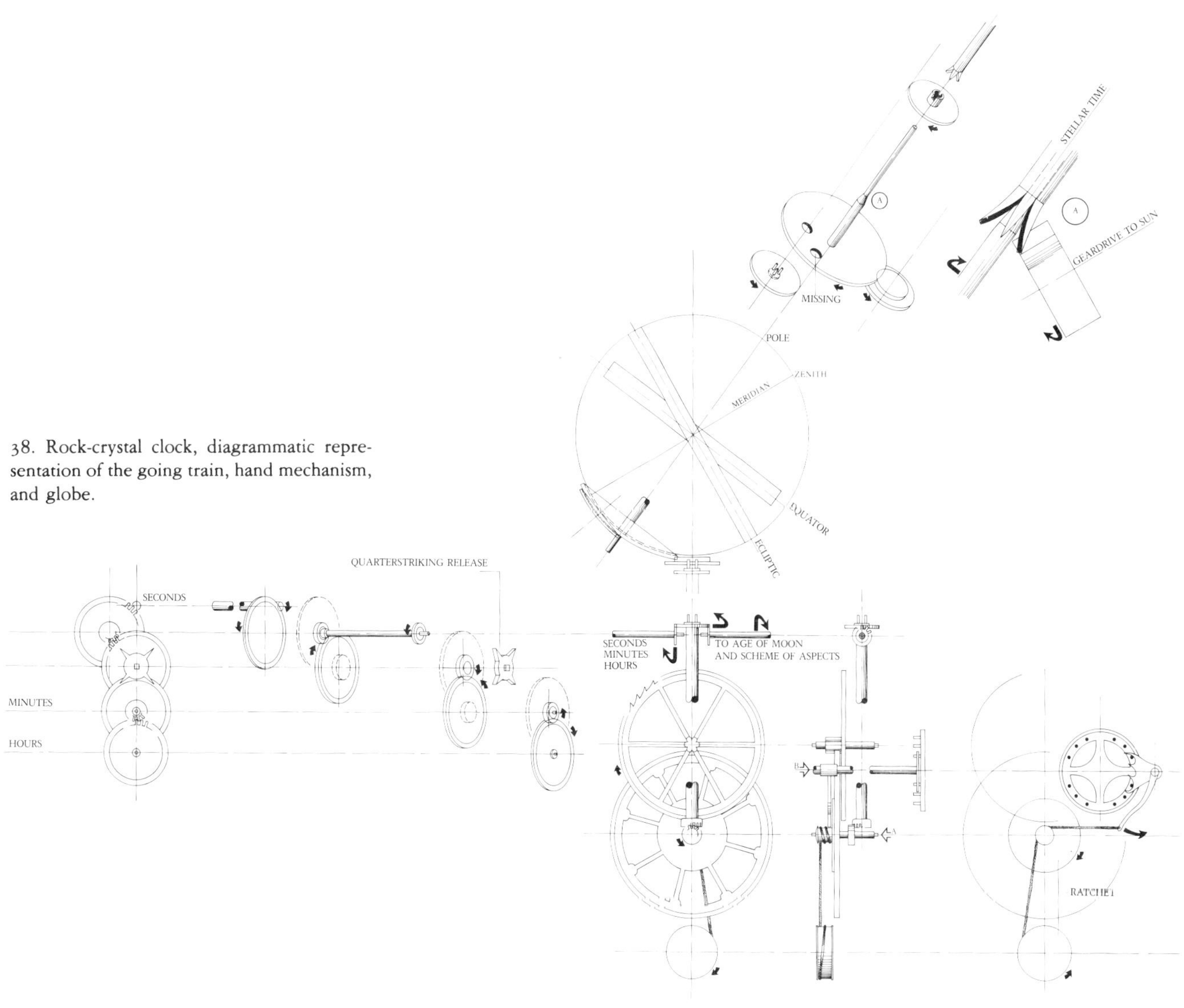

38. Rock-crystal clock, diagrammatic representation of the going train, hand mechanism, and globe.

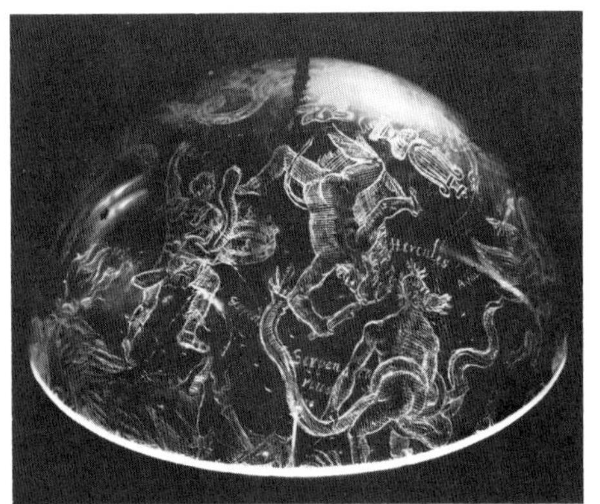

39. Rock-crystal clock, Northern Hemisphere of the terrestrial globe.

40. Rock-crystal clock, celestial globe.

41. Rock-crystal clock, steel structure inside the celestial globe.

compensation.[11] Bürgi solved this problem by repeatedly restoring the driving torque during the operating cycle; that is, at brief intervals he caused the mainspring to be rewound, or the weight to be drawn up. This winding action was induced by the running down of the drive mechanism, which itself was wound up again either through a separate mechanism or through the quarter-striking train, which the drive mechanism had set into motion. In this way a constant driving torque was assured for the going train of the clock. Bürgi coupled two separate processes—the drive for the clock and the drive for that drive—and achieved a cyclical operation which regulated itself.[12] Before Bürgi, the actions for the individual mechanisms (going and striking trains) had been connected sequentially. Bürgi then connected the separate functions into a control loop and thereby, for the first time in clockmaking, introduced a system of self-regulation under which the driving torque continuously kept itself at an optimum.

Bürgi, who was noted for his restraint and modesty, did not publish his mechanical designs. But he did sign one of his clocks—the rock-crystal clock shown in Figs. 36–50—and from this it has become possible to recognize his innovations and to ascribe his name to unsigned clocks incorporating the same principles of design.[13] Several mechanisms have

11   It should also be noted that the main cause for differing quantities of motive power was the variable friction produced as a result of imperfections in the size and profiles of the wheel teeth and pinion leaves.

12   For an historical introduction to this subject, see Otto Mayr, *Zur Frühgeschichte der technischen Regelung* (Munich, 1969) and Otto Mayr, *The Origins of Feedback Control* (Cambridge, Mass./ London, 1970).

13   Hans von Bertele was the first to establish that the cross-beat escapement and the *remontoire* were improvements introduced by Bürgi, on the basis of research on the rock-crystal clock. Hans von Bertele, "Jost Bürgis Beitrag zur Formentwicklung der Uhren," *Jahrbuch der Kunsthistorischen Sammlungen in Wien,* 1955, *51:* 170 ff. Hans von Bertele, "Eine mechan- ische Mondanomaliendarstellung auf der Basis der Kopernikanischen Sekundären Epicyklen in einer Räderuhr aus dem Ende des 16. Jahrhunderts," *Der Globusfreund,* Oct. 1973, pp. 162–168.

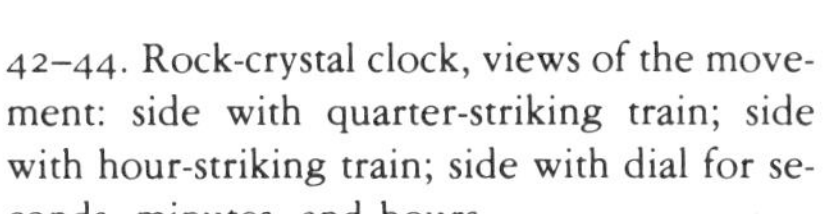

42–44. Rock-crystal clock, views of the movement: side with quarter-striking train; side with hour-striking train; side with dial for seconds, minutes, and hours.

been disassembled or restored for this exhibition. During this process it was possible to prepare various illustrations which should clarify many of the principles described above.

The rock-crystal clock, now at the Kunsthistorisches Museum in Vienna, is signed "Jost Bürgi." The monogram of its first owner, Prince Carl I of Liechtenstein, appearing together with the emblem of the Order of the Golden Fleece, narrows down the period in which the clock could have been built: after 1622, the year the Prince was made a member of the order, and before 1627, the year of his death.[14] The rock-crystal clock is constructed in a completely symmetrical fashion. The indications are laid out in a cruciform pattern: the subunit for telling the hours, minutes, and seconds is arranged opposite the one indicating the age of the moon, its phases, and the representation of its aspects; the quarter-striking train is opposite the hour-striking, which is adjustable for 12 or 24 hours. The cross-beat escapement has two balance wheels (not in the horizontal arms), one wheel installed above each of the striking trains. In this way the sides give the appearance of being in equilibrium.

When the hemispherical cap is removed from the cylindrical clock case, one finds beneath it a rotating globe of rock crystal (Figs. 39–41), its surface etched with the constellations (they follow figures that Eisenhoit had engraved upon other Bürgi globes).[15] The major stars—which were

14   No. 322 in the Catalog of the Sammlung für Plastik und Kunstgewerbe, Vol. II, Vienna, 1966.
15   Cf. our figures with the corresponding views appearing as Figs. 57 and 58 in Leopold and Pechstein, *Der kleine Himmelsglobus 1594 von Jost Bürgi.*

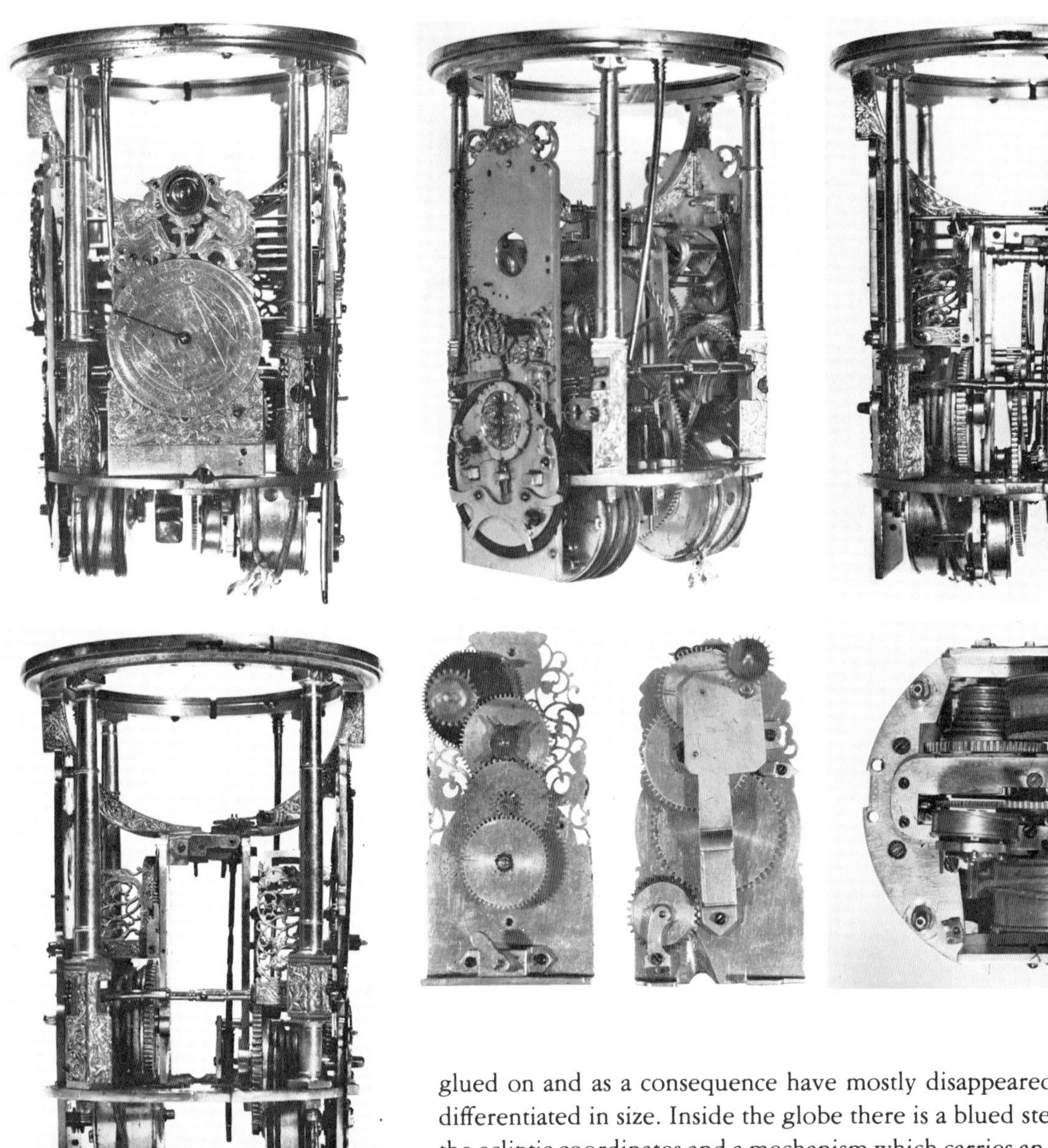

45–50. Rock-crystal clock, views of the movement: dial with lunar indication; escapement and dials removed. Going train disassembled. Back view of the structural elements of the dials for hours, minutes, and seconds and for the lunar indications. Lower side.

glued on and as a consequence have mostly disappeared by now—were differentiated in size. Inside the globe there is a blued steel framework of the ecliptic coordinates and a mechanism which carries an image of the sun through the zodiac during the course of the year (Fig. 41).

If one now draws out the whole movement contained in the base (Fig. 42), the arrangement of the three trains standing one behind the other becomes visible. The mainspring of the going train is wound up by a remontoire: every time the quarter hour strikes, the spring of the quarter-striking train winds up the mainspring of the going train.

The multitude of visual indications of time, from seconds to the annual course of the sun, is accompanied by an acoustical subdivision of the day. All of this Bürgi handled with a minimum of mechanics. From the first transmission member of the drive mechanism, the first wheel, he drives both time indicators and the globe by means of a single arbor. In this elegant way he achieves exact coordination of all the indications of the subassemblies (Fig. 38).

51–53. Views of the going train for the planetary clock by Jost Bürgi, Catalog No. 53.

Understanding the plan of construction of the rock-crystal clock makes it possible to ascribe the planetary clock of the same museum in Vienna (shown in Catalog No. 53) to Bürgi as well.[16] Its overall appearance is illustrated in the Catalog; let us describe here its construction elements and its functions on the basis of a side view (Figs. 51–53). In the center is a spring with a wooden fusee which draws up a weight (left). The weight then operates upon the actual clockwork mechanism via a toothed rack (Fig. 52). The going train is continuously regulated in its running-down by means of a vaned governor. To the right is a multi-layered subassembly holding the gearing for the planets (Fig. 54). The weight is drawn up hourly, actuated by a twelve-pointed star in the time-indicator mechanism. As winding-up takes place, the hand mechanisms for the two vertically arranged dials are advanced in hourly steps. In this clock, too, the drive system, the wheel train, and the indicator train are joined together for integrated action. The design of the remontoire is very simple: the wheel with the half-moon-shaped cam rotates through 90° with each hourly actuation (Fig. 52). In this process the wheel, via its cam, lifts a lever which draws the weight up. The mechanism of the remontoire for the weight is identical with that in the so-called first Kassel experimental clock (Fig. 56; Catalog No. 52).

16 Catalog No. 53 was ascribed by von Bertele first to Bürgi and subsequently to Christoph Margraf: "Der kaiserliche Kammeruhrmacher Christof Margraf und die Erfindung der Kugellauf-Uhr," *Jahrbuch der Kunsthistorischen Sammlungen in Wien,* 1963, 59:52, 73 f.

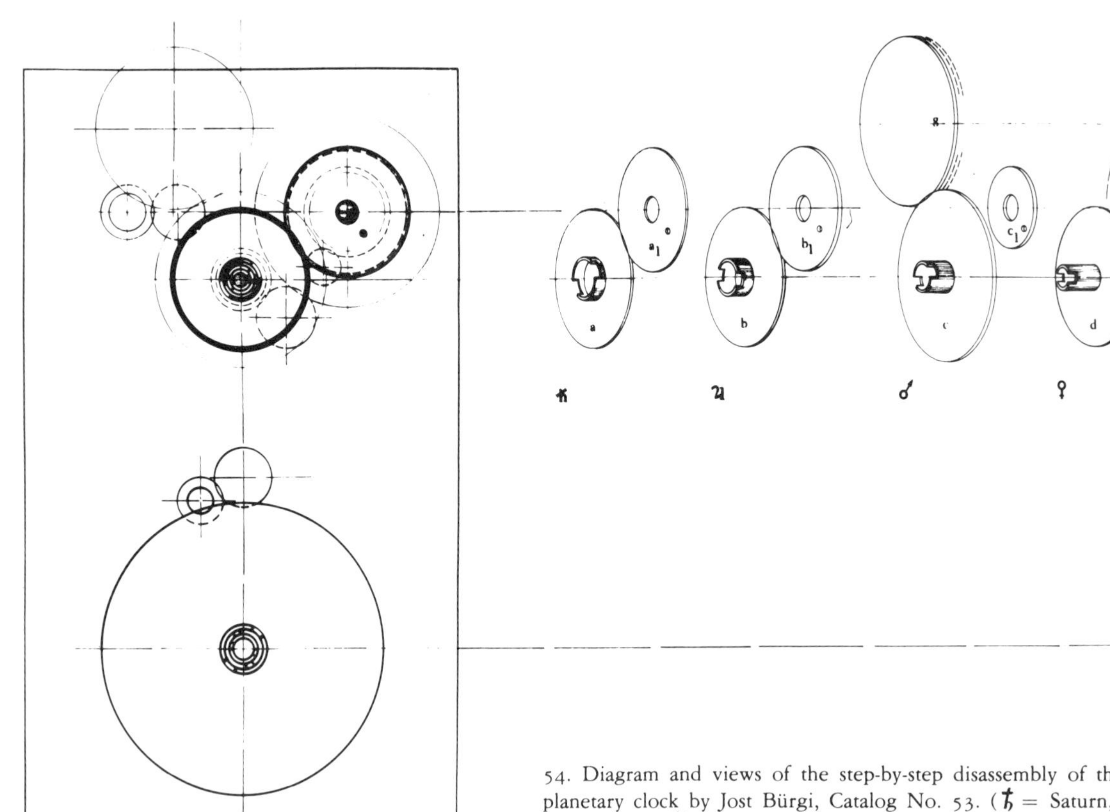

54. Diagram and views of the step-by-step disassembly of the hand mechanism of the planetary clock by Jost Bürgi, Catalog No. 53. (ħ = Saturn, 24 = Jupiter, ♂ = Mars, ♀ = Venus, ☿ = Mercury).

OUTSIDE VIEW

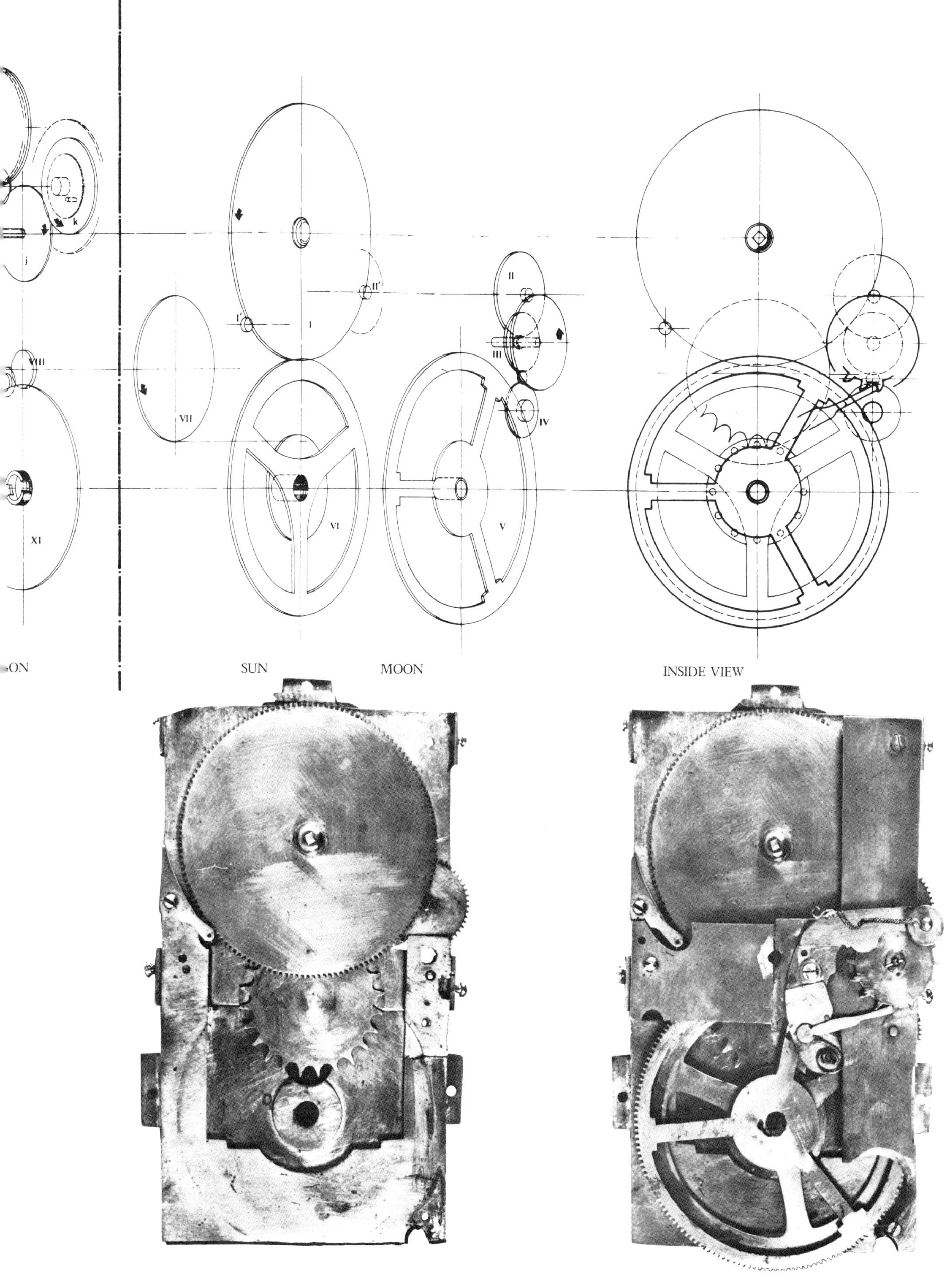

X
j
k
VIII
XI
I'
I
II'
II
III
IV
VII
VI
V
ON
SUN
MOON
INSIDE VIEW

55. Quarter-striking and hour-striking trains of the armillary sphere by Jost Bürgi, Catalog No. 119.

56. Movement of the experimental clock by Jost Bürgi, Catalog No. 52.

The armillary sphere from the Nordiska Museet in Stockholm (Catalog No. 119) is signed by Anton Eisenhoit. Bürgi was associated with this Westphalian goldsmith through collaborative work on apparatus and instruments. Restoration of the apparatus at the Bayerisches Nationalmuseum has revealed that in the sphere there are the same linkages as those of the little sphere in the rock-crystal clock signed by Bürgi. The clockwork is also made with uncommon precision; each wheel is fastened to an arbor by three screws in an absolutely uniform manner (Fig. 55). Such unusual precision of workmanship was an early indication of an outstanding master.

Another indication was the unique nature of this apparatus, in which the sphere and astrolabe are united. It is in fact a three-dimensional reproduction of the heavens with the fixed stars plus the apparent movement of the sun, combined with the astrolabe which shows the heavens in stereographic projection. I am acquainted with only one apparatus that has both these aspects. In Emperor Rudolf II's inventory of 1611, among "clocks and clockwork mechanisms," there is listed, in first place: "A large clockwork with an astrolabe and an annual indicator round about it, and upon this a sphere with its *circulis planetarum* and other appurtenances, and a written treatise with it; it stands on the table in the *Kunstkammer;* Jobst Bürgius made it, from the Duke of Braunschweig; presented to His Imperial Majesty."[17] Since this apparatus was presented to Rudolf II by Duke Heinrich Julius of Braunschweig, which lies at a latitude of 52°15′42″ N, and since the Stockholm apparatus is also set for a latitude of about 52° N,

17   Rotraud Bauer and Herbert Haupt, eds., "Das Kunstkammerinventar Kaiser Rudolfs II., 1607–1611," *Jahrbuch der Kunsthistorischen Sammlungen in Wien,* 1976, N.S. 36 (72): 110, No. 2138d.

this is a further argument for the identity of the two instruments,[18] since our sphere also has the "annual indicator round about it." But other evidence connecting the two is missing; thus ultimately missing is assurance that the two devices are identical. Yet the inventory in which Bürgi is named, and the instrument which bears the imprint of his technical fingerprints, do offer evidence in favor of an ascription to Bürgi.

The celestial globe which is further exhibited here (Catalog No. 116) contains no really "new" mechanisms. Its functioning has already been published in another connection, for which reason the description here is limited to the entry in the Catalog.

What chronometric development did Bürgi bring about? This very question, which presumes a linear evolution in technology, is tinged with views of progress which arose only at the end of the 17th century. Bürgi's mechanical genius lay first of all in his transcendence of traditional craftsmanship. Nonconservative and eager to experiment, he was building, not chronometers, but clocks which reflected all the cosmic movements and were at the same time synchronously functioning astrological diagrams. His works are those of an engineer, for with his application of the remontoire he introduced automatic control into mechanical constructions and therewith a new way of conceptualizing regulating techniques. "Thinking in terms of causal circles that stabilize themselves is a new and higher level of functional thinking."[19]

Bürgi's limited effect upon craft production has been explained on the basis of the turmoil created by the Thirty Years' War (1618–1648). This judgment was held even in Bürgi's own day, as expressed in this statement by his brother-in-law Benjamin Bramer: "and yet because the great unrest, which still persists throughout all Germany, sprang up at that time in Bohemia and had its start in that same place, all of this has been left lying."[20] Bürgi remained in Prague even after the imperial court had moved to Vienna after 1620, which meant that he was somewhat isolated for ten years.

Yet if one examines the numerous clocks that have survived, the remontoire and the cross-beat escapement were nevertheless more frequently used than has been believed. Hans Buschmann, Nikolaus Planckh, Matthäus Hallaicher, and Christoph Kreitzer in Augsburg, Nikolaus Radeloff in Schleswig (Catalog No. 60), Hager (?) at Arnstadt, Franziskus Schwarz at Brussels, and Georg Mayr at Munich all made the *Kreuzschlag*.[21] Johann Sayller at Ulm (Catalog No. 55) and Hans Buschmann at Augsburg (Catalog No. 57) used the remontoire. The dissemination of Bürgi's

18   The latitude had to be reconstructed; the lack of accuracy arose through the very great size of the hole bored through the hand.

19   Mayr, *Frühgeschichte der Regelung,* p. 123.

20   Benjamin Bramer, *Anleitung eines Berichts von M. Jobsten Bürgi geometrischen Triangular-Instruments* (Marburg, 1684), unpaginated foreword. (This account had appeared in the first edition of 1648.)

21   In Maurice, *Räderuhr,* Vol. I, all these works are illustrated; see p. 153.

improvements was thus greater than has been thought, but his influence was not. One might argue that there were only twenty-five years between Bürgi's death in 1632 and the invention and application of the pendulum as a clock regulator in 1657, and that the clock as a computing device was then replaced by the clock as a measuring instrument. But this consideration, too, is tinged somewhat by modern conceptions of progress.

An "old-fashioned" idea seems more convincing to me as an explanation of Bürgi's limited effect. Bürgi's was a talent of genius proportions. Two princes recognized and fostered his outstanding abilities. His constructions were quite up to the level of his considerable mathematical intelligence, and, above all, they worked. (Whenever Augsburg clockmakers built anything "out of the ordinary," the buyer usually experienced great difficulties in getting the clock to work; see Catalog No. 57.) Bürgi published, and he carried on an extensive correspondence with scientists. Who among the Augsburg craftsmen had his scientific colleagues, his opportunities, and especially his gifts? Burgi remained without influence not because craft traditions were too powerful, but because his genius was too great. This talent was something one could encourage and delight in, but not transmit.

Bruce Chandler and Clare Vincent

9 To Finance a Clock

An Example of Patronage in the 16th Century

Financing the dreams of men who work in science and technology has been a great trial throughout the ages. From the time of Plato's call to the city state to "join in supervising" and "take the lead in honouring" the study of solid geometry,[1] to today's scientific communities' search for government funding, the interaction between the practitioner and his source of support has been problematical. Of the many arrangements for support, most fall into three basic types: the entrepreneurial, the institutional, and the various forms of patronage. These classes are not necessarily distinct, and often they overlap. In Renaissance Germany clockmaking was one of the important products of science and technology, and the history of German clockmaking of the period invites study of all three types of support.

The guilds, the rigorously enforced regulations for commerce in individual cities, and the regularly scheduled fairs were central to the entrepreneurial side of the clockmaking trade in Germany. The support of an institution such as the church or the university seems at first glance not to have had any direct relationship to clockmakers, except in the usual sense of buying or commissioning clocks for utilitarian purposes. But we will see that there was, indeed, an indirect institutional support, which in certain cases proved to be very important. Patronage by individual rulers, the nobility, and the church was extensive in this period, especially for projects of any great complexity or cost. The distinctions between these types of support are not very rigid, and often several played a part in the production of any one project.

Insofar as they can be reconstructed, the fortunes of a 16th-century mathematician from Strasbourg, Philip Imser, and a clockmaker, Gerhard Emmoser von Rainen, provide an excellent opportunity to examine the interrelationships of all three types of support. Philip Imser, the older of the two, was professor of mathematics at Tübingen University. Born in Strasbourg,[2] he matriculated at Tübingen in 1526 and studied under the

---

1 *Republic.* Bk. VII, 528c, trans. Allan Bloom (New York, 1968).

2 The date of Imser's birth is not known. In his matriculation he signed himself *Philippus Ymser Argentinensis* according to Reinhold Rau, "Die Kunstuhr des Philipp Imser," *Tübinger Blätter,* 1962, *49:* 25. Elsewhere he repeatedly calls himself Philip Imser of Strasbourg, notably on the inscription on the astronomical clock he designed, which is now in the Technisches Museum in Vienna, and in the contract for building the clock, which is now in the Hauptstaatsarchiv Stuttgart (see n. 14). The title page of the autograph manuscript describing the functions of the clock is inscribed *Ausslegung vnd geprauch des newen Astronomischen Uren wercks Dariñ alle himlische leuff täglich vor augen stond. Durch Philippū Imserū von Strassburg beschribē. Anno Domini. 1560.* The MS is now in the Österreichische Nationalbibliothek,

57. Table clock from the workshop of Philip Imser. Tübingen, 1554. London, British Museum. (Courtesy of the Trustees of the British Museum.)

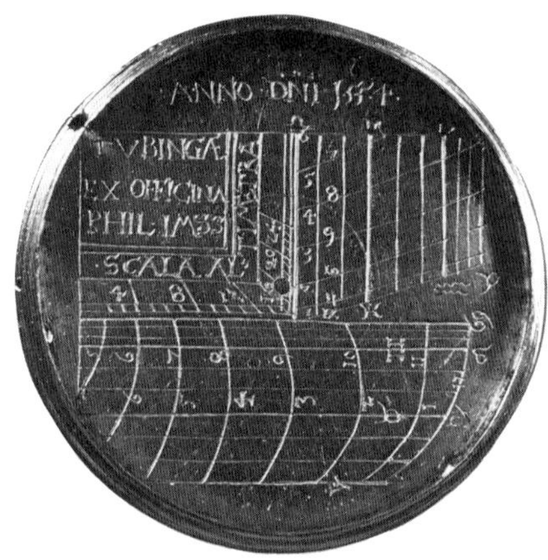

58. Sundial, signed by Philip Imser and dated 1554 on the interior side of the bottom cover of the table clock shown above.

mathematician Johannes Stöffler.[3] Upon Stöffler's death in 1531, Imser was appointed lecturer in astronomy at Tübingen, although he had not yet obtained his doctorate.[4] Just when he became professor of mathematics is not certain, but it was in either 1537 or 1538.[5] Imser finally obtained the doctorate at the University of Ingolstadt at some time between 1544 and 1546.[6] He remained on the Faculty of Arts at Tübingen until his resignation in January of 1557.[7]

During his tenure at Tübingen he was given money by the university senate to acquire astronomical instruments.[8] These were probably for demonstration purposes in the classroom—it was not unusual for universities to support such acquisitions. In Imser's case the tradition of university support for purchasing instruments probably had two important results. First, it may well have led Imser to involve himself in the actual production of the instruments. We have some indication that he did this, for there is a small spring-driven table clock, now in the British Museum, with an unusual sundial (Figs. 57 and 58) inscribed on the inside of its base TVBINGAE EX OFFICINA PHIL IMSS and dated 1554.[9]

The second is of more consequence. It probably encouraged Imser's dreams of building a grand astronomical clock. A curious drawing for an instrument to illustrate the theory of the sun's motion found in a manuscript of about 1538, now in the Bibliothèque Nationale in Paris, may be

Vienna, Codex 10,783 (see notes 21 and 23).

3 Rudolf Roth, *Urkunden zur Geschichte der Universität Tübingen aus den Jahren 1476 bis 1550* (Tübingen, 1877), p. 167, and Rau, "Imser," p. 25.

4 Tübingen Universitätsarchiv (referred to subsequently as TUV), Acta Universitatis, Liber Conductionum 5/13, fol. 61v, cited by Johannes Haller, *Die Anfänge der Universität Tübingen 1477–1537* (Stuttgart, 1927), Vol. II, p. 123, and by Rau, "Imser," p. 25.

5 Andreas Christoph Zeller, *Ausführliche Merkwordigkeiten der hochfurstl. Würtembergischen Universitaet und Stadt* (Tübingen, 1743), p. 495. Zeller says that Imser was made professor of mathematics and astronomy in 1537, the year he returned to the university after his departure over religious differences (see n. 6). In Bibliothèque Nationale MS Latin 7417, fol. 14r, Imser's undated treatise, *Compositio Theoritarum Planetarum per Philippum Imser Astronomice Pfessoram Tubingae . . . ,* is followed by an addition written by Nicolas Gugler and dated 1538 (see n. 10 below). Imser's treatise must therefore have been written in 1538 or earlier.

6 Zeller, *Ausführliche Merkwordigkeiten,* p. 495, and Karl Heinz Burmeister, *Georg Joachim Rhetikus 1514–1574,* Vol. I (Wiesbaden, 1967), p. 40. Burmeister gives the date as 1544. However, Zeller states only that Imser was working to obtain his doctorate at the University of Ingolstadt in 1544, for which he was reprimanded by Prince Ulrich of Württemberg. The payrolls in the Acta Universitatis volumes in the Tübingen Universitätsarchiv refer to Imser for the first time as Doctor with the quarterly payment of his salary in 1546. Prince Ulrich, the enthusiastic promulgator of the Protestant Reformation in Württemberg, was presumably upset by Imser's connection with Catholic Ingolstadt. It was not the first time Imser found himself in trouble with the authorities over his religion, for between 1534 and 1537 he left the Tübingen faculty rather than convert to Protestantism. (See Rau, "Imser," p. 25, and Haller, *Die Anfänge,* Vol. I, p. 336.)

7 TUV, 15/5 Facultas Philosophica G. Prof. Math. & Physics I, 1557–1700, pp. 1a–1b.

8 TUV, Acta Universitatis Jahresrechnung, 1538, fol. 86v; 1549, fol. 173r; 1554, fol. 143r.

9 London, British Museum 74,7-27, 1. See Ernst Zinner, *Deutsche und niederländische astronomische Instrumente des 11.–18. Jahrhunderts* (Munich, 1956), p. 397; P. G. Coole and E. Neumann, *The Orpheus Clocks* (London, 1972), pp. 48–49; and Klaus Maurice, *Die deutsche Räderuhr* (Munich, 1976), Vol. I, p. 62, Vol. II, Fig. 492 and p. 65, No. 492.

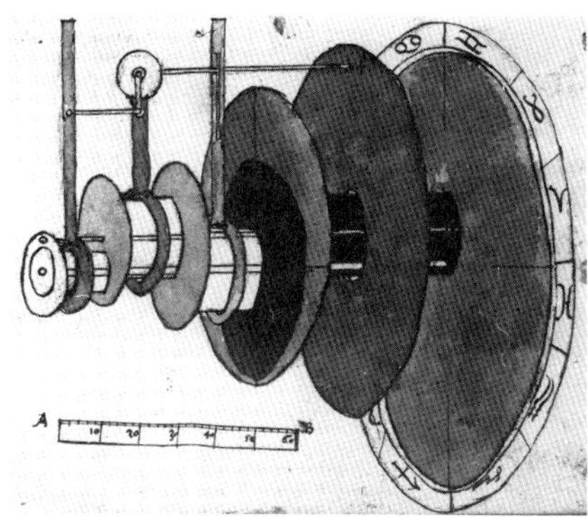

59. Plan for an instrument to illustrate the motion of the sun according to the theory of Georg Peuerbach's *Theorica nova planetarum,* probably drawn by Philip Imser around 1538. Paris, Bibliothèque Nationale, MS Latin 7417, fol. 30r.

a first step in the evolution of an idea for a gear train for a clock to illustrate the planetary motions (Fig. 59).[10] The idea for building such a clock must have finally taken shape in Imser's mind in 1554, when he convinced Pfalzgraf Ottheinrich to finance its construction.[11]

Ottheinrich was a likely patron for the undertaking, being a typical Renaissance prince who took pride in sponsoring many outstanding projects in the arts. Among the more ambitious were the well-known additions to Heidelberg Castle and the rebuilding of his own castle at Neuburg on the Donau. The rebuilding of Neuburg was begun in 1527 and contains some of the earliest importations of the Italian Renaissance style into German architecture. Ottheinrich's taste as a collector ranged from German paintings and Flemish tapestries to small sculptures, medals, coins, and decorative arts. He commissioned the Raphaelesque fresco cycle by Hans Bocksberger the elder in the chapel at Neuburg Castle as well as the impressive tapestry illustrating the genealogy of the house of Pfalz in the collection of the Bayerisches Nationalmuseum. Another of Ottheinrich's former possessions that is now in the Bayerisches Nationalmuseum is the drum-shaped table clock in this exhibition (Catalog No. 38).[12]

One of the most splendid manuscripts to survive from the Pfalzgraf's library is an illustrated astrological treatise (now in the Universitätsbibliothek, Heidelberg, Codex Ger. 832). Ottheinrich, like many another Renaissance prince, was also passionately interested in astrology, collecting books and manuscripts on the subject in addition to supporting astrolo-

10 Paris, Bibliothèque Nationale, MS Latin 7417. The drawing is on fol. 30r. The MS is a collection of various treatises. Following Imser's undated treatise (see n. 5) is an addition (fols. 27–29), *Structura Instrumenti Imaginatorii Theorica,* signed N. Guglero (Nicolas Gugler) and dated 1538. Following this, on fols. 30–35, is the instrument for illustrating the motion of the sun together with various diagrams for instruments illustrating Imser's treatise. These devices are similar to the one Imser drew in his 1560 treatise (see n. 2) and to those in the tables he compiled for Erasmus Oswald Schreckenfuchs' *Commentaria in Novas Theoricas Planetarum Georgii Purbachii* (Basel, 1556).

It is within the realm of possibility that the original idea for the gear train of the clock was Nicolas Gugler's. Although we know very little about Gugler, he was mentioned in an entry for July 5, 1540, TUV, 2/1a, Protocollum Senatus de an. 1524–1541, fols. 237v–238r. See also Bruce Chandler and Clare Vincent, "The Mathematician Philip Immser and the Clockmaker Gerhard Emmoser," *Proceedings of the XIIIth International Congress of the History of Science, Moscow, 1971,* Sec. XVI (Moscow, 1974), p. 362, and Maurice, *Deutsche Räderuhr,* Vol. I, p. 62, Vol. II, Fig. 214, and p. 37, No. 214.

11 Born in 1502 to Ruprecht, the third son of Palatine Elector Philip the Upright (r. 1476–1508), Ottheinrich became Prince of Pfalz-Neuburg in 1522. Upon the death of his uncle, Elector Ludwig V, in 1544, Ottheinrich was, according to the provisions of the Golden Bull, next in line to succeed as Elector, but in accordance with the wishes of his grandfather, Elector Philip, he stepped aside in favor of another uncle, who ruled as Friedrich II from 1544 until 1556. It was not until 1556, therefore, that Ottheinrich became Palatine Elector of the Holy Roman Empire. See Alexander von Reitzenstein, *Ottheinrich von der Pfalz* (Bremen/Berlin, 1938); Siegfried Reiche, *et al., Kurfürst Ottheinrich,* 1956.

12 For further information on Ottheinrich as patron and collector see Hans Rott, *Ottheinrich und die Kunst* (Heidelberg, 1905), *Ottheinrich: Gedenkschrift zur vierhundertjährigen Wiederkehr seiner Kurfürstenzeit in der Pfalz (1556–1559),* ed. Georg Poensgen (Heidelberg, 1956); and Walter Paatz, "Ottheinrich und die Kunst," *Mitteilungen der Vereinigung der Freunde der Studentenschaft der Universität Heidelberg e.V.,* 8th year, 1956, 20: 15–30.

gers.[13] Perhaps it was this interest which led him to commission the clock from Imser, for it would provide astrological as well as astronomical information. The contract for building the clock was concluded on June 12, 1554. Imser promised delivery within a year's time and guaranteed that if there were any fault found in the work, he would rectify it at his own expense. In return, Pfalzgraf Ottheinrich agreed to pay Imser 700 gulden plus an honorarium of 100 gulden for his services. Also Imser supplied a detailed description of the clock to be built.[14]

The clock itself survives (Figs. 60–63) and is now in the Technisches Museum in Vienna.[15] Its spring-driven, iron-framed movement is housed in a rectangular case of gilded and silvered copper which supports an observatory-like element filled with the automata common to astronomical clocks and surmounted by a going celestial globe. In all, the structure is 90 centimeters high.

The construction of the clock turned into a nightmare for Imser, as has been amply documented by letters in the Haupstaatsarchiv in Stuttgart and the Tübingen Universitätsarchiv. Imser's anguish is expressed in one of his last surviving letters, written near the end of November 1558, to Ottheinrich, who had by then become Kurfürst (Elector). The letter contains a plea for some compensation for his difficulties with the clock. Imser says that he has not only ruined his livelihood but also broken his health on account of the lengthy calculations, work, and cost connected with building the clock. He had hoped that he would be rewarded with consoling words from the Kurfürst which would "somewhat revive my melancholy head and deranged mind." But, he continues, the Kurfürst's last letter, which pressed Imser in the most ungracious terms to complete the final sections of the clock, has dashed his hopes, and as things now stand,

13 / G. F. Hartlaub, "Die Kunst und das magische Weltbild," in Poensgen, *Ottheinrich,* pp. 280–282.

14 A copy of this contract is in the Hauptstaatsarchiv Stuttgart (henceforth HSAS), Pfalz No. 115, CVI, 14, Bd. 9d, fols. 87–87b, published by Rott, "Ottheinrich," p. 220, and reprinted by Poensgen, *Ottheinrich,* p. 190. Rott quoted only Imser's part of the contract. The description of the clock is on fol. 99, published by Rott on pp. 117–118, n. 2, by Poensgen on p. 189, and by Rau in "Imser," p. 31, n. 20.

15 Vienna, Technisches Museum für Industrie und Gewerbe, Inv. No. 11939/22. See Rott, "Ottheinrich," pp. 117–120, 220–226. This material was reprinted with additions as "Die Planeten-Prunkuhr des Philipp Imser," in Poensgen, *Ottheinrich,* pp. 185–193. See also Heinrich Pfisterer, "Die astronomische Kunstuhr von Graz und deren Schöpfer Philipp Imser," *Blätter für Heimatkunde,* ed. Historischen Verein für Steiermark, 1941, *19:*9–12; Zinner, *Astronomische Instrumente,* pp. 27, 397–398; Alfred Chapuis and Edmond Droz, *Automata,* trans. Alec Reid (Neuchâtel, 1958), pp. 70–74; Alma Helfrich-Dörner, "Die Planeten-Prunkuhr des Philipp Imser für Ottheinrich von der Pfalz," *Die Uhr,* 1960, *11:* 43–44; Hans von Bertele, *Globes and Spheres* (Lausanne, 1961), p. 24; Rau, "Imser," pp. 25–32; H. Alan Lloyd, *The Collector's Dictionary of Clocks* (South Brunswick, N.J., 1964), pp. 111–112; *Vienne à Versailles: Les Grandes Collections Autrichiennes au Chateau de Versailles,* May 6–Oct. 11, 1964, pp. 116–117, No. 152 (entry by Erwin Neumann); Chandler and Vincent, "The Mathematician Immser," pp. 362–367; Richard Krcal, "Astronomisches Uhrwerk von Philippus Imsserus Anno Domini 1555 im Technischen Museum in Wien," *Blätter für Technikgeschichte,* 1974, *35:* 7–43; Maurice, *Deutsche Räderuhr,* Vol. I, pp. 60–62, Vol. II, pp. 36–37, No. 213 and Figs. 213a–c; H. C. King with J. R. Millburn, *Geared to the Stars: The Evolution of Planetariums, Orreries, and Astronomical Clocks* (Toronto, 1978), pp. 68–72.

60–63. The four views of the planetary clock designed by Philip Imser and made by Gerhard Emmoser. Tübingen and Weil-der-Stadt, 1554–1559. Vienna, Technisches Museum für Industrie und Gewerbe.

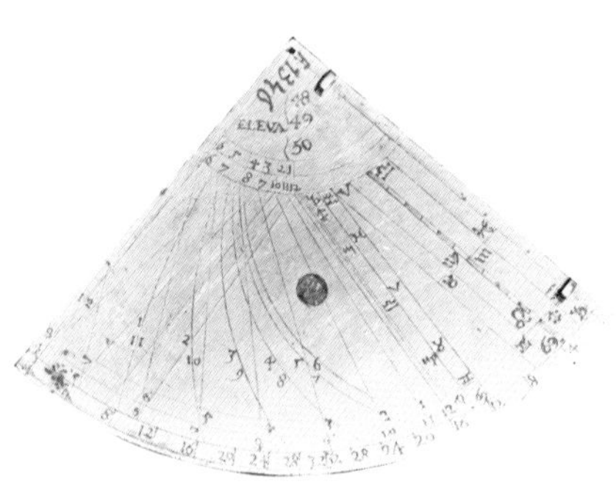

64, 65. Quadrant to be used as a sundial on one side, as a nocturnal on the other, by Gerhard Emmoser, 1558. Vienna, Österreichisches Museum für angewandte Kunst.

he can no longer apply his mind to making this clockwork run correctly. Therefore he has decided that he must forget all his troubles and have a complete rest. He would like to have 800 gulden, although he is aware that the clock is unfinished and not working quite right.[16]

In fact, it seems most unlikely that the clock would ever have been more than a shell had not Ottheinrich in 1556 sent the Heidelberg clockmaker Gerhard Emmoser von Rainen to Tübingen to carry out Imser's ideas.[17] The involved negotiations, and the complications which arose between the signing of the contract in 1554 and Ottheinrich's death in 1559, are too lengthy to include here; they have been well summarized by Hans Rau in an article in the *Tübinger Blätter.*[18]

We know nothing of Emmoser before 1556.[19] In 1558 he made a single instrument (Figs. 64, 65), a combination quadrant sundial and nocturnal, which he signed GERHART VON RAEINEN/VHRMACHER.[20] The collaboration between the clockmaker and the mathematician was far from amiable. Early in their relationship Emmoser was complaining to Ottheinrich that he needed more guidance from Imser in order to do his work, while at one point in 1558, when the clock was almost finished, the clockmaker took some parts of the movement and locked them into a trunk, claiming that he and not Imser was their inventor.[21]

16 HSAS, Pfalz No. 115, CVI, 14, Bd. 9d, fol. 99, published by Rott, "Ottheinrich," pp. 224–226: *"hab ich gentzlich verhofft, E. churf. gn. würde aus angeborner mülte mich betriebten man gnedigst bedacht und mit tröstlicher zusprechung mein melancholischen Kopf und verrückte sinn widerumb etwas erquickt haben."*

17 *Ibid.,* Bd. 9d, fol. 87c, published by Rott, "Ottheinrich," pp. 222–223. The realization that Gerhard von Rainen, the clockmaker who carried out Imser's plans for the astronomical clock, and Gerhard Emmoser, clockmaker to the Holy Roman Emperors Maximilian II and Rudolph II, were one and the same person was slow in coming. Both Zinner (*Astronomische Instrumente,* p. 27) and Neumann (*Vienne à Versailles,* p. 117) thought that there was a strong probability that such was the case, but neither offered definite proof. A number of contemporary documents do in fact furnish the proof by giving the clockmaker's full name, Gerhard Emmoser von Rainen. Among them are the Stadtarchiv Augsburg, Zunftbuch der Schmiede, No. 56, fol. 96b, Oct. 24, 1563, and the Hochzeitsamt Protokolle, No. 1 (1563–1569), fol. 36v, Oct. 24, 1563.

18 "Imser," pp. 26–29 (see n. 2).

19 It has not been possible to find a record of Emmoser's birth. The village of Rain, however, lies but a few kilometers from Neuburg a.d. Donau, the residence of Emmoser's first recorded patron Pfalzgraf Ottheinrich.

20 Vienna, Österreichisches Museum für angewandte Kunst, F. 1346. This instrument is wrongly listed by Zinner (*Astronomische Instrumente,* p. 242) as the work of an otherwise unknown maker, Gebhart von Baeinen, a misreading of the inscription on the instrument. (The statement, in Chandler and Vincent, "The Mathematician Immser," p. 365, that the instrument was attributed by Zinner to Gebhart von Gaeinen is a misprint.) The instrument is also incorrectly stated by Zinner to be in the collection of the Kunsthistorisches Museum in Vienna. The authors are indebted to Erwin Neumann for the discovery of its true location.

21 HSAS, Pfalz, No. 115, CVI, 14, Bd. 9d, fol. 87, Philip Imser to Stephan Chonberg in Tübingen, Weil-der-Stadt, July 13, 1558, published by Rott, "Ottheinrich," pp. 223–224, and cited by Rau, "Imser," p. 29. There is reason to believe that the case of the clock had been made before Emmoser entered into the project, for the case is dated 1555 in two conspicuous places. It is not clear whether the actual construction of the movement had begun before the end of 1555, but apparently it was either incomplete or nonexistent when Emmoser began to assist Imser. Thus, although it is clear that the calculations for the more complicated gearing of the planetary train of the clock must have been the work of the

The most important aspect of the clock, which made it a departure in the long tradition of European astronomical clocks, was the single dial which showed at once the position in the zodiac of each of the known planets and the sun, as well as the moon's position, age, nodes, and phases. All eight hands for this dial are operated by a single complicated gear train that turns eight concentric arbors.[22] Apparently Imser's clock was the first to have such an arrangement, and Imser was very proud of it. In fact he boasts in his manuscript describing the clock that "the positions of all the planets run unequally off a central axis (which no clockmaker has yet been able to understand how to make)."[23] The germ of the idea for this train is undoubtedly the diagram that appeared in the Paris manuscript mentioned earlier. In addition, the positions of each of the five known planets above and below the zodiac (celestial latitudes) are shown on the five dials on the left side of the clock (Fig. 62). They are apparently a mechanical transformation of the celestial latitude volvelles in Peter Apian's *Astronomicum Caesareum* (Ingolstadt, 1540). Such mechanical volvelles are not found on any other known clock of the 16th century.

Because the clock was designed to be used throughout Europe, maps showing the latitudes of a large number of European cities—necessary to the proper setting of both the clock and its globe—were included on the doors at the rear of the clock case (Fig. 63). With the exception of a later clock made in 1566 by Gerhard Emmoser alone, these maps are the only

---

mathematician, there is perhaps some justification for their later differences over the credit for the invention of parts of the clock. In the formal inscription on the case, the entire credit is claimed by Imser. However, on the inside of one panel of the case, the clockmaker scratched his name "Gerhard."

A lingering feeling of having been exploited may also explain the existence of a manuscript by Emmoser, *Ausslegung unnd von dem geprauch des Neuen Astronomischen Uhrenwerckhs darinn alle himlische Leuff Täglich vor Augen steen,* now in the Universitätsbibliothek in Graz (MS I, 151), which is an almost word-for-word copy of Imser's manuscript of 1560 (see notes 2 and 23). Emmoser's manuscript contains the theoretical chapter, as does Imser's, contrary to Zinner's statement (*Astronomische Instrumente,* p. 398), but more interesting is the fact that Emmoser's copy is signed "Gerhard Emmoser Ro. Kay. Mt. Uhrmacher beschriben." From the evidence of the signature the copy could not have been made before 1566, when Emmoser was indeed Clockmaker to the Holy Roman Emperor, and therefore it must be at least six years later in date than Imser's. The purpose of Emmoser's manuscript remains a puzzle, but its presence at Graz is explained by the later history of the clock, for by 1632 it had been placed in the Archducal Hofbibliothek in Graz. In 1752 it was part of the Museum Mathematicum of the Jesuit College in Graz, and in 1779 it was listed in an inventory of the Universitätsbibliothek in Graz (see Pfisterer, "Astronomische Kunstuhr," pp. 9–10, and Neumann, *Vienne à Versailles,* p. 117). The manuscript in the library at Stift Rein (Zinner, *Astronomische Instrumente,* p. 398, and Zinner, *Verzeichnis der astronomischen Handschriften des deutschen Kulturgebietes,* Munich, 1925, No. 2509) is an 18th-century copy of the Emmoser MS at Graz.

22 For detailed descriptions of the planetary train see Krcal, "Astronomisches Uhrwerk," pp. 7–43, and King, *Geared to the Stars,* pp. 68–72.

23 Philip Imser, *Ausslegung vnd gebrauch des newen Astronomischen Uren wercks,* Vienna, Österreichische Nationalbibliothek, Codex 10,783, fol. 5r: *"dem die stellung aller Planeten ungleichen leuff auff einer achs oder centro (welches bisher noch keine uhrmacher zich hat understanden auff die art zu machenn . . .)."* The MS describes the function and setting of the clock, and it also includes as the last chapter a short astronomical treatise. (See also notes 2 and 21.)

such representations on a 16th-century clock that are known. The main
dial of the fourth side registers the Nuremberg hours, as well as the equal
hours in the twice-twelve or Whole Clock system (Fig. 61). The shutters
for the Nuremberg hours are controlled by an ingenious mechanical
device which is possibly the invention of Imser or Emmoser. Also there
is a perpetual calendar with saints' days, an epact for finding the date of
Easter and the other movable feasts of the Church, and subsidiary dials for
an alarm and for setting the Nuremberg hour dial for use in the latitudes
42°, 45°, 48°, 51°, and 54°.

The octagonal structure at the top of the clock is divided into two tiers.
The upper is covered with representations of the human activities which
were thought to be governed by each of the planets. At the front there
is a sort of puppet theater for automata representing the appropriate
astrological planetary rulers of the hours. The lower tier depicts scholars
using various astronomical instruments and the ages of man: the child, the
youth, the older man, and death.[24] The last four figures are engraved on
the doors that open at the quarter hours to reveal automata that echo the
theme of the engravings on their respective doors. In the sequence of their
appearance they also symbolize the birth, aging, and death of the hour.
In addition, a maiden makes a complete circuit of the top of the clock once
an hour and indicates the minutes as she passes. Surmounting the observatory-like element is the celestial globe engraved with the constellations
derived from Albrecht Dürer's celestial maps of 1515.

It is commonly thought that most 16th-century patrons were more
delighted by automata than seriously interested in the accuracy of the
astronomical features of such instruments. Clearly this is not true, at least
in Ottheinrich's case. The automata must have played a part in tempting
Ottheinrich's commission, but the overriding concern was the accuracy of
the astronomical data that it would provide, for in fact the errors found
when the clock was examined in late 1558 or early 1559 were serious
enough to prevent Ottheinrich from accepting the clock.[25]

On February 2, 1559, an arrangement was made between Ottheinrich
and Imser whereby Imser would pay Ottheinrich 600 gulden within a
month and would then be free to sell the clock, although Ottheinrich
would retain the right of first refusal on the sale.[26] Almost immediately
(Feb. 12) Ottheinrich died, ending all hope Imser had of patronage from
the court of the Pfalzgraf.

This left Imser with a splendid, very expensive, and somewhat inaccurate clock on his hands. Moreover, he had resigned his position at the
university and could no longer count on support from that quarter.[27] The

---

24 It has been suggested that one figure in the group of scholars using instruments is Ottheinrich
    himself. See Poensgen, *Ottheinrich,* Plt. I, opposite p. 324.
25 HSAS, Pfalz No. 115, CVI, 14, Bd. 9d, fol. 110, which contains an extensive and detailed
    list of errors in the positions of the planets as shown by the clock from Dec. 1554 through
    Dec. 1558.
26 Rau, "Imser," p. 30.
27 We do not know whether Imser paid Ottheinrich's estate at the agreed time, but he did

story of what happened during the next two years is not known in all its details, but in a letter written by the Holy Roman Emperor Ferdinand I to his son, Maximilian, King of the Romans, on October 7, 1561, the Emperor agreed that he would pay Imser 2,400 Rheinish gulden for the clock, plus a yearly stipend of 200 gulden for life, as well as 100 gulden per year for his widow, upon his death. The Emperor added a postscript, however, that Imser must first make the clock work correctly.[28] In addition, we have the manuscript written by Imser in 1560, *Ausslegung vnd geprauch des newen Astronomischen Uren wercks,* describing the operation of the clock, beginning with an elaborate introduction and ending with a short astronomical treatise.[29] From these two documents the most reasonable conclusion one can draw is that Imser wrote the manuscript in an effort to sell the clock either to the Emperor or to his son.[30]

Furthermore, in a series of letters from the early part of 1562, Gerhard Emmoser, who signed himself "Clockmaker to the Pfalzgraf," pleads with the Emperor to pay Imser his stipend of 200 gulden and requests that the Emperor write to Ottheinrich's successor as Pfalzgraf, Friedrich III, to excuse the clockmaker's long absence and to grant an extension allowing him to stay longer at the court of the Emperor.[31] In the same bundle of archival documents, immediately after the Emmoser letters, there is an accounting of the expenses for a trip with a clock from Weil-der-Stadt to Ulm and down the Danube to Vienna. Since Weil-der-Stadt was the town where Imser took the clock after his retirement from Tübingen University, it seems certain that this account refers to the transportation of Imser's clock to the court of the Holy Roman Emperor. It is probably safe to conclude, therefore, that Imser and Emmoser settled their differences and combined their efforts to sell the clock. In this they succeeded, and Imser received the Emperor's permission for his stipend and at least one payment in February 1562.[32] With the sale of the clock to the Emperor, Imser

---

borrow 400 gulden from Tübingen University in May of the same year, and still later 200 gulden, perhaps to pay the debt *(ibid.).* As the result of this last agreement between Ottheinrich and Imser, the provision of Ottheinrich's testament (Karlsruhe, Generallandesarchiv, Pfälz., Kopialb. 845, fol. 125, and Munich, Hausarchiv, Kasten 15, Lade 1, No. 3007, Akt 976) bequeathing the clock to the library at Heidelberg University must have been considered void. (See also Poensgen, *Ottheinrich,* p. 187.)

28 Franz Kreyczi, "Urkunden und Registen aus dem K.u K. Reichs-Finanz-Archiv," *Jahrbuch der Kunsthistorischen Sammlungen des Allerhöchsten Kaiserhauses,* 1887, 5: lxxxviii, No. 4314. See also p. lxxxviii, No. 4315.

29 See notes 2, 21, and 23.

30 For Maximilian's interest in clocks and automata see Alphons Lhotsky, *Geschichte der Sammlungen (Festschrift des Kunsthistorischen Museums in Wien, 1891–1941.* Zweiter Teil, Erste Hälfte) (Vienna, 1941–1945), pp. 177–178. See also Hugh Trevor-Roper, *Princes and Artists: Patronage and Ideology at Four Habsburg Courts, 1517–1633* (London, 1976), pp. 85–95.

31 Hofkammer- und Finanzarchiv, Vienna (hereafter abbreviated HKA), uncatalogued material, 1562, Hoffinanz note No. 11.

32 HKA, Gedenkbuch, 90, fols. 39r–39v, published by Kreyczi, "Urkunden," p. lxxxix, No. 4221. HKA, Finanz Register, R 254, Feb. 18, 1562. There are no records of later payments to Imser in the volumes of the Finanz Register that are preserved in the Hofkammer- und Finanzarchiv, a fact which seems at first to indicate that the mathematician must have died in that year. However, Emperor Ferdinand I had promised both in the letter to Maximilian

111

assured himself of an honorable place in the history of German clockmaking, and Emmoser's career took a decisive upward turn. It proved to be the beginning of Emmoser's rise from clockmaker at a provincial German court to posts in the courts of two of the great imperial patrons of art and science in the 16th century, Maximilian II and Rudolph II.

On July 15, 1562, Emperor Ferdinand I excused Emmoser's debts to Pfalzgraf Friedrich III; on February 4, 1563, the Emperor installed the clockmaker in the imperial city of Augsburg, then the major center of European clockmaking; three weeks later he was made a burgher of that city.[33] During the course of the next three years Emmoser married[34] and settled in Augsburg, working to complete the requirements for the position of master clockmaker.[35]

In 1564 Maximilian became Holy Roman Emperor. Although it was Maximilian's father who paid for the Imser-Emmoser clock, the surviving letters about the purchase show that it was clearly Maximilian who had been instrumental in its purchase. Maximilian must have remained in contact with the clockmaker, for in February of 1566 he interceded with the Augsburg officials and asked that Emmoser be given three months more in which to complete his masterpiece.[36] Furthermore, on October 14 of that year the Emperor wrote to the city council that "we want to have your burgher Gerhardtus Emmoser, clockmaker, with his clock and handwork, to work for a period in our Imperial Court."[37] The letter goes on to explain that because of the Emperor's wishes, Emmoser had come with his wife and child to the court, but Emmoser's Augsburg citizenship should not be affected by his absence. Thus in 1566 Emmoser became Clockmaker to the Holy Roman Emperor, a title he was to keep until his death in 1584.[38]

---

and in the Gedenkbuch entry cited above to pay 100 gulden per year to Imser's widow, and no such payments can be found. Imser's death date is given as Oct. 20, 1570, by Martin Crusius, *Schwäbische Chronik* (Frankfurt, 1733), Vol. II, p. 322. See also Rau, "Imser," p. 31.

33 HKA, Finanz Register, R 254, July 15, 1562; Stadtarchiv Augsburg (hereafter STAA), Ratsprotokolle, 32, 1563, fol. 10v, Feb. 4, fol. 17v, Feb. 25.

34 STAA, Hochzeitsamts Protokolle, No. 1, 1563–1569, fol. 26v, No. 268. Emmoser married Eufronsina Weinolt or Weinold, the daughter of the Augsburg goldsmith Michael Weinold, in 1563.

35 STAA, Ratserkentnissen, Vol. XXXIII, 1564–1565, fol. 25v, Oct. 16, 1565; Vol. XXXIV, 1566–1567, fol. 18r, Feb. 23, 1566.

36 *Ibid.*, fol. 19r, Feb. 28, 1566.

37 *Ibid.*, Augsburg and Kaiser, Fasc. 12, Maximilian II, 1558–1576, Produkt vom 4.10.1566, *"wir euer Burger Gerhardtus Emmoser urmacher mit seinen uhr und handwerk an unserem Kaiserlichen Hof in unserem Diensten ein Zeitlang zu gebrauchen vorhaben sein."* The authors would like to thank Dr. Friedrich Blendinger, Archivdirektor, Stadtarchiv Augsburg, for locating the document and Dr. Gudrun Becker for her invaluable help in transcribing this almost unreadable manuscript.

38 HKA, Hofzahlamtsbuch, 21, 1566, fol. 277v, Dec. 31, 1566, published in part by Wendelin Boeheim, "Urkunden und Registen aus der K. K. Hofbibliothek," *Jahrbuch der Kunsthistorischen Sammlungen des Allerhöchsten Kaiserhauses*, 1888, 7:cxxiv, No. 5065. As Clockmaker to the Emperor *(Hofuhrmacher)*, Emmoser received 12 gulden per month. He was given back pay to August of 1566. Hofzahlamtsbuch, No. 42, 1591, fol. 237r, Feb. 15, 1591, states that Emmoser died in November of 1584. Published in part by Boeheim, p. ccxxiii, No. 5509.

There is no record that Emmoser ever became a master clockmaker at Augsburg. But a spring-driven astronomical table clock (Fig. 66), which was formerly in the Imperial collections and is now in Budapest,[39] signed GERHART EMMOSER/VHRMACHER. IN AUGS/PVRG and dated 1566, is, because of its unique construction, probably the clock that Emmoser was making as his masterpiece. Because the imperial arms are on the pierced housing for the bell—an element that could be easily and quickly replaced—and not on the main body of the clock, it is tempting to believe that the Emperor purchased this clock before it could be presented in Augsburg as Emmoser's masterpiece.[40]

The Hofkammer- und Finanz Archiv in Vienna provides records of many payments for his work throughout the rest of his life, but only two instruments from this period are known to have survived. One, now in the collection of the Germanisches Nationalmuseum in Nuremberg, is an unusually elaborate sundial and nocturnal, dated 1571.[41] The other, made for Emperor Rudolph II and now in the Metropolitan Museum of Art in New York, is Emmoser's masterpiece, the magnificent celestial globe with clockwork shown in Catalog No. 113.

66. Table clock by Gerhard Emmoser. Augsburg, 1566. Budapest, Iparmüvészeti Múzeum.

39 Budapest, Iparmüvészeti Múzeum. See Ernst von Bassermann-Jordan, "Die Uhr des Augsburgers Gerhard Emmoser für Kaiser Maximilian II," *Die Uhrmacherwoche,* Dec. 22, 1928, 35:847–888; Zinner, *Astronomische Instrumente,* p. 304; Erwin Neumann, "Die Tischuhr des Jeremias Metzger von 1564 und ihre nächsten Verwandten. Bemerkungen zur Formengeschichte einer Gruppe süddeutschen Stutzuhren der Renaissance," *Jahrbuch der Kunsthistorischen Sammlungen der Allerhöchsten Kaiserhauses,* 1961, 57:113 and Fig. 129; Maurice, *Deutsche Räderuhr,* Vol. I, p. 62, and Vol. II, p. 29, No. 156 and Figs. 156a,b.

40 HKA, Finanz Register, R271, 1566, fol. 225r, Oct. 6, 1566: Emmoser was to be paid 500 gulden for his work. The Hofzahlamtsbuch, No. 21, 1566, fol. 604r in the same archive records a payment of an additional 50 gulden for a clock *(aines Uhrwercks)* for the Emperor in 1566. In addition, on July 10, 1567, Emmoser was paid an installment of 300 gulden for a large clock *(aines . . . grossen uhrwerck)* (see Kreyczi, "Urkunden," p. cvi, No. 4408). On Aug. 20, 1567, Emmoser received another 100 gulden for the same clock (see Boeheim, "Urkunden," p. cxxv, No. 5081). Also in the letter of Emperor Maximilian II to the Augsburg council, the Emperor says that Emmoser has brought his clock with him to the court (see n. 37).

41 Nuremberg, Germanisches Nationalmuseum, WI 1803. See Zinner, *Astronomische Instrumente,* p. 304, and Plt. 38, Fig. 1.

The authors wish to acknowledge their debt to their former colleague and friend, the late Dr. Erwin Neumann, Leiter der Sammlung für Plastik und Kunstgewerbe, at the Kunsthistorisches Museum in Vienna, whose enthusiastic encouragement prompted their initial study of the collaboration between Philip Imser and Gerhard Emmoser.

Peter Scott Honig

10 History and Mathematical Analysis of the Fusee

The application of the spiral spring as a power source gave birth to a new breed of clocks as the freedom from dependence on gravity permitted movement and orientation in three dimensions. Portability and size reduction gave even more leeway to the clockmaker in experimenting in mechanical design. Experimentation was indeed necessary, for this new power source had a great problem associated with it: the more the clock was wound, the faster it ran. Something had to be done to make the driving force constant.

Germany experimented with a device called a stackfreed, a springy piece of iron fitted with a roller that pressed against a cam. The cam was shaped so that the stackfreed produced a retarding force on the barrel when wound and aided the driving force when the spring was running down. Although the mechanism helped somewhat, it was a far cry from achieving constant power.

The solution was developed in the form of the combination of a cord or chain (first reference in 1539) with a sort of cone called a fusee. Qualitatively it works in the following manner. When the clock is unwound, the chain or cord is wrapped on the outside of the mainspring barrel with one end of the cord anchored in a slot in its side; the other end is anchored in the fusee. As the timepiece is wound, the chain becomes wrapped along the spiral grooves of the fusee as it is unwrapped from the spring barrel. While this is happening, the tension in the chain increases as a result of the growth in spring force. Since the fusee is the first member of the gear train, the torque or twist which it develops must be constant if timing is to be accurate. As tension in the chain increases, the radius of the fusee where the chain makes contact decreases to compensate theoretically for force changes. A lot of force acting on a small radius is equivalent to a little force on a large radius. That is why the fusee possesses its unusual shape (see Fig. 67).

The answer to the question of when or by whom the first fusee was made has long been shrouded by time; however, many years of research have yielded some pieces of information that suggest various dates in horological history. Drawings by Leonardo da Vinci around 1493–1500 induced horologists to credit him with the fusee mechanism's invention, until the discovery of a manuscript in the Bibliothèque Royale of Brussels which established that the device existed by the mid-fifteenth century.[1] Another early document is the Almanus manuscript from the last quarter of the 15th century, where we find detailed mechanical descriptions of

67. A wooden fusee from the Minerva table carriage, Catalog No. 108.

---

1 Cedric Jagger, *The World's Great Clocks and Watches* (London, 1977), p. 91.

some fusee clocks. The first surviving example is the so-called Burgundian clock, circa 1430 (in the Germanisches Nationalmuseum, Nuremberg).

It is evident that the craftsmen of the period did not have the knowledge to make the appropriate mathematical calculations, so the development of the form for a well-working fusee came about by empirical methods. To generate a fusee the clockmaker used two basic properties: the principle of torque transmission and the fact that a fully wound spring exerts much more force than one that is run down. Therefore a logically intuitive guess for a fusee shape would be a truncated regular cone, and I hypothesize that the first fusee was indeed such a cone, without any curvature.

An example of this is found in a clock by Jost Bürgi, circa 1600 (Fig. 51, Catalog No. 53). The profile of the fusee is a cone as previously described, and it is made of wood. Because of its high workability, wood is an excellent material to use where repetitive adjustment of the shape is necessary, and with the early spring-driven clocks this was certainly the case. Curvature in the profile was no doubt the result of design modification through the study of torque generation and timing-rate errors.

If the making of the fusee were to be recreated, we would begin with a truncated regular cone that would be inserted into the movement. It would have a spiral groove like a fusee, and with the chain or cord in its most unwrapped position the initial force of the mainspring would be set up. A very simple but important tool called an adjusting rod would then be placed on the fusee winding square and the weight on the rod adjusted until it counteracted the torque.[2] With the clock slightly wound up, the rod (with its weight in the same position) would again be placed on the square to see whether the torques still balanced. If the torque due to the spring exceeded that of the rod, the fusee radius had to be reduced along that region. If it was less, the radius needed to be increased, which unfortunately could not be done practically. That is why the clockmaker would begin with a wide cone, so there would be enough material to avoid having to increase the radius at any point. After much testing and applying the observations and indications, the reworked cone would appear in its highly recognizable shape. Since the necessity of obtaining steady torque was such a great task, it is no wonder that we find some clocks with wooden fusees.

Perhaps the method of trial and error was not the most accurate, but it yielded a shape that gravitated toward the ideal. Considering the errors in escapements that still remained, and the fact that a time-regulated society had not yet developed, the early fusees performed their task sufficiently well.

Another problem in achieveing constant torque was the lack of uniformity and consistency in the quality of mainsprings. Variations in thickness and uneven tempering resulted in unpredictable changes in force, thus preventing any theoretical analysis from being successfully put into practice.

2   F. J. Britten, *Watch and Clockmakers' Handbook* (London, 1955), p. 3.

One of the early analyses to be performed on the fusee from a mathematical point of view was by Pierre Varignon in 1702.[3] Using the differential calculus developed by Leibniz during the last quarter of the 17th century, Varignon developed a differential equation of the form

$$dx = -\frac{dy}{m}\left(\frac{b}{y}\right)^{\alpha/2}$$

where y = fusee radius
  x = fusee height
  m = spring force and $\alpha = (4m + 2)/m$
  b = constant

Varignon first developed an expression for the spring force m on his own, as well as giving two other solutions developed by Johann Bernoulli, and then proceeded to the final solution:

$$xy^{(2m + 1)/m} = \frac{2a^{(m + 1)/m}\,b^{1/m}}{2mc + c}$$

where c is a constant. The method of obtaining the solution was not stated, for analysis of differential equations during Varignon's time was often accomplished by trial and error, unlike today's methodical procedures.

An attempt was made by the English mathematician Benjamin Martin in 1764 to solve the fusee problem without the difficulties of a differential equation. After a good qualitative description of the equalization principles of the fusee, Martin formulated equations using a graphic superimposition of the spring force upon the device's profile.[4]

Martin formed a relationship between the force and radius without a differential equation or even any calculus. His result by means of algebra was an equilateral hyperbola, but he was unfortunately incorrect, for the profile is actually a more complex form of a hyperbola (see the comparison shown in Fig. 68). His serious error was made at the beginning of his analysis when he assumed that the spring force varies linearly with the fusee height. Although it is true that force varies linearly with the length of chain pulled off the spring barrel (for a linear spring), the relationship is lost when it is wound upon the fusee, for the decrease in the radius causes less length of chain to be wrapped up per revolution as the timepiece is wound. Without integrating along the surface of the fusee it is not possible to determine accurately the amount of chain or cord that has been utilized by the fusee for any given position.

It seems strange that the various individuals who incorrectly attempted to formulate fusee equations did not notice their errors. The craftsmen who used the method of observing errors in the fusee and then reworking

3 Pierre Varignon, "De la figure ou curvité des fusées des horloges à ressort," *Mémoires de l'Académie Royale des Sciences, Paris,* 1702, p. 195.
4 Benjamin Martin, *System of Mathematical Institutions* (London, 1764), Vol. II, p. 366.

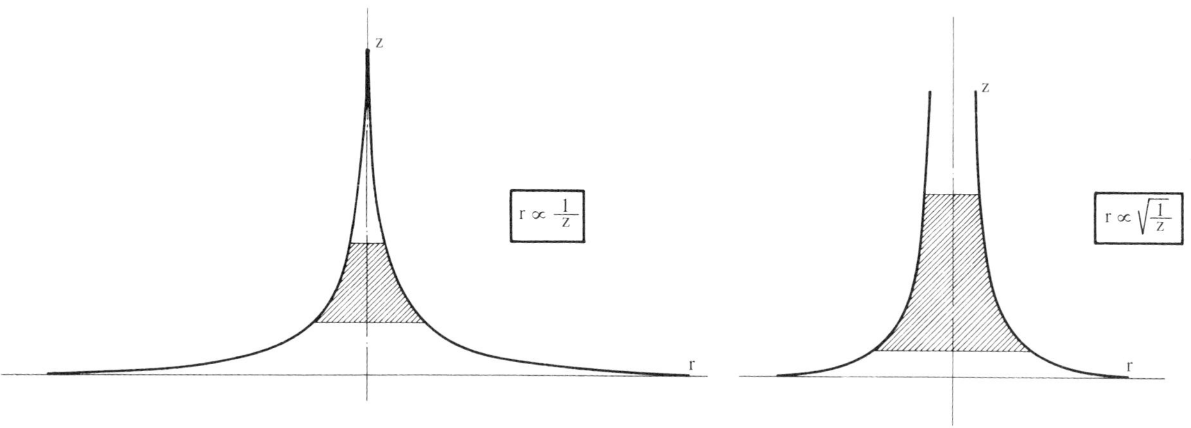

68. Equilateral hyperbola, where r = radius of the fusee and Z = height of the fusee (this was an incorrect depiction of the relationship). More complicated hyperbola (correct relationship).

the profile achieved the correct form for the device. The majority of the theoreticians, however, continued to obtain false results, while believing them to be correct. The fact that theory and practice differed by an easily observed amount tends to indicate that those involved with theory had nothing to do with actual production, and the craftsmen who worked empirically had nothing to do with theory. No doubt there was a great dichotomy between scientists and craftsmen of the period.

Using calculus and some algebraic relationships, I have proposed the equation for the ideal fusee. The following mathematical derivations demonstrate how the solution can be accomplished successfully while avoiding the complications of a differential equation. This method, however, is only applicable to linear springs or springs of higher order, and for more complex force functions a differential equation must be formulated to obtain a solution.

I: Formulation of the mathematical theory

Given that the spring force $f(x)$ is a function of the distance x (or chain extension), it is required that the spring force times the fusee radius r be a constant value of torque T. That is, $f(x)r = T = $ constant.

It is now necessary to specify the wrap spacing t. In other words, specify the distance t, which the chain moves axially for one revolution of the fusee; note that it may not be less than the chain or cord diameter. Using the point where the chain is anchored to the fusee base as an angle reference, we can say: If the angle $\theta = 0$ when height $Z = 0$, then

$$\frac{\theta}{2\pi} = \frac{Z}{t} \tag{1}$$

where θ is in radians. Let x be the length of the chain wrapped on the fusee.

117

Now

$$x = \int_0^\theta r \, d\theta \tag{2}$$

where $r = r(Z)$, meaning the radius r is a function of height Z. Also spring force $\times$ radius $=$ torque, or

$$f(x) \, r = T \tag{3}$$

Substituting Eq. (2) in Eq. (3) yields

$$rf\left(\int_0^\theta r \, d\theta\right) = T \tag{4}$$

Rearranging Eq. (1) gives

$$\theta = \frac{2\pi}{t} Z \tag{5}$$

Now taking the derivative produces

$$d\theta = \frac{2\pi}{t} \, dZ \tag{6}$$

Inserting Eq. (6) into Eq. (4) yields

$$rf\left(\int_0^z r \, \frac{2\pi}{t} dZ\right) = T \tag{7}$$

Or, rearranging this, we obtain the *basic working equation:*

$$rf \, \frac{2\pi}{t} \int_0^z r \, dZ = T$$

where $\quad T = $ torque
$\quad\quad\;\; t \; = $ wrap spacing
$\quad\quad\;\; f \; = $ spring force
$\quad\quad\;\; Z = $ axial distance or height
$\quad\quad\;\; r \; = $ fusee radius

II: Fusee for a linear spring

A linear spring creates a force which is proportional to the extension of the chain. For example, with a cord wrapped around the barrel, we make a reference indicator at the point where the spring just begins to pull. A 1-inch extension might yield a force of 1/2 pound; then a 2-inch extension will yield a force of 3/4 pound, and 3 inches yields 1 pound, and so on.

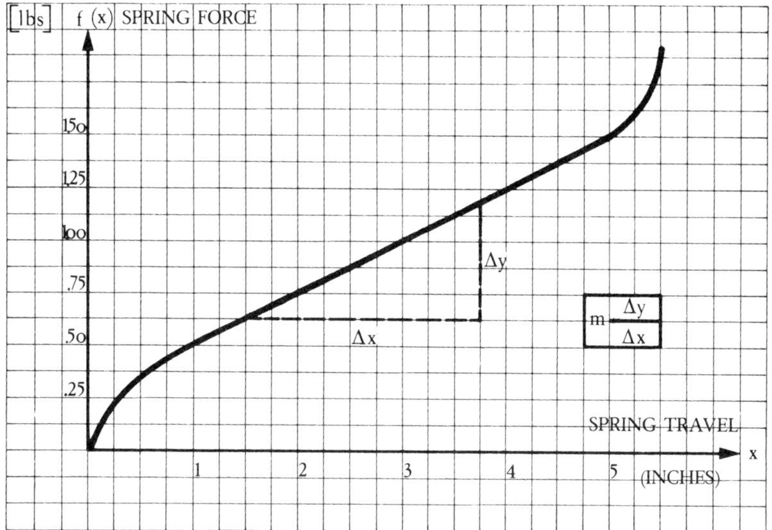

Thus for every inch of extension 1/4 pound of force is generated, or the spring constant is 1/4 pound per inch. Using the letter symbol of the following equations, we can say $m = 0.25$. A graphic representation of the above example is shown in Fig. 69.

Let us take the simple case where the spring force is completely linear. Then the equation for the force can be written as $f(x) = mx$ where m is the spring constant. Substituting this expression into the basic equation derived in Part I yields

$$rm \; \frac{2\pi}{t} \int_0^z r \, dZ = T \tag{8}$$

Since we are interested in the fusee profile, it may be represented by the general equation:

$$r = AZ^n \tag{9}$$

where A and n are constants to be found. Inserting Eq. (9) into (8) produces

$$AZ^n m \; \frac{2\pi}{t} \int_0^z AZ^n \, dZ = T \tag{10}$$

Integrating yields

$$AZ^n m \; \frac{2\pi}{t} A \; \frac{Z^{n+1}}{n+1} = T \tag{11}$$

Or in simplified form

$$A^2 m \; \frac{2\pi}{t} \; \frac{Z^{2n+1}}{n+1} = T \tag{12}$$

The purpose of the fusee is to maintain constant torque. For the left side of the equation to be constant, it is required that

$$Z^{2n+1} = \text{constant} \tag{13}$$

Now taking the derivative of Eq. (13) implies

$$2n + 1 = 0 \tag{14}$$

Therefore,

$$n = -1/2 \tag{15}$$

Inserting the value for n into Eq. (12) we get

$$A^2 m \; \frac{2\pi}{t} \; \frac{1}{1 - 1/2} = T \tag{16}$$

or

$$A^2 = \frac{Tt}{4\pi m} \tag{17}$$

Since $n = -1/2$, Eq. (9) now becomes

$$r = AZ^{-1/2} \quad \text{or} \quad Z = \frac{A^2}{r^2} \tag{18}$$

Inserting Eq. (17) into Eq. (18) produces the *linear spring fusee equation:*

$$Z = \left(\frac{Tt}{4\pi m} \right) \frac{1}{r^2} \tag{19}$$

or, in radius-solving form,

$$r = \sqrt{\frac{Tt}{4\pi m} \; \frac{1}{Z}} \tag{20}$$

where again
$$
\begin{aligned}
T &= \text{torque} \\
t &= \text{wrap spacing} \\
m &= \text{spring constant} \\
Z &= \text{axial distance or height} \\
r &= \text{fusee radius}
\end{aligned}
$$

With but a small amount of insight, almost anyone practicing chronometry will find that the equations of this analysis can readily be employed to produce a perfect fusee. Unlike those of the past, today's craftsmen should have no difficulty in achieving harmony between theory and practice. In fact, the existence of high-precision computerized lathes makes it possible to automatically produce a complete fusee by merely using the given equation and stating a few parameters. It is unfortunate that now that the requisite technology has finally arrived, the production of fusees has all but disappeared.

Georg Himmelheber

11   The Clock and Its Base

As articles of unusual value, clocks have been more subject to modification than other works of craftsmanship. Household articles such as furniture and dishes are usually discarded when they have outlived their use, but clocks are modernized and technical improvements are introduced into the works. The cases which house the clockwork have been subject to alteration as well.

The earliest German table and wall clocks that have survived have almost no cases at all. Their works are suspended in a frame that is made of the same material, and with the same care, as the mechanism itself (see Catalog Nos. 2 and 3). Even display clocks, such as the so-called clock of Philip the Good in the Germanisches Nationalmuseum at Nuremberg, were built with a major part of the mechanism visible. It was only at the end of the Gothic period that the mechanism came to be disguised—hidden in a case.

Since the clockmaker was used to working in metal, most of the early cases were metal and were probably constructed with the help of other artisans such as the brass-founder or the goldsmith. In the hundred years from 1550 to 1650, however, clockmakers also regularly collaborated with cabinetmakers, although primarily for only one part of the case—the base.[1] These bases were uniformly black—made from ebony or from stained hardwood—and thus presented an effective contrast to the gold color of the brass case.

A survey of the clocks surviving from the 1550–1650 period seems to show that those having wooden bases constitute a minority. But this impression is a false one. In order to elucidate what was presumably the original proportion between metal and wooden bases we must first consider a variant of the table clock—the figure clock.

In the first half of the 17th century these figure clocks—featuring moving horses, camels, elephants, lions, birds, human figures, Madonnas—enjoyed great popularity. Very rarely were their mechanisms installed in the body of the figure itself; usually the works were in the base of the clock. In a remarkable number of instances the bases of figure clocks are of wood. To draw the conclusion that the wooden base is part of the figure-clock class is, however, erroneous; presumably just as many figure clocks were made with metal bases as with wooden ones. The suprising preponderance of wooden bases is merely a question of preservation. The wooden bases of figure clocks could not be removed without ruining the

---

1   Much less numerous than metal cases, there were occasionally wooden cases which housed the clockwork. These are the immediate predecessors of the clock cases of the late 17th, 18th, and 19th centuries.

whole clock, for the bases contained the mechanism.[2]

If one examines the surviving wooden bases of other table clocks, one perceives a certain evolution. In the third quarter of the 16th century they were low, consisting of a single concave molding delicately contoured on the outside. They sat directly, without feet, upon their bottom surfaces. In the 1570s the base gradually became higher, and there might be a double concave molding. In the last quarter of the century the base for the most part had vertical walls instead of the molding, and now the walls were often used to hold a drawer at one side. The key, or instructions for using the clock, could be put here for safekeeping. The requirement for such a drawer was among the reasons to commission a cabinetmaker to make the base, for the drawer was an invention of these craftsmen—only in wood could it be made to slide smoothly and to close tightly.

In the first quarter of the 17th century the base scarcely changed; its vertical sides became somewhat higher, and they might also be fitted with metal ornaments, usually of silver. But the closer the century approached its midpoint, the more varied and abundant did the shapes become. To the vertical surfaces there was added an upright concave molding, or a convex molded base, or an ogee molding. Yet feet, of whatever shape, are only rarely encountered in these wooden bases.[3] This situation changed radically around 1650: now the wooden bases always have metal feet, and the bases become more lively through the addition of scrollwork and more colorful through the use of tortoiseshell. The stylistic evolution of the metal bases corresponds exactly with the changes in the wooden bases, with a single exception: metal bases, even before 1650, almost always have feet. Sometimes these feet have developed to such considerable size that they take the place of a regular base.

Yet there remain a surprisingly large number of surviving clocks that stand neither upon a base nor upon feet.[4] There can be no doubt that these are incomplete clocks—the bases have been removed. Since only wooden bases were easily removed, we can safely conclude that this large number of baseless clocks actually had wooden bases when they were originally built. Given this assumption, we arrive at a proportion of wooden-based to metal-based clocks that corresponds completely to the proportion in the case of figure clocks.

How important the base was for the clock is something we learn from the fact that there were even clocks with fully evolved metal bases that were additionally mounted upon a wooden base (see Catalog Nos. 25 and

2　The dromedary, Catalog No. 81, from the period of around 1600, shows that the base was sometimes modernized in figure clocks as well; it has a base of the rococo period.

3　From time to time feet were added as an afterthought. This sort of addition is easy to recognize, for they usually had to be stuck on at unsuitable places.

4　As examples we may cite the following figures from Klaus Maurice, *Die deutsche Räderuhr,* Vol. II (Munich, 1976). Third quarter of the 16th century: 94–96, 101, 103, 115, 119, 146, 149, 150, 154, 157–159, 161–163, 167, 182, 228. Fourth quarter of the 16th century: 123–125, 170, 172, 226, 252, 256, 257. First quarter of the 17th century: 233, 255, 292. See also Catalog Nos. 21 and 26.

29). Bases apparently could also be acquired subsequent to purchase of the clock, as can be seen in an Augsburg broadside showing samples from the period around 1670–1680.[5]

The collaboration of the cabinetmakers was especially important in the construction of bases for figure clocks, for these bases served to house the mechanism itself. These bases were almost always elongated octagonal boxes with projecting molded borders at the top and bottom. The wall surfaces consisted of framing and panelling, which sometimes bore metal ornaments. Metal lattices or panes of glass might also be used in place of panelling, to render the mechanism visible.

The polygonal shape (metal bases for figure clocks were almost always oval or round in cross-section) arose from the techniques of the cabinetmaker, who made his product out of flat boards or strips. Enrichment of the form was achieved through frequent use of miter joints, strips of beading, ornamental moldings with a wave-like pattern, and volutes. In the construction itself, the cases were all alike. They were presumably made by cabinetmakers specializing in clock bases and cases, who had the proper bands and strips in stock and put them together as the need arose.[6] Evidently individual parts were even standardized to the extent that cases were often built that were not tall enough to contain the mechanism of the clock, forcing the clockmaker to add metal feet that thoroughly spoiled the overall impression conveyed by the base.[7] There were also instances of finished cases being heightened by the addition of wooden elements.

For more expensive clocks, of course, cases were also custom made. But even on the automated figures that moved across table tops one observes a certain standardization.[8] In the Augsburg Minerva from Vienna (Catalog No. 108) it is quite obvious that stocked, prefabricated parts were used which caused certain problems. The joints did not match, the edging of the foot was not compatible with the volute, and a molded profile was attached without miters. The little cabinet at the back was also one of those prefabrications, definitely not made to measure. In order to attach it, the bottom of the housing had to be cut.

The panels of the case could be inlaid with silver or could be replaced with metal lattice ornaments, and additional metalwork was decidedly common. For instance, expensive clocks frequently had the edges of their cases fitted with little metal columns. Then there were clocks in which woodwork predominated. Some of these were chests or portable cabinets with all the appropriate fittings such as tableware and game boards, topped with a clock.[9] Others were the type of table clock indistinguishable in

---

5 Maurice, *Deutsche Räderuhr.* Vol. I, p. 193.
6 Catalog Nos. 67, 84, and 85 show clock bases that are identical in construction; only the details of molding are different.
7 For conspicuously incongruous feet see Maurice, *Deutsche Räderuhr.* Vol. II, Figs. 308, 309, 315, 318, 321, 324, 325, 332. Cf. also Catalog Nos. 62, 64, 82.
8 Maurice, *Deutsche Räderuhr.* Vol. II, Figs. 280–284; Catalog Nos. 107 and 109.
9 Maurice, *Deutsche Räderuhr.* Vol. II, Fig. 549.

70. Augsburg cabinetmakers' marks for ebony, along with the Augsburg hallmark, the pine cone (*Pyr*), transcribed from Catalog Nos. 108, 84, 29, and 86.

shape from precious little domestic altars, which were fine enough to ornament princely chapels or *Kunstkammern*.[10]

The specializations of cabinetmakers, above all in the 16th and 17th centuries, is well known. Christof Weigel wrote in 1698 that the type of work done by the cabinetmakers is "variegated beyond all measure." There were cabinetmakers who "make nothing but commonplace pieces out of poor wood." Others specialized in "black-stained work and they excel to the extent that their work is viewed with wonderment. One also finds those who produce incomparably painstaking decorative work by using tortoiseshell. . . . Others, again, work mainly in good wood, such as ebony, cypress, olive, brazilwood, Indian rosewood, mahogany [from sugar crates]," and other woods. He also mentioned the cabinetmakers who covered their furniture with silver and finally those who inlaid mother-of-pearl, stone, and glass into wood.[11]

Because the cabinetmakers often made the turned parts for their furniture themselves, there were frequently contentions with the wood-turners. In fact, in Augsburg in 1589, cabinetmakers were expressly permitted to turn parts, but only those of "ebony alone."[12] Also conceded to the Augsburg cabinetmakers, in 1613 and 1617, was fretwork in gold, silver, ivory, and tortoiseshell.[13]

In addition to ebony, bases were frequently made of stained pearwood, which is comparable to ebony in its dense texture. In order to prevent fraud, it was decreed at Augsburg in 1625 and again in 1690 that true ebony should be marked with the city hallmark and the word EBEN (see Fig. 70).[14] Some cabinetmakers also placed their initials on the cases (see Catalog Nos. 29 and 74).

Little is known regarding collaboration between clockmakers and cabinetmakers. Jurisdictional disputes were probably quite common at the border areas. Stetten reported on the cabinetmakers Heinrich Eichler and Christoph Ellrich who allegedly also manufactured "clock- and organwork."[15] Stetten further reported that in 1589 the clockmaker Georg Roll had made "a very artistic clock" for the Turkish court, "for which Jacob Schaur, an ingenious cabinetmaker, has made the housing."[16] In 1624 the Augsburg cabinetmaker Philipp Mayr delivered an organ case for the clockmaker Matthäus Runggel.[17] Aside from these few exceptions, the makers of wooden cases and bases for clocks—specialists who developed a rational division of labor—remain anonymous for us.

10 *Ibid.*, Figs. 659–670; Catalog No. 47.
11 Christof Weigel, *Abbildung der Gemein-Nützlichen Hauptstände* (Regensburg, 1698), p. 433.
12 Stadtarchiv Augsburg, Handwerksakten Kistler, 24, fol. 30r. I thank Dr. Eva Groiss for this source. See also Fritz Hellwag, *Die Geschichte des Deutschen Tischlerhandwerks* (Berlin, 1924), p. 90.
13 Hellwag, *Tischlerhandwerks*, p. 94.
14 *Ibid.*, pp. 335, 461.
15 Paul von Stetten, *Kunst-, Gewerb-, und Handwerks-Geschichte der Reichsstadt Augsburg* (Augsburg, 1779), p. 116.
16 *Ibid.*, p. 185.
17 Hellwag, *Tischlerhandwerks*, pp. 86, 348.

Eva Groiss

12 Automatic Music

The Bidermann-Langenbucher Lawsuit

71. Program drum of the musical work of Catalog No. 108, a table carriage with Minerva, persumably by Achilles Langenbucher.

According to the account of the master wood-turner Balthasar Langenbucher,[1] the essence of "self-playing musical works" consists of a programmer recording a piece of music on a drum or cylinder by means of various studs which produce a whole note, a half, a quarter, or a sixteenth note, just as a composer writes musical notes on paper. This drum he can then apply to not only organ works but also violin and other musical works. Such devices were often, perhaps mostly, part of clocks and automata (see Fig. 71 and Catalog Nos. 8, 11, 108, 111).

In 1647 the brothers Samuel and Daniel Bidermann, both of them organ and instrument makers, brought suit against Balthasar and Achilles Langenbucher for infringement upon their craft. The Langenbuchers, they claimed, had become guilty of *Stimplerei*[2] through the unauthorized making of "self-playing instruments" and through the studding of programmed drums. The brothers Bidermann appealed to family tradition as the basis of their claim that only instrument makers were authorized to make these musical works.[3] Both their father, Samuel Bidermann, and an older brother had long before learned this art from the renowned organist and composed Hans Leo Hassler (1564–1612) and had passed it on to the two younger brothers. Their father had supplied works for Meersburg on Lake Constance and, as early as 1586, for the collegiate monastery of Saint Stephen at Constance.

After the death of both the father and the oldest brother in 1622, the surviving brothers, at that time nineteen and twenty-two, "learned the skill even more properly" from the cabinetmaker Conrad Eisenburger. Samuel had worked and studied with Eisenburger as a journeyman until the latter's death in 1625, and subsequently he had obtained and completed numerous commissions from Moravia, Innsbruck, and elsewhere.[4] Even today Balthasar would be unable to stud any programmed drum, continued the Bidermanns' suit, if a composer or an experienced organist were not to provide him with a prospect or "design." The Langenbuchers

---

1 Also called a "silver-turner." Where not otherwise specified, the Bidermann-Langenbucher lawsuit described here is taken, along with other facts and dates, from Stadtarchiv Augsburg, Handwerksakten Kistler, Fasc. VI, fols. 69–157 (Aug. 19, 1647–Nov. 23, 1651).

2 *Stimplerei* was work performed by a craftsman who was not properly licensed by a guild and whose work was therefore presumed to be incompetent or shoddy.

3 Samuel Bidermann the elder is named in the documents as an instrument maker from 1569 onward. He died in 1622. Samuel Bidermann the younger, born around 1600, studied the cabinetmaker's craft up to 1625. Daniel Bidermann was born in 1603.

4 After his work for the archducal court at Innsbruck he had a commission from the Augsburg Jesuits in 1644. It is emphasized at a number of points that the Bidermann family is Catholic and the Langenbucher family Protestant.

were interfering with not only the musical instrument makers but other craftsmen, such as the clockmakers, whose guild had recently confiscated half-finished clocks and tools from the Langenbuchers. Balthasar Langenbucher had two brothers who also "lie ill in the sick-house of *Stimplerei,*" and all must be fined and prevented from taking on further work which they are not competent to perform.

The defense of the Langenbuchers, for the most part advanced by Balthasar, was full of indignation and was so convincingly presented that it strongly elicits both belief and sympathy. First Balthasar unfolded the origin of self-playing musical instruments, characterizing as a myth the story that Hans Leo Hassler, admittedly highly renowned as a composer and organist, had discovered this art. Quite the contrary, it could be shown that an Augsburg weaver named Georg Hainle had received a patent for this invention from the Emperor Rudolf, along with a gold chain and a cloak of velvet, as "inventor and author of this art."

This last part of Langenbucher's history is not correct. Hainle did in fact receive a reward of 50 talers from the Emperor when he journeyed to Prague in 1601 with Hassler and installed an organ "automaton" there. But the greater part of the financial and personal "honors," as well as the patent, fell to Hassler. Hainle could neither read nor write, but according to tradition he had extremely skillful hands and an inventive mind. Hassler took advantage of these talents rather unscrupulously. The weaver Hainle died a poor man burdened with debt, his whole life spent in the shadow of Hassler, who financed his work and inventions and then grandly brought them forth as his own. Hassler's great readiness to help Hainle, which expressed itself through the provision of medicines and physicians' services during the illness of the gifted weaver, certainly arose in great part from concern for his own reputation. In fact, the completion of a major work commissioned by the Emperor was at stake.

During his stay in Prague Hassler had publicly boasted that he was in a position to build an even more perfect organ mechanism, and he had received a commission to do this. The execution commenced at Augsburg with the participation of various collaborating craftsmen. "The master who was to fit the members together, to cut the teeth on the gears and mesh them, to make the bellows function automatically, to apply the studs in the programmed drums, and to bring life into the whole was Heinlein [i.e., Hainle], the weaver and 'maker of wooden clocks,' whereas Hassler merely provided general guidance and composed 'the piece or song that was to come onto the drum.' "[5]

At the same time Hainle had to continue working as a weaver in order to earn his daily bread, and plainly his health was not up to this double burden. Despite all Hassler's endeavors, Hainle did not arise from his sickbed; he died in September 1603. After his death a number of credi-

---

5 F. Roth, "Der grosse Spieluhrprozess Hans Leo Hassler's von 1603–1611," *Sammelbände der Internationalen Musik-Gesellschaft,* 1912, *14*(1): 34–49. All our facts and dates bearing on this matter are taken from this passage.

tors, including Hassler, made claims on the estate. In the subsequent inventory of the contents of Hainle's little workshop there are listed two design sketches of clocks, a "small workbench with all sorts of turners' and locksmiths' tools . . . more than the better half of a clock that belongs to Doctor Schissler," in addition to the large, uncompleted organ mechanism. A seven-year trial ensued, during which Hassler was exposed to considerable ridicule, as reflected in such statements as this: "Seeing, however, that Hassler is so artful a master that he can bring forth an imperial patent in his own name while suppressing those of the persons who do the most, indeed the essential part, of the work on this sort of self-playing organ mechanism. . . ."

What finally happened to the long-disputed unfinished organ work is not apparent from the documents of the first great musical clock trial at Augsburg in 1603–1611. Perhaps it came into the possession of Achilles Langenbucher, conceivably through the mediation of the cabinetmaker Conrad Eisenburger, to whom, as a close collaborator, Hainle had entrusted it on the condition that he "neither reveal nor use it," as long as the weaver was alive. One cannot determine whether after Hainle's death Eisenburger transmitted this knowledge to the Langenbuchers or the Bidermanns.

In any event, Achilles Langenbucher did not rest content. By taking apart the works left behind by Hainle—perhaps including the unfinished work on the imperial commission—he determined the principles of their technical construction. It seems that Achilles Langenbucher was fairly brilliant; in addition to being a goldsmith and wax embosser, he was also an inventor. He had produced so many unusual articles that he had been honored with the gift of Augsburg citizenship in 1611.[6] He brought with him from Giengen his wife Agatha Blaserin of Oberammergau and four sons: Christoph, Hieronymus, Michael, and Hans; subsequently a son Gottfried was born in Augsburg.

The head of the other branch of the family was Veit Langenbucher, whose origins and specific relationship to Achilles cannot be determined.[7] In any event, he was Achilles' pupil and, upon payment of 50 gulden, was introduced to the art of setting studs in programmed drums. Veit Langenbucher in turn transmitted his knowledge to his sons Gottfried, Melchior, Balthasar, Caspar, and Christian.[8] Veit pursued the art of self-playing mechanisms about twenty years, made deliveries throughout Germany and other countries, and at his death left his art to his sons as "a treasure" —something that as late as 1647 Balthasar was still able to demonstrate through living witnesses and "artistic sketches," that is, construction draw-

6 The entry in the Augsburg book of burghers names Giengen as his place of origin. G. Nebinger, "Das Augsburger Bürgerbuch 1557–1680," *Blätter des Bayerischen Landesvereins für Familienkunde,* 1974, *12*(9/10): 323.
7 Balthasar twice characterizes Achilles as his *Vetter,* but according to modern usage ("cousin"), this can hardly be right.
8 Melchior, the oldest son, died young. Caspar's principal occupation was clockmaking. Balthasar and Christian were silver-turners.

ings. The Bidermanns, according to the Langenbuchers' defense, had then "filched" this science from the Langenbucher family by various insidious tricks and were comporting themselves like so many "peacocks . . . that have taken all sorts of fine feathers from other praiseworthy birds . . . and have decked themselves out in them."

It was claimed that Daniel and Samuel Bidermann had stolen the method of making programmed drums from the sons of old Achilles behind their father's back. Veit's widow had always warned that "the Bidermann boys should never be allowed to see anything," and she had once chased them out of the house. They had "snatched what was best and most distinguished" from among the drawings left behind by Veit Langenbucher and had also gleaned information from his eldest son Gottfried. Mrs. Langenbucher had "let herself be persuaded by the Bidermanns to let Gottfried live in their house, on the pretext that they should work together for an equal share of profits."

The mother, described in this context as simple-minded, subsequently proved herself to be decidedly businesslike on being deserted by her second husband: she took advantage of the skills of her sons and had them install clockwork in dolls she had made. This was the occasion for a writ of complaint "according to which they, the clockmakers, had found my unmarried brothers sitting at a vise . . . and they were fain to do so, for our stepfather, that Drefler [Tobias Treffler] who is now with the Prince at Wolfenbüttel, had at that time in most unkindly fashion thrust our mother from him. . . ." With this exception, according to Balthasar Langenbucher, no trespasses into the fields of other craftsmen had occurred, and the complaints of the Bidermanns were completely unwarranted.

Although no one could deny Balthasar the right to teach the art of setting programmed drum studs, he had in fact never done this, he claimed. Old Achilles, however, had transmitted the method to Cornelius Zwizel and Zacharias of Kürch (a wool merchant and a cloth cutter); Zacharias in turn had taught young Schröckh and the maker of large clocks Caspar Marquart. Finally, Marquart had instructed Hans Schmidt, a young wax embosser. Balthasar Langenbucher proposed to the magistrate that in order to avoid poor quality, all work should be inspected by "us who have experience." But which of the two parties had senior and better experience in the construction of automatic musical instruments had not yet been fought out in the courtroom.

The Bidermann brothers slowly lost ground in this trial, and eventually they adduced arguments which lay outside the field of the original complaint: Veit Langenbucher was alleged to have cheated, some twenty years before, the goldsmith Claudius Pontier, whose son Elias Pontier was currently being brought in as a witness. At that time a large automaton with mechanical beasts, which in themselves had cost some hundreds of gulden, had been fitted out with beautiful choral music and with automatic violin music by Veit Langenbucher. In Balthasar's opinion the goldsmith might have sold this automaton at that time for 900 gulden and got a good profit. The fact that he had instead held on to it and that in the meanwhile tastes had changed in musical as well as artistic respects, and that the piece, now old-fashioned, had lost its value, were things his father had been unable to do anything about.

Samuel and Daniel Bidermann, it seems, also sought to advance their campaign by buying influence in other cities. A bribery attempt of theirs was revealed by the son of a Viennese organ and instrument maker, Valentin Zeissler. The son came to Augsburg and reported before witnesses that the Bidermanns sent word to his father asking whether he could influence the Emperor into declaring that only organ and instrument makers might pursue the art of setting studs in programmed drums. To which his father had quickly replied that he would have nothing to do with this and had no wish to forbid anything to anyone; furthermore, if they had a lot of money to waste, let them send a good sackful of it to Vienna.

The Bidermanns did succeed in presenting a list of cities where the studding of drums for automatic music mechanisms was allegedly incorporated into the craft of organ and instrument makers: in Nuremberg, Regensburg, Strasbourg, and other "distinguished Imperial Cities." But a document from Nuremberg attached by the Langenbucher party asserts the exact opposite with the instrument maker Paulus Wissener's statement that the studding of programmed drums was a free art everywhere and had always been, and that he further certified the good reputation of the Langenbuchers.

The question of invention recedes more and more into the background: it is simply held "demonstrable that a weaver named Hainlin had brought it forth, prior to which it may doubtless have also been discovered elsewhere and yet have remained hidden a while."

A particular offer of help came from within the family itself: Christoph Langenbucher, who in the meantime had become "grotto master" to the Prince Elector, sought to intervene by means of a written communication from Munich. He wrote that Samuel Bidermann had lent money to Hieronymus Langenbucher, had then drunk it away with him, and had finally taken him into his house secretly—without the father or brothers knowing about it. While there, Hieronymus had made a self-playing instrument, and in that way Samuel Bidermann had caught on to it. According to Hieronymus, Samuel had admitted himself that he "would have been unable to make the mechanism, and would have come to grief over it if I had not helped him out of his fix. . . . He would have been unable to do it alone and he would sing my praises his life long."

Balthasar Langenbucher also presented an example which had obviously become known in Augsburg:

> . . . I completed Schlotthaimer's mechanism, which the Bidermanns had in their house for a long time and could do nothing with, so successfully after Mr. Schluderspacher and Mr. Hippen had entrusted it to me, that demand for my work came to be addressed to these gentlemen, just as it did to Mr. Zihnlin and other merchants who had availed themselves of my services, which is something that displeases the Bidermanns most of all. . . .

What displeased the Bidermanns doubtless even more was the fact that they gained nothing through their charges against the Langenbuchers, despite two extensions of the proceedings and an expensive law suit lasting four years. The Senate decree of November 1651, which concludes the matter, allows their opponents to continue building self-playing instruments as long as they do not encroach upon other crafts. The Bidermanns are bound to carry out dutifully any commissions the Langenbuchers may give to them. We may assume, however, that the Langenbuchers proceeded to seek out other instrument makers to place their commissions with.

13 Telling Time without a Clock

Um Mitternacht.
Gelassen stieg die Nacht ans Land,
Lehnt träumend an der Berge Wand,
Ihr Auge sieht die goldne Waage nun
Der Zeit in gleichen Schalen stille ruhn.

[At midnight.
Night steps serenely ashore,
Leans dreaming against the mountain face,
Contemplating the golden balance of time
As it evenly comes to rest.]

—Eduard Mörike

To characterize "midnight," Mörike uses the image of a balance whose pans hang suspended at the same level: for him the middle of the night is marked by the transitory equilibrium of the measuring instrument. Metaphors to indicate a point or an interval in time are decidedly common, although not always in so poetic a form.[1]

"At cockcrow," we used to say without thinking much about it, and by this we meant the time of early rising, even though few were still being awakened by a cock. In Europe today, "the time it takes to smoke a cigarette" or "to stay for a cup of tea" is more suitable. Time "runs out" or "flows by" are concepts minted by the hourglass, which we revisualize with some effort. Just as early measurements of length were derived from parts of the human body, so have physiological functions enriched our vocabulary with expressions which are intended to designate a very brief span of time: that required for "a heartbeat," "the wink of an eye," "the drawing of a breath," "the turning over of one's hand." But natural phenomena—"brief as a summer shower"—can likewise describe intervals of time.

Up to the decline of the Middle Ages people found no occasion to designate points and periods in time exactly, except in court proceedings and in a few market or fair regulations. This need arose only with the development of ordered community life, in which every member was called upon to perform specific duties at prescribed times. For example, in the cloisters and also in the cities, established hours of day or night called for community services; hours for their rounds and for their cries were prescribed for the nightwatchmen. In medieval cloisters the day was divided into seven *horae canonicae* which determined the times of prayer.[2]

Purely verbal subdivisions of the day and the night are a commonplace in all languages, but as the exactitude of the mechanical measurement of time has increased, the use of these concepts has declined. Designations such as "lauds" or "vespers" to subdivide the working day are no longer common.[3]

Substantially more differentiated subdivisions of the day are found among the Slavic peoples. For example, in the neighborhood of Cracow nineteen different temporal concepts are known: predawn, dawn, sunrise,

---

1   See Otto Weinreich, "Phöbus, Aurora, Kalender und Uhr" in *Schriften und Vorträge der Württembergischen Gesellschaft der Wissenschaften* (Stuttgart, 1937).
2   1 Matins = first crow of the cock, first mass; 2 Prime = sunrise, first hour of the day; 3 Tierce = mid-morning, end of the third hour of the day; 4 Sext = mid-day, end of the sixth hour; 5 Nones = middle of the afternoon, end of the ninth hour; 6 Vespers = one hour before sunset, the close of the day; 7 Complin = the end of the twelfth hour.
3   Dietmar Wünschmann, *Die Tageszeiten. Ihre Bezeichnung im Deutschen* (Marburg, 1966).

breakfast time, forenoon, cattle time (i.e., when the livestock return home in the morning), noon, afternoon, pre-evening, mid-evening, lower evening, sunset, post-sunset, twilight, evening, after supper, before midnight, midnight, after midnight.[4]

Among the German-speaking people of Transylvania—the so-called Saxons—there is again a multitude of expressions for specific times of day and night.[5] Day breaks: "The chariot of heaven stands on the pole" (i.e., the Big Dipper stands on its handle); "The morning star has appeared"; "Day has caught fire"; "One no longer hears the rattle of the mills." Evening falls: "The sun goes home"; "The people come from the field"; "The herds return."

In Hungary, too, different segments of the day are associated with certain concepts.[6] Morning gray, break of day, half-noon, noon, and afternoon are current designations; but here the phase of "making light," "lighting lamps," becomes associated with evening twilight. Twilight is also logically termed "the time before one makes light." These divisions apply to household activities. Outside the domicile the time of dawn is known as before and after "feeding the stock," before and after "giving them to drink." The "break of day" is the time one breaks out into the fields. From sunrise to breakfast one has the first spell of field work; thereafter there is a time "before breakfast" and a span "after breakfast." The second spell of field work lasts till "half-noon." Noon divides the whole day into two parts, and in the same way "vespers" partitions the afternoon. "Evening twilight" marks the end of field work.

Among primitive peoples there are innumerable variants on this sort of subdivision of the day and night; they all resemble each other in the use of certain concepts and in the selection of the times to be designated.[7] An African example may serve for many: the Baganda divide the day into night, midnight, cockcrow, early graying, morning, little sun (from 6 to 9 A.M.), full daylight (from 9 A.M. to 2 P.M.), midday, afternoon, evening.

A real awareness of time evolved in the West only at the end of the Middle Ages. But this did not arise solely from a single-minded desire for more exact subdivisions of life on earth. Becoming aware of time went hand in hand with the painfully growing awareness of the problem of eternity. The tortuous and perplexing endeavor to find answers to the question "how long is eternity" has given rise to as many consoling legends as has the wish for closeness to those who have already departed this world. The pattern of tales about the meeting of the three living with the three dead who intervene helpfully in earthly events is an expression

4   Kazimierz Moszyński, *Kultura ludowa Słowian* (Warsaw, 1967), Vol. II, Pt. I, #94. I thank Jiřina Lehmann of Munich for the translation.
5   Karl G. Frommann, "Bildliche Redensarten . . . der siebenbürgisch-sächsischen Volkssprache," in *Die deutschen Mundarten. Vierteljahresschrift für Dichtung, Forschung und Kritik,* 1858, 5:328; further examples there.
6   Edit Fél and Tamás Hofer, *Bäuerliche Denkweisen in Wirtschaft und Haushalt* (Göttingen, 1972), pp. 418 f.
7   Martin P. Nilsson, *Primitive Time-Reckoning* (Lund, 1920), pp. 25 ff.

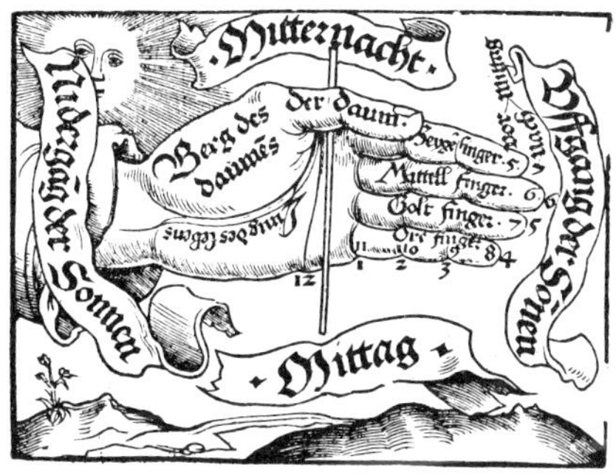

72. Using the hand as a sundial. Jacob Koebel, *Eyn künstliche sonn Uhr inn eynes jeden menschen Lincken handt* (Mainz, 1532).

of this desire for correlation of time with eternity.[8]

Despite the existence of the pocket watch, which has become a matter of course, and despite the presence of innumerable church towers, certain skills in measuring time solely on the basis of natural phenomena have held their ground through the last two centuries, in part even into our own. In Europe it is above all the farmers and herdsmen who have used their wide range of knowledge in telling time. (Pocket watches are a part of Sunday dress and are not taken into the fields.) The time determinations of primitive peoples exhibit striking agreement with those of the Europeans working in field and forest. For that reason we shall make no distinction between European and non-European ways of telling time in a simple milieu.

Setting the beginning and the end of work, the times for feeding stock and for milking, calls for the determination of a point in time. The even course of the constellations has always helped mankind in establishing termini. Sun, moon, and stars mark time periods for a properly informed person, even in the absence of artificial aids. The sun first of all does this quite simply through its appearance in the sky. In the *Peregrinatio aetheriae* there is notice of "the hour at which a man can begin to recognize one of his fellowmen."[9] In Islam the time of first morning prayer is set at the moment when one can distinguish a thread of white wool from a black one.[10] Until very recently the monks in Burma rose "when it was bright enough to see the veins on one's hand."[11]

These are approximate determinations of the graying of morning, the indicator of which is the sun. We need not labor the point that this star has always served man for the measurement of time; the positioning of megaliths and ancient obelisks convinces us of this.[12] Determination of time by means of the sun, but without instruments, is described by Jacob Koebel in 1532 in a woodcut which shows a human hand and bears the caption "an artificial sundial in every man's left hand." One had need only of a straw to measure the time on the fingers (see Fig. 72). There are numerous indications that this method was still current in the 19th century and indeed even into the 20th.[13]

8 See Leopold Kretzenbacher, "Versöhnung im Jenseits. Zur Widerspiegelung des Apokatastasis-Denkens in Glaube, Hochdichtung und Legende," in *Bayerische Akademie der Wissenschaften. Philosophisch-Historische Klasse. Sitzungsberichte Jg. 1971.* H. 7, Munich 1971; and W. Rotzler, *Die Begegnung der drei Lebenden und der drei Toten. Ein Beitrag zur Forschung über die mittelalterlichen Vergänglichkeitsdarstellungen* (Winterthur, 1961).

9 *Peregrinatio aetheriae* XXV, 6. Quoted in Nilsson, *Time-Reckoning.* p. 20.

10 Communication from Öszcan Karataş, Istanbul.

11 E. P. Thompson, "Zeit, Arbeitsdisziplin und Industriekapitalismus" in *Gesellschaft in der industriellen Revolution.* Rudolf Braun, Wolfram Fischer, Helmut Grosskreutz, and Heinrich Volkmann, eds. (Cologne, 1973), p. 83.

12 See Georg Innerebner, "Sonnenlauf und Zeitbestimmung im Leben der Urzeitvölker," Heft 2 of *Germanien. Monatshefte für Germanienkunde. Zeitschrift aller Freunde germanischer Vorgeschichte* (Berlin-Dahlem, 1942). Christian Caminada, *Die verzauberten Täler. Kulte und Bräuche im alten Rätien* (3rd ed., Olten/Freiburg, 1970), as well as some studies by Ulrich and Greti Büchi, mainly in the *Vierteljahresschrift der Naturforschenden Gesellschaft in Zürich.*

13 References from Styria (Knittelfeld, Schwanberg) in the Ferk Archive, Steirisches Volks-

To tell when it was noon there were the phrases "the sun stands over-head" and "the sun shines directly upon one's head." Members of various primitive tribes define a given point in time after sunrise by raising an arm: "when the sun stands this high."[14] The position of the sun just above the horizon shows farmers the time for stopping work.[15] The farmer's wife at home, on the other hand, could only take her directions from the sun if she had a kitchen facing south. It has been demonstrated for Poland that there were appropriate markings on the windowpane or the window frame which made it possible to read off the time during the summer months.[16] There are similar reports of African tribes whose houses were always built "facing the sun."[17] Finally, getting an idea of what time it is can also be managed by observing the position of the sun between two buildings in a town or in a farmyard.

The height of the sun produces changing lengths of shadows which have always been familiar measures of time. "When your shadow is sixteen feet long, Berenice, Amasis will be waiting for you in the olive grove," the young suitor of the princess writes in the Egyptian tale, on a tablet which he secretly passes to her.[18] In the Slavic woodlands herdsmen determined the shortest lengths of shadows by means of notches on their whip staffs, according to a method no longer fully understood, and so did huntsmen, with notches on their gunstocks. Frequently this shortest shadow length is also expressed in inches (thumbs-breadths) or finger-breadths and recorded in memory, so that at an appropriate time one can compare that length with the length prevailing at the moment. This shortest length of shadows measured, for example, at the beginning of May holds good with slight alteration until the end of August—that is, a quarter of a year; the necessary corrections are introduced as measured by eye.[19]

A determination of time reported from Styria, but one which might be used anywhere, depends on the form taken by the shadows: "Children, come in," a farm woman of west Styria is supposed to have cried out each

kundemuseum, Graz. Communication from Dr. Elfriede Grabner. On the same subject cf. the journal *Deutsche Gaue* (Kaufbeuren, 1938), commencing with No. 99; here there was a lively reader response to the communication by a Landsberg pastor who had written about measuring time by means of a straw or a blade of grass. In subsequent issues a number of readers confirmed the fact that they knew of this method.

14 Nilsson, *Time-Reckoning.* pp. 18 f.

15 In *Weistum von Lindscheid im Taunus,* quoted by K. Bücher in *Arbeit und Rhythmus* (5th ed., Leipzig, 1919), p. 336, we read: "When the farmers cut their grain they should have a piper to pipe to the reapers, and once the sun is at the height of a tree they should dance until night falls." In his autobiography the farmer Robert Rave of Moorhusen, Kreis Steinburg, describes late afternoon in the field: "At that time one held up a hand between sun and horizon and when the distance came to only the breadth of three fingers, one thrust the sickleblade through a knapsack loop, hung both over a shoulder, and went home." Robert Rave, *Das Leben auf einem Bauernhof in der Kollmar-Marsch um die Jahrhundertwende* (Moorhusen, 1966), p. 26.

16 Moszyński, *Kultura.* #94.

17 Nilsson, *Time-Reckoning.* p. 21.

18 Helga Pohl, *Wenn dein Schatten sechzehn Fuss misst, Berenike. Das Geheimnis der Zeitmessung* (Munich/Vienna, 1955), p. 9.

19 Moszyński, *Kultura.* #96.

evening, "the grass is getting pointed." This was the way she characterized the shadows as they grew long at evening.[20]

A particular variant of the measurement of shadows is reserved for the Alpine regions—"the sundial of the mountain peaks." As early as the Middle Ages these peaks served to mark time. In Wallis one had to start community farm work when the sun gleamed over the highest mountain tops. The original document concerning the Neuwerk aqueduct at Ausserberg, dated 1381, states: "Let it be known that the said laborers working in the said aqueduct are to come very early in the morning, when the sun irradiates the peaks of the mountains, and if anyone should come afterward let him be fined for that day as is set forth above."[21]

Depending on the place, certain hours of the day were determined by the sun as it stood above individual peaks, which correspondingly received such names as Neuner (Niner), Zehner (Tenner), Elfer (Elevener), Mittagsspitze (Noon Peak), and Einser (One-er). This is the way the naming of the familiar Sexten Sundial came about: the denizens of the village of Moos in the Sexten Valley in the Dolomites—and only they alone—were advised of the times of day by the Niner, the Elevener, the Twelve-er, and the One-er peaks.[22] These mountains naturally had other names on the other side of the valley. The Midday Pass (Mittaglücke) in the Gorwetsch Ridge is at the same time the Tenner Pass (Zehnerlücke), and the Midday Massif (Mittagstock) of Gwüest is the Niner Massif (Nünistock) for the people of the Göschen.[23] But modern cartography has frozen once and for all these popular designations of mountains that were valid for only one point in a valley, so that today numerous Midday Peaks and Twelve-er Summits—as well as those called Peak Ten (Cima Dieci) and Peak Eleven (Cima Undici) in Italian—bear names that are meaningless, for one side of a valley at least.

Mountain peaks with names like Twelve-er Summit (Zwölferkogel) have also occasionally given individual fields names that depend upon the fall of shadow upon the land. The field names "Sun-Turn," and "at the Sun-Turn" (Sonnwenden, auf der Sonnwend), like the early form of the same, Sunibenten, are examples of this; at the winter solstice the shadow of the Zwölferkogel—in one case from Bad Goisern in Upper Austria—reached the Sonnwenden field at noon and covered it almost entirely.[24]

20 Communication from Prof. Leopold Kretzenbacher, who received this information from Prof. Victor von Geramb, Graz.

21 Quoted in St. Schmid, "Die Wasserleitungen am Bischofsberg," *Blätter der Walliser Geschichte.* 1928, 6:452.

22 See Hermann Mang, "Volkstümliche Zeitbestimmung in Südtirol," *Beiträge zur Volkskunde Südtirols. Festschrift für Hermann Wopfner.* Part 2 (Innsbruck, 1948), pp. 253–261 (= Schlern-Schriften, 53); and Harro Heinz Kühnelt, "Die Sonnenuhren in Nordtirol," in *Tiroler Heimat. Jahrbuch für Geschichte und Volkskunde.* 1952, 16:131.

23 See Cueni in *Schweizerische Zeitschrift für Vermessungswesen.* 1943, p. 295 For precise computation of the position of the sun above specific peaks see Georg Innerebner, "Bergspitzensonnenuhren," *Der Schlern.* 1947, 21:204–210.

24 Cf. Franz Laimer, "Die Bergsonnenuhren von Goisern," *Heimatland. Wort und Bild aus Oberösterreich.* 1955, 2:20 f.

In addition, at certain times of the day or year sunlight striking rock formations such as clefts or holes in the mountain can serve for timetelling. There are examples in Switzerland in the Felsgrat of the Bodenhorn or on the eastern slope of the Eiger, Martin's Hole (Martinsloch). At certain dates in the winter and early spring the sun appears precisely in this hole carved out by nature, serving both as a heavenly spectacle and an indication of time for the local population.[25]

The Untersberg between Berchtesgaden and Salzburg, which has been known since Baroque times as a magic mountain, a place of the departed, a mountain of prophecy and of those from the nether world, and as the throne of Emperor Frederick waiting for his return, owes its name to the position of the sun and not to connotations of the Underworld. The latter interpretation finds its origin in a treatise of Lazarus Gizner of 1557 or 1558, which was disseminated in the 17th century in countless transcripts;[26] it placed the word *unter* in conjunction with *unterirdisch*. Actually, the naming of the mountain is connected with the setting of the sun as seen from a particular angle.[27]

Since the decline of the Middle Ages, work procedures have been divided in accordance with the fall of shadow upon certain fields. This was the case in the 15th century on the Lasser Höhe in the South Tyrol: the times for watering at the Gadria stream for six neighboring villages were set in accordance with the position of the sun, that is, with the fall of shadows: The first turn begins with the singing of the birds and lasts till 'die Sonne af d'Etsch,' that is, when the sun comes to the Etsch river, and so on. The division of turns ends with the Angelus.[28] This call to evening prayer itself depended, to the end of the 19th century, upon the position of the sun, not upon the time of the clock. Thus it shifted with the season of the year. For example, at Meransen at the entrance to the Puster Valley the call to prayer came when the last rays of the sun lay upon the Gopprater Kaser, an isolated mountain farm above Rodenegg.[29]

In certain areas of the Alps, where farmyards frequently lie hours apart, even in the last century the reckoning of time in accordance with an individual event was not uncommon. Thus from the 1870s it was reported, again from a mountain village in the South Tyrol, that in wintertime the ringing of the Angelus commenced only when a certain farmhand came into sight at a certain point on his way to church from a pasture high in the mountains. Similar stories are reported from Afing and from Mölten, also in the South Tyrol: the call to mass on feast days came only when the

<hr>

25 E. Friedli, *Bärndütsch als Spiegel bernischen Volkstums* (Bern, n.d.), Vol. II, Grindelwald, p. 132.

26 Cf. Wilhelm Herzog, "Die Untersbergsage nach den Handschriften untersucht und herausgegeben, in *Veröffentlichungen des historischen Seminars der Universität Graz.* Vol. VI (Graz/Vienna/Leipzig, 1929).

27 Julius Miedel, "Ortsnamen und Besiedelung des Berchtesgadener Landes," *Altbayerische Monatsschrift.* 1913–1914, *12:*75. See also Johann Andreas Schmeller, *Baierisches Wörterbuch* (Munich, 1872), Vol. I, col. 116.

28 Mang, "Volkstümliche Zeitbestimmung in Südtirol," pp. 253 f.

29 *Ibid.,* p. 254.

farmer whose land lay farthest from the parish came into sight at a certain high point mounted upon his horse.[30]

Now let us return from divisions of time which are of an individual character, even though they may be binding on a whole community, to the similar measurement of time that is prescribed by the heavenly bodies. All through the Middle Ages court sessions were determined by the position of the sun, commencing at sunrise and concluding with sunset. One who had not by then appeared in response to a summons might be held guilty.[31] Pawned objects had to be redeemed before sunset or "at the time of the Ave Maria." In the regulations of the village of Villanders, in South Tyrol, we read, for example, that if the items taken in pawn by the parish clerk have not been redeemed, "they are lost at the time of vespers."[32] According to the Safier regional code (Landsatzung) from the middle of the 16th century, in parts of Switzerland the appraisal of objects in pawn should take place "before the sun or sunshine disappears behind the mountain tops."[33] In many places the community baking oven was no longer to be fired up once a shadow fell upon it. Domestic chores had to be finished by sundown. "One may spin until the sun disappears behind the mountains" was the rule at Grindelwald in Switzerland.[34]

On the heels of the lengthening shadows there came the night, which interfered with telling time. The moon has not been popular as a timekeeper. True, it is reported from Hungary that formerly men lodged in the stalls went "to look at the moon" just as they later drew their watches from their pockets to determine the time.[35] But the position of the moon could lead one astray; concepts like "the treacherous moon" or "luna mendax" bear upon the unreliability of the moon as an indicator of time.

The stars are better suited than the moon for indicating the time of year and the hour. Among the most celebrated are the Pleiades or Seven Sisters, a cluster of stars in the constellation of Taurus.[36] Because of their importance the group is called *Zegar*, or clock, in Serbo-Croatian (= German *Zeiger*).[37]

The cock as morning alarm clock is the most common of the "natural measures of time," but for it, too, the graying of dawn sets a subdivision of time. "Before the cock will crow, thou shalt deny me thrice," Jesus says to Peter and thus expresses the conclusion that his fate will be sealed before sunrise. Cicero writes: "In time of peace the day begins with the first cry of the cock, in time of war with the first trumpet blast."[38] In military encampments as in cloisters the cock has for centuries played the

---

30 Cf. the "Mitteilungen" in *Der Schlern,* 1924, 5:293, and 1925, 6:31, 67.
31 Communication from Prof. Hellmut Rosenfeld, Munich.
32 Mang, "Volkstümliche Zeitbestimmung in Südtirol," p. 255.
33 Quoted in Paul Zinsli, *Grund und Grat. Die Bergwelt im Spiegel der schweizerischen Alpenmundarten* (Bern, n.d. [c. 1945]), p. 179.
34 *Ibid.*
35 Fél and Hofer, *Bäuerliche Denkweisen,* p. 414.
36 See *ibid.,* p. 415 and Moszyński, *Kultura.* # 15.
37 Moszyński, *Kultura.* #99.
38 Quoted in Hans Denis, "Zeitmesser ohne Räderwerk," *Alte Uhren,* 1978, 1:379.

part of the awakener. In the *Canterbury Tales* it is eternalized, and here there is a plain reference to the sun and the Pleiades which in turn give the time to the cock:

> . . . caste up his eyen to the brighte sonne,
> That in the signe of Taurus hadde yronne
> Twenty degrees and oon and somwhat moore,
> And knew by kynde, and by noon oother loore,
> That it was pryme, and crew with blisful stevene.[39]

In Shakespeare *(Love's Labour's Lost)* the larks as they awaken announce the start of work: "And merry larks are ploughmen's clocks." Juliet calms Romeo at morning cockcrow with the words: "it is the nightingale, and not the lark." We have already cited above the first songs of birds as indicating the beginning of the watering of cattle at the Gadria stream.[40]

Gradually encroaching twilight was recognized by Slavic herdsmen from the color of the fire burning in the open or inside the house. Throughout the day the flame was red; as day declined the flame became brighter.[41] For cloudy days the Huzul shepherds had a further determination of time, although one that was only approximate: they knew that the pupils of sheeps' eyes were oval throughout the day but became circular at just about the hour they were driven home to their folds. Polish farmers used the eyes of their cats in the same way to measure time.[42] Even inexplicable heavenly phenomena which recurred regularly were suited for subdividing time. "If in the summertime one looks from the vineyards of Unterdöbling, Heiligenstadt, or Nussdorf toward St. Stephen's [in Vienna], there appears at four o'clock every afternoon by the tower a white apparition that has the shape of a woman. In earlier times, when the vineyard workers had no watches, they simply looked for the 'white lady' or 'mother of heaven,' and when she appeared they knew that it was vespers."[43]

The human body too can serve as an instrument for measuring time if it has a chance to follow a regular daily schedule. " 'When did you commit the theft?' I asked a suspect some years ago," the administrator of a state domain in Styria wrote early in the 19th century. " 'Somewhere around the second visit,' the miscreant replied," and the writer added in his notes, "Further inquiry instructed me that the *visit* every night meant when the countryman gets up to pass water. The first *visit* is at eleven o'clock at night, and the second at three in the morning."[44] Physiological determina-

39 *Canterbury Tales,* quoted in Thompson (n. 11 above), p. 81.
40 For morning awakening through sunrise and birdsong see Fél and Hofer, *Bäuerliche Denkweisen,* p. 415; Ernst Finder, *Die Vierlande, Beiträge zur Geschichte, Landes- und Volkskunde Niedersachsens* (Hamburg, 1922), p. 158.
41 Moszyński, *Kultura,* #17.
42 *Ibid.*
43 Gustav Gugitz, *Die Sagen und Legenden der Stadt Wien* (Vienna, 1952), p. 68.
44 Johann Nepomuk Felix Knaffl, *Die Mundart des obersteyermärkischen Landmanns im cammeralischen Bezirke Frohnsdorf. Ein kleiner Beytrag zur Volkskunde und zur Statistik der oberen Steyermark* (Judenburg, Oct. 13, 1813—date of the dedication to Archduke Johann), MS 940 of the

tion of time is also known as the theme or a well-known picaresque tale and as the motif of a fairy tale. The tale has it that a noble lady and her maidservant had fallen prisoner and refused to say who was the lady and who was the maid. In order to determine which was which, they were asked how they could tell, in a darkened room, that day was coming. One of them replied that she would know because she would simply have to "go out"; the other, who was then recognized as the lady, said that for her the sign of dawning was when her pearl necklace felt cold on her breast.[45]

The imaginary flower-clock of the Swedish botanist Carl Linnaeus can be ascribed to the charming whim of a scholar.[46] His notion depended upon the fact that blossoms open at different times which are absolutely binding for their particular species. In his *horologium florae* the wheel was the sun and the hands were various types of flowers which bloomed in succession. (The primitive inhabitants of the Cross River region actually use a plant about 50 centimeters high as a measure of time; it has violet-white flowers that open gradually after sunrise, are wide open at noon, and close again step by step as evening approaches.[47]) The flower clock of Linnaeus aroused the ridicule of Jean Paul, who described "a clock made out of men" in his *Siebenkäs* (1797). He described his room, one window looking out over a garden while another opened toward the marketplace, thus giving him the opportunity of contemplating the succession of visitors to the market as they relate to the flowers coming into bloom.[48]

If it was maids and bakers that gave Jean Paul the time, it can also be a regularly returning vehicle in the vicinity of a roadway. "The express among the horse-cars was the post, which drove at a trot over the Mass Meadow daily at 9:30 A.M. with its passengers in the trap. One had no need to look at the clock—the farmer would say to his wife, 'get to your cooking, it is half-past nine.' "[49] While some years earlier it had been a matter of course for the farmer to read the time from the position of the sun, in the meanwhile it had become just as natural for him to use the coach as a measure of time. It seems to have involved little difference if it were the crowing cock, the church bell, or the spanking new horse-drawn coaches that gave him the time; it was just looking at the clock that he was not used to.

Aids in reckoning a span of time are much rarer than are devices for identifying a point in time. The moon, not much help with determining

Steiermärkisches Landesarchiv at Graz, p. 13v. I am grateful to Dr. Sepp Walter of Graz for the reference to this passage.

45 Cf. Kurt Ranke, "Schwank und Witz als Schwundstufe," in *Festschrift für Will-Erich Peuckert* (Berlin, 1955), pp. 47 f., where there are still more passages relating to this theme.

46 Quoted in Denis, "Zeitmesser ohne Räderwerk," p. 380.

47 Nilsson, *Time-Reckoning,* p. 19.

48 Jean Paul, *Blumen-, Frucht- und Dornenstücke oder Ehestand und Hochzeit des Armenadvokaten F. St. Siebenkäs,* edition of Hanser Verlag (Munich, 1959), Vol. II, pp. 381 f.

49 I am grateful to Dr. G. R. Schroubek of Munich for the opportunity to peruse this memoir of the 80-year-old farmer N. Jaksch of Leopoldschlag-Dorf in Upper Austria.

exact time, provides a measure of duration in the tides. Since an entire ebb and flow period runs 12 hours and 24 minutes, it is basically unsuited for fixing a point in time, yet the Greenland Eskimos reckoned time by tides into the 20th century. They divided each day into ebb and flow, with a computation that was valid only for that single day; since two full periods were almost an hour longer than a day, they had to calculate afresh each morning and recommence their count. At the beginning of our century rudiments of this kind of reckoning were still to be encountered among some Polynesian tribes.[50]

A very pretty Mohammedan rule of thumb introduces the wind as a measure of duration of time: an argument may last only so long as a wet cloth hung out on a line has time to dry.[51]

From the domain of cloister life one is familiar with the "psalm clock": the brother on watch would run through a fixed number of psalms and in this way measure the sleeping time of his confreres. In A.D. 575 Gregory of Tours is supposed to have attempted to set down a psalm clock in writing. He introduced into a table on the risings of the stars the number of those psalms which ought to be recited prior to the *officium nocturnum* during the hours as they changed over the year; in this connection the duration of a psalm was set at 4 to 5 minutes.[52]

Prayers as measures of time were also much favored in worldly usage; they even made their way into cookbooks. "Then you sprinkle parsley, onion, spinach leaves, . . . on the roast hare which is already done, and you put it again in the oven for as long as it takes to say two Our Fathers."[53] It takes three Our Fathers for eggs to become hardboiled, something which is current even today among older housewives, principally in the country. The Austrian poet Karl Heinrich Waggerl was confronted in his childhood with the timing of kitchen intervals by means of prayers: Advent brought for him "the painful hours of stirring dough: four Our Fathers for the lard, three for the eggs, a whole rosary for the sugar and flour."[54]

Even in connection with revolutionary technological achievements, traditional methods of measuring time continued to be relied upon. At the Munich Jakobi Fair of 1841 the "frenzied portraitist" Johann Baptist Isenring set up his "Daguerreotype Salon" and turned out portrait photographs with an exposure time of three Our Fathers.[55]

Natural disasters were also measured by prayers: Chilean tradition tells of an earthquake in 1647 that lasted "two Credos long."[56] In a courtroom at the beginning of the 19th century a notorious murderer of Lower Bavaria speaks of his speed: "no matter how fast a person can pray, he

50 Nilsson, *Time-Reckoning.* p. 38.
51 Communication from Frau Dr. Neriman Ülkümen.
52 Pohl, *Wenn dein Schatten.* pp. 108 f.
53 *Altadeliges Bayer'sches Koch- und Konfektbuch für alle Stände* (Munich, 1837).
54 Karl Heinrich Waggerl, *Das ist die stillste Zeit im Jahr* (Salzburg, 1956), p. 14.
55 Heinz Gebhardt, *Königlichbayerische Photographie 1838–1918* (Munich, 1978), p. 9.
56 Thompson (n. 11 above), p. 83.

cannot recite a rosary or ten Ave Marias" as quickly as I can do my deeds.[57]

Georg Christoph Lichtenberg, a physicist in the 18th century, explored the idea of reciting word combinations of established length as an aid to astronomical observations. "If one learned to pronounce a word of five or six syllables so that it lasts just one second, or if one tried to see whether the ordinal numerals would perform this task, then one could subdivide time in astronomy: the astronomer who is observing, for example, an immersion or an emersion need only recite this instead of having to count seconds; then, after he has noted the moment of the occurrence, he can continue counting seconds and look at his clock, whereupon he can easily find his proper time through subtraction."[58]

Measuring time in verses or free texts is a tradition of long standing. The earliest example seems to be a passage in William Langland's *The Vision of Piers Plowman* of 1377 where "pater noster while" is mentioned.[59] In the *New English Dictionary* one finds further early passages within the Anglo-Saxon heritage, such as "misererewhile" (1450) and "speche while" (1430). There is also reference here to physiological concepts in the measurement of extremely short time spans (a heartbeat, a breath, the winking of an eye). In 1593 the concept "breathing while" appears, and even (after 1553) "pissing while," which is transformed in 1676 by the playwright Wycherley into "making water while."[60]

Primitive societies have not only measured cooking times by means of the length of prayers, they have also in the reverse sense expressed durations of events in terms of the cooking times of familiar foods. In Madagascar "rice cooking" meant about half an hour, "the scalding of a grasshopper" an instant. "In a time shorter than it takes maize to roast" means in less than 15 minutes. "The time in which one can boil a handfull of greens" means about an hour.[61] Even in Europe these references to cooking time are decidedly common. In Brandenburg, when a woman's labor pangs commenced, peas were put on the fire; the birth ought then to follow as soon as the peas were cooked.[62]

Distances —or the time it takes one to walk them—are also occasionally used as measures of time. In 1500 the Nuremberg master Erhard Etzlaub measured miles in terms of prayer lengths on his map of the pilgrimage to Rome.[63] In the cookbook of the Würzburg Codex one reads: "To make

57 Anselm Ritter von Feuerbach, *Merkwürdige Verbrechen,* selections published by Wilhelm von Scholz, Vol. I (Munich/Leipzig, 1913), p. 95.
58 Georg Christoph Lichtenberg, *Schriften und Briefe,* Vol. II (Darmstadt, 1971), p. 32.
59 William Langland, *The Vision of Piers Plowman,* 1377, line 348 (B-Text).
60 A compilation of these ideas is found in Gösta Langenfeldt, "The Historic Origin of the Eight Hours Day: Studies in English Traditionalism," *Kungl. Vitterhets Historie och Antikvitets Akademiens Handlingar,* 1954, 87:7.
61 Nilsson, *Time-Reckoning,* pp. 42 f.
62 H. Prahn, "Glaube und Brauch in der Mark Brandenburg," in *Zeitschrift des Vereins für Volkskunde,* 1891, 1:183.
63 Cf. H. Krüger, "Des Nürnberger Meisters Erhard Etzlaub älteste Strassenkarte von Deutschland," *Jahrbuch für fränkische Landesforschung,* 1958, 18:1–286. Communication from Prof. Dr. Wolfgang von Stromer, Nuremberg.

mead. . . . then boil the herb itself a field's length across and back. . . . then boil it with the herb for half a mile. . . ."[64] Similar measurements of time in miles and fractions thereof are also known from the English-speaking areas, "one mile" meaning approximately 20 minutes.

Japanese swordsmiths recite specific ritual formulas that determine the length of time the heated blade should remain in the quenching bath before it is removed to cool completely in the atmosphere; the texts serve as a measure of time and simultaneously as a conjuration. This method, however, is by no means found in Asia alone: Alfons Maissen reports that a farmer in Domat (Graubünden) sang all sixty stanzas of a particular song (the Plazidus-Lied) while sharpening his scythe. "This was his measure of time for a good whetting job, but at the same time also an edification and a laudation to God."[65] The same song, when it was sung antiphonally by the various groups in the procession, provided the measure of time for the pilgrimage from Somvix to the monastery of Disentis, which is about 7 kilometers away.[66]

Neither prayer nor religious song in serving as a measure of time has merely that profane function that could today be taken over by any clock. Prayer and song fill the working day with thoughts of God and his saints and imbue everyday life with Christian formulas as statuary symbols do. Time is not only measured and subdivided but is also utilized for prayer and edification.

Almost all examples of the simple designation of time without mechanical aids arise from the domain of farm life, especially in the last two centuries. Here the general dissemination of the mechanical clock had to wait a long time. When a clock did enter a house, it was, to begin with, not portable. The pocket watch could be dispensed with, since simple and natural means for calculating time were well known. Precision was scarcely sought; an approximate measurement sufficed. There were only a few points in time that must be observed, and prominent among these was the hour of the beginning of divine services on Sundays. The breaking down of the time available for a given procedure appears to have presented problems well into the present age. As late as 1900 farmers in Westphalia were still setting their watches so that time for the walk to church was already accounted for: a farmer who needed 20 minutes to get to church would set his watch 20 minutes fast. If mass began at 9 A.M., he then left the house when his watch showed 9 A.M.[67]

Farmers, woodsmen, and herdsmen are reluctant to let the course of their day be prescribed by a small mechanism constructed by the hand of man. Nor can they by any means, by the nature of their work, let themselves be bound to a working day that is fixed by the clock. For them

---

64 Schmeller, *Baierisches Wörterbuch.* Vol. II, col. 1161.
65 Alfons Maissen and Werner Wehrli, eds. *Rätoromanische Volkslieder, Die Lieder der Consolazium dell'olma devoziusa.* Part 1: *Melodien,* p. xxxvi, Melody in the same volume, p. 188, text in the second part, p. 135.
66 Communication from Prof. Alfons Maissen, Chur.
67 Communication from Dr. Barbara Heller, Munich.

sunrise signifies the beginning of the day; with sunrise the livestock awake and demand attention. Twilight and sunset mark the end of work for them; at least outdoor work is no longer possible. The time of sunrise and sunset varies over the year. He who lives by the sun and allows it to divide up his working time experiences much more consciously not only each day in its natural extent but also the changes of the seasons. The fact that the sun, as spring advances, rises each day 2 minutes and 30 seconds earlier than it had in the preceding week is of no significance to him, for there is another way of counting the length of the increasing day: "At Three Kings the crow of a cock, on St. Sebastian's day the spring of a stag, at Candlemas a whole hour."[68] This progression is fixed—it cannot be influenced by the hand of man.

The farmer, whose harvest is dependent upon the growth-promoting rain, on the bloom-forcing sun, on the richness of the soil, down to our time acknowledges first of all the heavenly bodies, especially the sun, as a measure of time. He thereby finds himself in bluntest contradiction to society, which is characterized by the exact determination of time, yet he quite naturally subjects himself to it in appropriate situations. The working day, however, is set by the "movable measure of time" provided by nature, attuned to the requirements of the season. For this determination of time no mechanical clockwork is required, merely certain traditional skills.[69]

68 Similar formulations can be referred to in numerous German dialects; their origin cannot be fully deduced.

69 This contribution could not have been written without the help of the following folklore experts, to whom I should like once more to offer my heartiest thanks: Prof. Dr. Hermann Bausinger, Tübingen; Prof. Dr. Rolf W. Brednich, Freiburg; Dr. Bernward Deneke, Nuremberg; Mr. Alexander Fenton, Edinburgh; Prof. Dr. Milovan Gavazzi, Zagreb; Dr. Elfriede Grabner, Graz; Dr. Edgar Harvolk, Munich; Dr. Alfred Höck, Marburg; Dr. Günther Kapfhammer and Prof. Dr. Leopold Kretzenbach, Munich; Dr. Arnold Lühning, Schleswig; Prof. Dr. Alfons Maissen, Chur; Mr. Fritz Markmiller, Dingolfing; Dr. Georg R. Schroubek, Munich; Dr. Sepp Walter, Graz; Prof. Dr. Günter Wiegelmann, Münster; Dr. Robert Wildhaber, Basel.

Sigrid and Klaus Maurice

14 Counting the Hours in Community Life of the 16th and 17th Centuries

The Prince, for his part, questioned a reliable astrologer regarding a time which would be propitious for the founding of the city. The calculation revealed that such a time would fall on Monday, April 15, 1460, at 10:21 A.M.; for at that moment a fixed star which was to have a bearing upon the earth, and which was ruled by Venus at sunrise, would rise under the auspicious sway of the moon. . . .[1]

73. Portrait of a marriage (detail), from the *Histoire de Renaud de Montauban.* French, 1467–1470. Paris, Bibliothèque de l'Arsenal, MS 5073.

The city in question was Sforzinda, the ideal utopian community of the Italian architect Filarete, founded in harmony with the cosmos, which required knowing the precise point in time at which a favorable astrological constellation would fall. Registering time was a matter of course in the community, and not just in astrological calculations; it was independent of the way the day was divided into sections or of the way in which time was measured.[2] In Europe time had always been a philosophical and a psychological problem,[3] one not linked with the physical problem of measurability. Ways of setting hours in the community were a part of the culture of the court, cloisters, and cities, and as an element introducing order they remained independent of the philosophy and the psychology of time. Ways of counting time appear in the rules of religious orders, in municipal law, in the regulations of crafts and the mining industry. There are specifications of time in the Bible, in journals, and in chronicles. A review of these sources turns up a relationship between time and one's experience with the surrounding world: the broader that experience becomes, the more various and frequent are the ways of counting the hours. The selection of examples for this study resembles part of a mosaic. It proposes to offer only those that show that during the period with which the exhibition is concerned—1550–1650—specifications of time were taken as a matter of course and were universally encountered.

The astrologically propitious moment for the founding of a city was calculated the way our first quotation shows; but for the laying of the cornerstone it had to be put in terms of an hour. In the same way, the favorable hour for conception and for birth were set. In the *Histoire de Renaud de Montauban,* approximately contemporaneous with the date for the foundation of Sforzinda, we find a miniature marriage portrait (Fig. 73). The newly wedded pair sits upon a bed, and a mirror hangs upon the back wall surface supporting the canopy over the bed.[4] The purpose of the mirror is to reflect the appropriate constellation of stars as this comes into sight through a window, and thus to signal the ideal moment for concep-

1 Antonio Averlino Filarete, *Tractat über die Baukunst* (c. 1464) (German ed., Vienna, 1896), p. 130. Cf. Peter Tigler, *Die Architekturtheorie des Filarete* (Berlin, 1963).
2 Gustav Bilfinger, *Die mittelalterlichen Horen und die modernen Stunden* (Stuttgart, 1892).
3 Werner Gent, *Die Philosophie des Raumes und der Zeit* (1926; reprint, Hildesheim, 1971). Richard Glasser, *Studien zur Geschichte des französischen Zeitbegriffs* (Munich, 1936). Erwin Stürzl, *Der Zeitbegriff in der Elisabethanischen Literatur* (Vienna, 1965). Ricardo J. Quiñones, *The Renaissance Discovery of Time* (Cambridge, Mass., 1972).
4 Madeleine Pelner Cosman, "Machaut's Medical Musical World," *Annals of the New York Academy of Sciences,* 1978, *314:* 1–36.

tion, as determined by an astrologer. Dates can thus be indicated in twofold fashion, by hour and by the prevailing constellation as is seen in the journals of the Augsburg citizen Lucas Rem, written from 1484 to 1541: "Berchtold, my son, was born Monday, January 11, 1529, at a quarter past two of the afternoon. . . . upon the day before, the new moon had been in the Ram at 9 hours 2 minutes," or "he died on October 14, 1530, at 9 in the evening. The moon was at its last quarter the day before at 2 hours 44 minutes of the afternoon, in Leo 12, the Sun in Scorpius."[5] Time specifications of other births from the diary run, for example: "one-third after 7 . . . during the day, just as it struck 2 . . . in the evening, half-past 6 . . . before it had struck 2 after midnight." They indicate that the associated astronomical constellations indicated in terms of minutes had been figured out and that the time of day had been registered through a clock which struck.

But let us first recall that despite the use of a standard clock, up to the 19th century the hours were sectioned in different ways depending on the city and country. Just as we figure in time zones and calculate the time of another zone in terms of our own, so the biblical specifications of time in accordance with the Jewish method of counting the hours were restated in "our time": "In this way, then, Jesus Christ too, savior and redeemer of ourselves and of the whole world, was himself condemned to death, taken out, and crucified by Pontius Pilate the Roman legate and criminal judge at Jerusalem, around the sixth hour of the day; which according to our time is to be placed at twelve o'clock. And at the ninth hour, which is at about half-past three in the afternoon, he gave up the ghost."[6]

This conversion into local time makes the life and passion of Christ more vividly present. In the 13th century the whole Gospel was put into strict chronological sequence: in the *Lucidarius* the time Adam was in Paradise is precisely set forth: "barely seven hours."[7] Why no longer? Because the woman, created only after six of these hours, proved false and gave Adam the forbidden fruit, and the Lord drove them out of Paradise at the ninth hour of the day.[8] Again, such specifications of time serve to make more immediately present events which were in a sense historical. In the same way the *Horologium devotionis,* the devotional tract of the Dominican monk Berthold, also served to intensify the Gospel. Commencing with the 14th century, this "Pious Timepiece of the Life and Passion of Christ, arranged according to the twenty-four hours" (Fig. 74) has assigned, in its many editions, appropriate prayers for each hour of the day and night.

But it is when one realizes how many ways of counting time coexisted up to the 19th century that one perceives above all the common sense and practical necessity of a means of converting the different statements of

74. Title page of the *Zeitglöcklein* of Berthold von Freiburg (Nuremberg, 1493).

5 *Tagebuch des Lucas Rem aus den Jahren 1494–1541.* ed. B. Greiff (Augsburg, 1861), p. 68.
6 Johann Geyger, *Horologium politicum. Das ist: Einfältiger Discurs und kurtze Betrachtung dess politischen materialischen selbst gehenden vnnd schlagenden Uhrwercks* . . . (Nuremberg, 1621), p. 38.
7 Felix Heidlauf, ed , *Lucidarius aus der Berliner Handschrift* (Berlin, 1915), p. 7.
8 Jacques Le Goff, *Kultur des europäischen Mittelalters* (Munich, 1970), p. 295.

145

time. Generally speaking, in Germany people reckoned by the Small Clock (1 to 12, commencing at midnight) and by the Whole Clock (1 to 24, commencing at midnight); in Italy according to Italian time (1 to 24, commencing at sundown), in the Bohemian lands according to Bohemian time (1 to 24, commencing at sunrise). In south Germany some imperial cities like Nuremberg, Regensburg, and Rothenburg ob der Tauber counted not only by the Small and the Whole Clock but also by the Great Clock, in which the hours of the day and night were separated into two different series, commencing either with sunrise or with sunset. Hence in the summertime the imperial cities reckoned with 16 daylight hours and 8 nighttime hours, in winter with 8 hours during the day and 16 at night. The multiplicity of counting methods, constantly shifting with the seasons as sunrise and sunset varied throughout the year, required endless conversions and synchronizations.[9] For this purpose tables were compiled with the hours printed in columns that could be compared against a sliding scale (Fig. 75). Simplest of all was conversion by means of mechanical clocks which automatically compared a number of counting methods and displayed the results (see p. 173).

Within the same community the day could commence at different hours. Grimmelshausen wrote in his perpetual calendar in 1670, "So the day also starts in different fashions with people in different branches of their daily operations. Thus in celebration of the holy offices one begins the day the preceding evening, but in an armistice during the course of a war, with sunrise."[10]

To begin with, the course of the day was regulated through the striking of a bell. Pope Sebastian (604–605) had already decreed that "churches should strike the hours of the day."[11] The rules of the monastic orders, beginning with the Benedictine, also indicated precise hours for the offices of day and night. These canonical hours in part continued to follow the subdivision of the day in classical antiquity: matins in the third quarter of the night, prime at sunrise, tierce at the middle of the forenoon, sext at noon, nones at mid-afternoon, vespers an hour before sundown, and complin at the end of the day. Somewhere around the 13th century this

9 As shown in the following schedule for the working hours of miners, sometime before 1584: "I. shift on the small- and the whole clock. In St. Joachimsthal and at other places were the small-clock is used in mining operations one shall strike early at 3 hours in the winter and summer times and at 4 one shall enter the shafts. . . . But at those places and at those mine-heads where the whole or Bohemian clock is used, there we desire that in the summertime, or from St. George's Day until St. Gall's, the first shift shall be sounded at 7 and the miners go in at 8, the second shift to be sounded at 15 and go in at 16, the third shift sounded at 23 and go in at 24. In the wintertime, however, from St. Gall's day till St. George's, the first shift shall be sounded at 9 and go in at 10, the second shift sounded at 17 and go in at 18; the third shift, however, at 1 in the night and go in at 2." From Sebastian Span, *Speculum iuris metallici oder Berg-Rechts-Spiegel . . . aus denen vom Kayser Ferdinando I. anno 1584 publicirten . . . wie auch bey Zeiten Kayser Rudolphi II. zu Pappier gesetzten Bergwercks-Ordnungen . . .* (Dresden, 1698), p. 232.
10 Hans Jakob Christoph von Grimmelshausen, *Ewigwährender Kalender* (1670) (Munich, 1925), p. 274.
11 Bilfinger, *Die mittelalterlichen Horen.* p. 3.

75. Tables of regional hour counts and of the ruling planets for the hours. Woodcut by Jost Amman, Nuremberg, 1568.

breakdown was shifted, since one might eat only after nones during fasts, and a deferral of eating until late afternoon was not so desirable. Nones crept gradually toward noon and became the midday office, after which the midday meal accordingly followed. The words *noon* and *afternoon* are reminders of this evolution.[12]

The ringing of bells did not indicate definite hours but signals for the start and finish of religious activities, and eventually also lay activities. In the Nuremberg books of regulations compiled between 1315 and 1360 we find numerous examples of this: "It has also been decreed that no baker shall have any bread on sale at home in his house or at his store, nor shall he sell any bread, on any day from the first early mass that is sounded at the chapel until complin is sounded at St. Egidius." The start of work determined by the call to early mass also held good for butchers and smiths: "no butcher shall open his shop on any feast day before the day's mass"; "no smiths shall rise to set about their work before the call to parish

12 *Ibid.,* pp. 59 ff.

147

mass, and they shall also work no longer than until the curfew bells are sounded."[13] Conversely, a public bell could also sound the call to prayer. Originally the evening signal, the *ignitegium,* or "fire-covering," was the striking of the curfew bell; when it sounded, open fires must be extinguished. It was probably William the Conqueror who introduced this rule. Pope John XXII (1316–1334) subsequently prescribed that when this bell rang, the Ave Maria should be said.[14]

The following bells ruled secular life: the communal bell, which convened the city delegates; the work bell, which sounded the start and the finish of the working day; and the curfew, which indicated the end of the burgher's day. In Nuremberg there was a fine for anyone who should "have music or gaming in his house by night after curfew" or who should "serve drinks after curfew."[15] In other cities curfew was sounded by the beer bell or the wine bell. In the municipal law of Munich in 1374 this bell signaled the hour when "my lords shall sit watch by night, and no one shall serve any sort of drink after the beer-bell"; it was also "forbidden that anyone shall walk about the streets without a light after the beer-bell."[16] There were other bells whose names indicated their functions: watch bell, toll bell, broom bell, porridge bell (Basel), market bell, grain bell. Frequently one bell served various functions. For example, in 1355 the magistrate of Aire (Flanders) hung in the watchtower a bell which not only struck the hours for the start and finish of the working day for the local textile workers, but also called the city delegates together on many days,

> . . . and seeing that the town in question is governed by the craft of drapery and other crafts where it is convenient that a number of workmen proceed to and from their work at specified hours, and also that the said mayor and councilmen and sundry of their citizens should be present and should come into the hall to do justice and make law as is customary for several days in each week, it will be necessary that there be bells in the said bell-tower to strike the hours for them.[17]

As the hours of daylight changed in length, so were the bells rung at different times. Nuremberg, for example, had its own bell schedule. From St. Gall's Day (October 16) onward the fire bell (curfew) rang at one-half hour after one at night according to the Great Clock, or according to the Small Clock, one and one-half hours after sundown; from St. Martin's Day (November 11), at two at night, thus two hours after sundown; from St. Anthony's day (January 17), one and one-half hours after sundown; and from the Monday after Estomihi (the seventh Sunday before Easter) until

13   *Satzsungsbücher und Satzungen der Reichsstadt Nürnberg aus dem 14. Jahrhundert* (Nuremberg, 1965), pp. 84, 87, 126.

14   Franz Rühl, *Chronologie des Mittelalters und der Neuzeit* (Berlin, 1897), p. 210.

15   *Satzungsbücher,* p. 69.

16   Franz Auer, *Das Stadtrecht von München nach bisher ungedruckten Handschriften* (Munich, 1840), p. 190, 133.

17   G. Espinas and H. Pirenne, *Recueil de documents relatifs à l'histoire de l'industrie drapière en Flandre.* Vol. I (Brussels, 1906), p. 6.

St. Gall's Day, at one hour after sundown.[18]

Whereas 14th-century ordinances set time restrictions corresponding to such signals as the call to early mass, in later regulations particular hours are specified. This abstraction did not appear suddenly, not even in the chronicles or decrees of a single community. In every city, in every market town, striking clocks were set up beginning with the 14th century, though specific dates cannot always be determined. Gradually confidence was bestowed on the striking of the hour, which was always the same. By the date 1555 we read that the Munich town council permitted bakers to shorten their baking period by three hours "in such fashion that from now on they shall bake no longer than from five o'clock in the evening until two after midday of the next following day."[19] From the year 1563 we have a Freiberg regulation for taverns: "And as soon as it shall have struck nine, the servant of the drinking premises shall bring forth no more beer," and "the tavern being regulated shall open when the hand stands at four and continue thereafter, drinking coming to an end when the bell has struck eight."[20]

The course of the public day and that of private life, the "mundane condition" and the "domestic condition,"[21] are ruled by the striking of a clock, no longer by the sounding of a bell. Just as the sounding of the curfew shifted with the seasons of the year, so did the indication of the hours—the hours delimited change with the changes in the light of day. Consider this sample of "music and gaming" hours. In 1616 the Bavarian provincial laws set times for dancing: "from Pentecost until St. Michael's day up to the fourth hour, from St. Michael's day, however, back to Pentecost once more up to the third hour after noon, and at these times all dances shall cease on feasts of the church."[22]

In mining operations, especially, everything was closely regulated by the clock. In a treatise on metallurgy (1556) Agricola precisely notes the times of shifts: "the first shift commences at 4 hours in the morning and lasts till 11 hours; the second begins at 12 hours and ends at 7 hours; these two are the day shifts. . . . The third is the night shift; it has its beginning at the eighth hour of evening and it ends at 3 hours of the night."[23] In Freiberg mining shifts of six hours are reported as early as the middle of the 15th century.[24] The mining regulations of Duke George of Saxony in 1499 prescribe eight-hour shifts which commence at four in the morning

18 Bilfinger, *Die mittelalterlichen Horen,* p. 56.

19 Anton Schlichthörle, *Die Gewerbebefugnisse in der K. Haupt- und Residenzstadt München.* Vol. I (Erlangen, 1844), p. 308.

20 Dr. Bursian, "Ordnung der Trinkstuben zu Freiberg," *Zeitschrift für deutsche Kulturgeschichte.* 1859, 4: 512, 514.

21 Geyger, *Horologium politicum.*

22 *Landrecht Policey. Gerichts- Malefitz- und andere Ordnungen der Fürstenthumben Obern und Nidern Bayrn* (Munich, 1616), Bk. 3, Art. 8.

23 Georg Agricola, *Zwölf Bücher vom Berg- und Hüttenwesen.* translation of the Latin original of 1556 (Munich, 1977), p. 77

24 Hubert Ermisch, *Das sächsische Bergrecht des Mittelalters* (Leipzig, 1887), pp. xc f.

and twelve noon.[25] Every mine had to contribute half a groschen weekly for the "hand-setter," that is, a clockmaker. Thus a man was provided simply for looking after the clocks, but at the same time also the mine bell. Warning was to be given of each shift an hour beforehand, "whereby the miners may make themselves ready and have so much the less excuse for their negligence" (1589).[26] In reading the mining regulations one discovers a parallel with the present day: the less the dependence on daylight, the greater the need for a regulation of time by means of a clock.

Indication of time is also a criterion for regulations within the sovereign's court. In the *Teutsche Hofrecht* of 1754 timeliness was characterized as a "regulatory principle":

> With regard to the service itself, punctuality and order in and among all offices and persons pertaining to the court is one of those general duties which not only give outsiders a good or a poor impression of the court, but also bring substantial benefit to the regent himself. A court without order is a building without its roof: it falls in of itself. This order finds its expression in respect, in obedience, . . . and further in precise observance of the time which is set for each principal undertaking.[27]

One principal activity regulated by the clock was the taking of meals by the prince and his retinue, and even the feeding of the animals. The Braunschweig court regulations of 1510–1520 mention the following hours: "The kitchen shall always be made ready, when no fasting is to be done, at 9 hours and in the evening at 4 hours; when there is a fast, then at 11 hours of midday. . . ." In 1547–1548, at the same court, "In the summertime the feed stableman and the grain warden shall feed and issue at two hours of the afternoon, and in the wintertime at the stroke of one . . . ." At the court of Anhalt in 1546 the opening of the wine cellar, too, was subject to rule:

> With our cellar one shall handle it in such fashion that in the morning it be available until the stroke of seven, . . . and from mealtime onward, until we and also the last shall have eaten, the cellar shall again stand open, thereafter being closed until the stroke of two, from two once more opened for perhaps a quarter of an hour, so that each may drink according to the requirement of his thirst, yet not to excess.

Finally, the cellar was opened once more in the evening for the nightcap.[28]

Gargantuan appetites could always make light of these mealtime regulations. In the *Zimmerische Chronik* we find under 1552:

27   Friedrich Carl von Moser, *Teutsches Hofrecht.* Vol. I (Leipzig, 1754), p. 129.
28   *Deutsche Hofordnungen des 16. und 17. Jahrhunderts.* ed. Arthur Kern (Berlin, 1905), pp. 6, 14, 25.
25   *Ibid..* p. 131.
26   *Bergk Ordnung des Durchlauchtigsten Hochgebornen Fürsten und Herrn Christianen Hertzogen zu Sachssen . . .* (n.p., 1589), p. 37.

> In the mornings it might be a mere seven of the clock or the stroke of eight at the latest, and he [Gottfried Werner Graf von Zimmern] would want to have a little snack. And if no one was feeling gay, then to make him happy one had to eat. . . . At night around nine hours and thereafter, everyone had to eat heartily. Then at the time when one ought to sleep and rest, one commenced to really steam. . . . It was in such disorder as this that the summer and also the following autumn were for the most part spent.[29]

But then, following rigidly prescribed dining hours might simply run counter to an aristocratic sense of style, as in 17th-century France. In *De la vraie honnêteté* the Chevalier de Méré (1607?–1684) considers it a bourgeois turn of phrase to say of a princess that she dallied a while in a garden "while awaiting the supper hour; for neither princesses nor persons of high society have the least need of precise times for sitting down to meals."[30]

Like kitchen and cellar doors, house and court gates were also opened and closed by the clock: "The Swiss guards will close the door at six hours in the wintertime, and will hand over the keys to my doorkeeper at ten, and in the morning they will receive the keys when day breaks; and one of them shall stand watch every day at the door; and in the summertime they will close at eight hours of the evening"—this from the court regulations of Prince Conti (1668).[31] The presence at court of the prince's retinue was similarly regulated, for example for pages at Paris and for cavaliers at Gotha: "They [the pages] shall rise at six hours from Easter till All Saints, and after All Saints and until Easter, at six hours and a half."[32] And "first, one Cavalier after another shall wait upon our orders outside our bedroom from seven in the morning by summer, and at eight hours by winter, until curfewbell in the evening; the rest shall present themselves all together at the same place in the mornings by 10 hours and in the afternoons by 5 hours."[33]

The education of the prince also followed a fixed plan of hours. We need only mention in passing that Oxford University was already fixing hours for students as early as 1412.[34] In the instructions for the steward the times for rising, praying, eating, studying, recreation, recapitulation, supping, edification, and going to bed are precisely prescribed. Panokrates, too, sets up Gargantua's plan of studies in such a fashion that no hour of the day is lost; quite the contrary, the entire time is applied to learning the sciences and acquiring useful knowledge.[35] From the end of

29 *Zimmerische Chronik.* ed. Karl August Barack (Freiburg, 1881–1882), Vol. IV, p. 64.

30 Antoine Gombaud de Méré, *Oeuvres complètes.* ed. Ch.-H. Boudhors (Paris, 1930), Vol. III, p. 131.

31 Claude Fleury, *Les devoirs des maîtres et des domestiques* (Paris, 1668), pp. 80 f.

32 *Ibid.,* p. 83.

33 "Herzog Ernst zu Sachsen-Gotha, Hofordnung 1, Februar 1648," in Moser, *Teutsches Hofrecht.* p. 25.

34 Bilfinger, *Die mittelalterlichen Horen.* p. 40.

35 François Rabelais, *Gargantua . . . et Pantagruel* (Brussels, 1659), I, 23.

the 16th century to the 17th, all courses of instruction show increased
regulation, both in Catholic and in Protestant education; there is a shift
from a description of the way the day is divided up to a curt, military table
of hours. Two examples must suffice. In 1584 Duke Wilhelm V of Bavaria
instituted the following instructions for the education of his son Maximilian:

> . . . the regulation hour for rising in the morning is six or half-past six
> . . . for dressing, washing, and the like, three-quarters of an hour. . . . prayer
> in his oratorium. . . . from seven until eight hours there shall be study of
> grammar . . . at eight hours is the time for a bit of breakfast to be taken.
> . . . whereafter a mass should be heard. . . . after the mass there shall be study
> once more for a while. . . . until about half an hour before the midday
> mealtime. . . . after the meal there shall be approximately a couple of hours
> free for recreation . . . at two hours study is resumed (Latin and German)
> . . . until about a half or a whole hour before the evening meal, the time shall
> be passed in the practice of music. . . . the evening meal has its own regula-
> tion. . . . at eight hours he shall be off to bed. . . .[36]

The table of hours according to which Leibniz in 1673 organized the
education of Philipp Wilhelm von Boineburg at Mainz is the following:

> 5 1/2: Rise, dress, and take care of prayers.
> 6–7: Review what the language teacher has expounded or assigned the day
> before, in order to be quite prepared when he comes.
> 7–8: Language teacher, with pronunciation and orthography having first
> priority, and after that translations from Latin into French, also at times from
> French into Latin, to be set as exercises. In addition he may from time to time
> tell Herr von Boineburg a story in French, and have it told back to him.
> 8–9: Mathematics, with instruction being given above all in the fundamen-
> tals of arithmetic and elementary geometry.
> 9–10: Mass and homily, including a sermon.
> 10–11 and 11–12: Exercise, that is to say, dancing master and fencing
> instructor.
> 12 noon: Mealtime.
> 1–2: Rest, or conversation with Mr. Heissen and his beloved [Mother],
> after conclusion of the meal.
> 2–3 and 3–4: History and geography, with a view to his understanding
> both the sequence of events of universal history and the locations and bound-
> aries of countries; and so that he shall also from time to time pick up some-
> thing of chronology, genealogy, and heraldry or blazonry.
> 4–5: The language teacher may come in again.
> 5–6: Guitar teacher.
> 6–7: At his own disposal; and to read some book at the same time both
> useful and agreeable.
> *Nota bene:* At times the hours from 5 to 7 may be devoted to attendance
> at a comedy.
> 7–8: Evening meal.

---

36   Friedrich Schmidt, *Geschichte der Erziehung der Bayerischen Wittelsbacher von den frühesten Zeiten
     bis 1750* (Berlin, 1892), pp. 34 ff. Change in the regulations up to the 18th century becomes
     manifest here.

8–10, etc.: To be used in conversation, in review of what has been absorbed, or in doing the homework that the above-mentioned language and other teachers may have assigned; or as occasion may offer, in the reading of a diverting and at the same time useful book.[37]

Such an inviolable organization of the day now strikes us as stiffly ceremonial. As Saint-Simon was able to say of Louis XIV, "if one had a calendar and a watch, one could tell at three hundred leagues from there [i.e., the palace of Versailles] just what he would be doing."[38] This planning of the hours of the day also took place in the life of the ordinary citizen. As early as 1582 the alchemist Leonhard Thurneyesser of Basel wrote to his wife:

Don't let the children sleep any later in the morning than six o'clock. Bring them up to be clean, first washing the mouth, hands, and eyes with fresh water and cleaning them, then to prayers, and finally to work. Hold their meals at the right time, about 10 hours. In the mornings one should give the children and the household a good strong soup, every day, for they are young and must have something to eat. On this account one should provide them with a sup at eight hours, at ten the midday meal, and in the evening at three their supper; at five one gives them their night meal.[39]

The time for meals shifted over the course of the centuries, and it also changed from one estate of society to another. In 1577 meals were taken at the following times at London: "the nobility, gentry and students doe ordinarily go to dinner at aleaven before noone and to supper at five, or betweene five and sixe at afternoon. The marchaunts dine and suppe seldome before 12 at noone and sixe at night . . . the husbandmen dine also at highe-noone as they call it and sup at seaven or eyght; but out of the terme in our Universities the schoolers dine at tenne."[40] Rabelais was acquainted with the following catch verse for mealtimes: "breakfast at five, dinner at nine, supper at five, bed at nine."[41] Shakespeare's Henry VI spoke longingly of how a shepherd divided his day and his life from minute to hour, from day to year:

To carve out dials quaintly, point by point,
Thereby to see the minutes how they run,
How many make the hour full complete;
How many hours bring about the day;
How many days will finish up the year;
How many years a mortal man may live.

---

37   Gottfried Wilhelm Leibniz, *Allgemeiner politischer und historischer Briefwechsel.* Vol. I (Darmstadt, 1923), p. 322. Even more detailed is the plan of studies which Wolfgang Dietrich von Beichtling reported to his father at about the same time: cf. E. Bülau, "Studentenbriefe aus dem siebzehnten Jahrhundert," *Zeitschrift für deutsche Kulturgeschichte.* 1859, 4:460 f.

38   Glasser, *Studien zur Geschichte des französischen Zeitbegriffs.* p. 177.

39   Karl Seifart, "Zur Sittengeschichte, des 16. Jahrhunderts," *Zeitschrift für deutsche Kulturgeschichte.* 1859, 4:769. The sequence of events in this quotation has been altered somewhat.

40   Bilfinger, *Die mittelalterlichen Horen.* pp. 90 ff. and 47.

41   Rabelais, *Gargantua et Pantagruel.* IV, 64.

When this is known, then to divide the times:
So many hours must I tend my flock;
So many hours must I take my rest;
So many hours must I contemplate;
So many hours must I sport myself;[42]

The breakdown of the hours for waking and for sleeping is already set forth in an ancient mnemonic verse, which we may quote here in the version of an Englishman. The jurist Sir Edward Coke (1552–1634) translated it: "six hours in sleep, in law's grave study six, four spend in prayer, the rest on Nature fix."[43] Comenius divided the day into three parts (according to tradition, dating from the days of Alfred the Great): eight hours for sleep and twice eight hours for work and prayer. Just as in his *Didactica magna* of 1657 Comenius sets apart time for work and for rest, so a measuring out of time in hours appeals even to the utopians, commencing with six hours in Thomas More's *Utopia* (1516), in Campanella's *Civitas solis* (1600) four hours. In Huxley's *Brave New World* three and a half hours are first chosen, but at length a decision for seven and a half is reached.[44]

In the literature of domestic husbandry, hours are indicated for every occupation in the house and garden: for example, the best time for sowing savoy cabbage was set at ten in the morning at Shrovetide.[45] In addition, the planetary hours are also influential: the seven planets rule every single hour, one after the other in turn (Fig. 75). All of them influence growth and the rest of the activities of man as well: "it is good to write letters, send couriers, and visit others in their houses at the hour of Mercury."[46]

Before the days of the newspaper and the post, news was brought by couriers. By the 8th century recipients were already indicating the hour of arrival for the communications brought by couriers wandering from cloister to cloister, so that a sort of time check came into being.[47] The Teutonic Order likewise noted the hour of receipt and of dispatch on the letters that they exchanged with one another.[48] According to the Augsburg courier regulations brought up to date in 1555, a letter sent from Augsburg took a week to reach Venice and one from Venice a week to reach Augsburg; the post had to be at Augsburg by eight hours on Saturdays, and at Venice at twelve hours.[49] At that time there were also, naturally enough, established times for closing the post to other cities.

---

42 *King Henry VI. Third Part.* II, v, 24–34.
43 Gösta Langenfelt, "The Historic Origin of the Eight Hours Day," *Kungl. Vitterhets Historie och Antikvitets Akademiens Handlingar.* Lund, 1954, 87:80.
44 *Ibid.*
45 Balthasar Schnurr, *Kunst-. Hauss-. und Wunder-Buch* (Frankfurt, 1664).
46 Balthasar Schnurr, *Calendarium oeconomicum & perpetuum . . .* (Hanau, 1619).
47 Ernst Kiesskalt, *Die Entstehung der deutschen Post und ihre Entwickelung bis zum Jahre 1932* (Erlangen, 1938), p. 32.
48 *Ibid..* p. 37.
49 *Tagebuch des Lucas Rem.* p. 77.

The letters which the agents of the commercial house of Fugger sent to Augsburg from various countries and cities between 1568 and 1605 make up the "Fugger newsletters."[50] In these, all stories and anecdotes regarded as important to the firm were recounted explicitly, including exact times: On June 4, 1568, the counts Egmont and Hoorn were taken from Ghent to Brussels, where they arrived at three in the afternoon, and at seven in the evening their sentences were read to them. At eleven on the following day Egmont was led to the scaffold, spoke "the time of a pater noster" with the bishop, and was beheaded, being followed by Hoorn. Their heads, stuck on iron spikes, were exposed till three in the afternoon. The times of the deaths of Don Carlos, Queen Isabel of Spain, and other dignitaries were precisely indicated, likewise how many hours the battle of Lepanto lasted and when the tremors of a great earthquake of February 18, 1600, in Peru commenced. Just as the Fugger newsletters report the "strange events and happenings that come to pass in the world" and name the hours for these incidents, so one naturally finds in all the other "relations" that now appear in their train, or in the broadsheets, the naming of the hour at which battles, tempests, celestial phenomena, conflagrations, earthquakes, robberies, and murders took place.

The imprecision of the clock in the specification of times was also taken into account: the code of juridical procedure made it incumbent on the parties to indicate to the judge the (church) clock, in accordance with which all other times having to do with the proceedings should then be arranged.[51] Instead of court terms, let us cite here a literary example. The English poet Robert Greene in 1594 describes in a court scene a treacherous usurer who seeks with his clock, which is running fast, to prevent the payment of a bill of exchange on time:

Usurer: .... for here is the obligation, to be paid between three and four in afternoon, and the clock struck four before he offered it, and the words be between three and four, therefore to be tendered before four.

Thras: Sir, I was there before four, and he held me with brabbling till the clock strook, and then for the breach of a minute he refused my money and kept the recognizance of my land for so small a trifle. Good Signor Mizaldo, speak what is law. . . .

Alc.:   Faith, sir judge, I pray you let me be the gentleman's counsellor, for I can say thus much in his defence, that the usurer's clock is the swiftest clock in all the town' 'tis, sir, like a woman's tongue, it goes ever half an hour before the time; for when we were gone from him, other clocks in the town strook four.[52]

"What is a'clock?"—this question is frequently asked in Shakespeare;

50   *Fugger-Zeitungen,* ed. Victor Klarwill (Vienna, 1923), pp. 3 ff., 14 ff., 226 ff.

51   *Erläuterungen und Verbesserungen der Chur-Fürstlich-Sächsischen Process- und Gerichts-Ordnung. 1724;* quoted in Johann Heinrich Zedler, *Grosses vollständiges Universallexikon aller Wissenschaften und Künsten* (Halle/Leipzig, 1732–54), Vol. 48, col. 501.

52   Robert Greene, *Dramatic and Poetical Works* (London, 1861), p. 82.

specifications of time are numerous in his works. "By seven a'clock I'll get you such a ladder. . . . It hath strook ten a'clock. . . . let him be sent for tomorrow, eight a'clock. . . . is it four a'clock? . . . . let Claudio be executed by four of the clock. . . . sure, Luciana, it is two a'clock. . . . be ready at the farthest by five of the clock. . . . I pray you, what is't a'clock? . . . . by two a'clock I will be with thee again. . . ."[53]

The hour as indicated by the clock regulated the day for all classes of society in the 16th and 17th centuries. The hour as given by the mechanical clock brought the preacher to the pulpit, married betrothed couples, and set baptisms and burials. Instructors at schools and universities, councilors, municipal board members, and judges, watchmen and the proprietors of inns, commenced their operations at firmly set hours. In 1621 Geyger wrote:

> Ein richtig Vhrwerck in der Stadt
> Zeigt an, dass da ein weiser Raht
> Ein richtigs Regiment fuehr eben
> Auch gute Policey darneben
> Die Burger regier mit Weissheit
> Ertheil nach Grechtigkeit die Bscheid.[54]

Yet it was precisely Geyger's emphasis on regimen and on regulation by means of the clock—the order and the consigning of activities to fixed hours which he praised so much—that limited the freedom of action of the individual and caused him to resist. The outcry of Brother Jean to Gargantua, "the hours are made for man and not man for the hours,"[55] shows us how much the day had already been divided up in 1534 when Rabelais wrote his *Gargantua.* People were already rebelling against the compulsion of the clock before the period our exhibition covers—1550–1650. It was for this reason that Gargantua did not set up a clock in the Abbey of Thelema:

> And since in the cloisters of this world everything is elsewhere circumscribed, divided up, and regulated by the hour, it was decreed that there no sort of clock nor dial should be present, but that all activities should be undertaken as circumstances and convenience might indicate. For, says Gargantua, the only true waste of time is that of counting hours. What does one get out of it, anyway? And it is the most unreasonable thing in the world to regulate oneself by the stroke of the clock's bell and not by what your common sense and insight tell you.[56]

53   Martin Spevack, *The Harvard Concordance of Shakespeare* (Cambridge, Mass., 1973), passim.
54   Geyger, *Horologium politicum.* p. 39.
55   Rabelais, *Gargantua et Pantagruel.* I, 41.
56   *Ibid..* I, 52.

Catalog

The introductory texts I–V have been written by Otto Mayr and Klaus Maurice, the catalog texts by Klaus Maurice. The archival citations regarding the Augsburg clockmakers are the work of Eva Groiss, thanks to a grant from the Volkswagen Plant Foundation. General information about the activities of specific clockmakers was taken from Jürgen Abeler, *Meister der Uhrmacherkunst, über 14000 Uhrmacher aus dem deutschen Sprachgebiet* (Wuppertal, 1977).

Abbreviations of archival sources and references:
EMA = Evangelisches Matrikelamt Augsburg
HAP = Hochzeitsamtprotokolle
HKA = Hofkammerarchiv Vienna
HZAB = Hofzahlamtsbücher Vienna
RPR = Ratsprotokolle
STAA = Stadtarchiv Augsburg
SZB = Zunftbuch der Schmiede

D I, II, . . . = Handwerksakten Drechsler, Fasc. I, II
G I, II, . . . = Handwerksakten Goldschmiede, Fasc. I, II
K I, II, . . . = Handwerksakten Kistler, Fasc. I, II
S I, II, . . . = Handwerksakten Schmiede, Fasc. I, II
U I, II, . . . = Handwerksakten Uhrmacher, Fasc. I, II

Maurice I, II = Klaus Maurice, *Die deutsche Räderuhr* (Munich, 1976), Vol. I, II

Goldsmiths' marks:
Rosenberg[3] = Marc Rosenberg, *Der Goldschmiede Merkzeichen* (3rd ed., Frankfurt, 1922)
Seling = Helmut Seling, *Die Kunst der Augsburger Goldschmiede 1529–1868. Geschichte, Werke, Marken* (Munich, in press)

I The Clock as an End in Itself: The Ideal of a Perfect Machine

From the sources known to us we cannot say by whom or in what country the mechanical clock was invented, but we find the first literary evidences of mechanical clocks at the end of the 13th century. The clock then spread throughout Europe with astonishing speed. A wealth of inventive fancy was devoted to it, many technical solutions competed with each other, and by the middle of the 16th century a number of mature design configurations had emerged. The technical characteristics of the clock, as well as the styles and specialities of the production centers, were differentiated geographically.

During the century between approximately 1550 and 1650 European clock technology reached a culmination in the free imperial cities of southern Germany. Easy access to raw materials, highly developed crafts skills, combined with operation on a division-of-labor basis under the rule of the guilds, made cities such as Augsburg and Nuremburg the centers of clockmaking.

Only part of the production of clocks was directed to the needs of everyday life; much effort was devoted to ends that served no practical or economic purposes. Astronomical and calendrical indications plus rich ornamentation outweighed the mundane function of timetelling. The clock became the expression of a fascination which reflected the thoughts and dreams of an age but did not bear upon practical utility.

Clocks and automata can be classified by function and by design configuration. Their mechanisms can perform three functions: telling time, playing music, and animating mechanical figures. The visual indications measure time intervals ranging from minutes (the seconds hand became common only in the 19th century) to planetary revolutions of several solar years. The acoustical indications break down the day into hours and, frequently, quarter hours. The movements for visual indications (the going train) and acoustical ones (the striking train) are distinctly separate within the clock.

The going train consists of

- drive: weight or spring
- gearing: toothed wheels or cord transmission
- escapement: verge escapement with foliot or balance wheel; crossbeat escapement; rolling-ball regulator (the use of a pendulum as a regulator begins just after the time period of our exhibit).

The striking train consists of the same elements as the going train, but it has in addition an ingenious program control feature, the count wheel, which governs the number of strikes required.

The musical mechanism is similar to a striking train, but its program

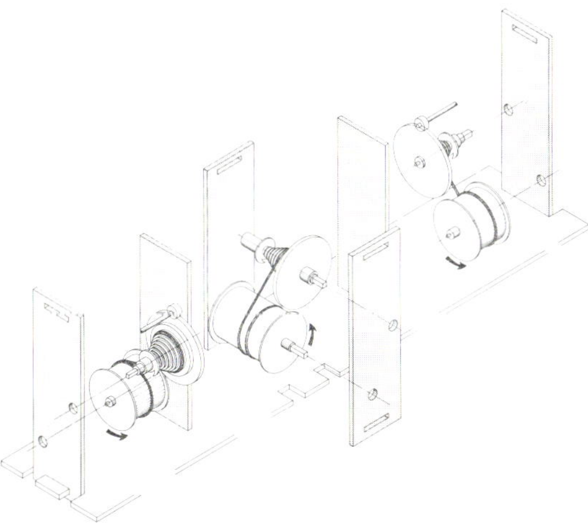

76. Layout of a cruciform arrangement of going, quarter-striking, and hour-striking trains.

77

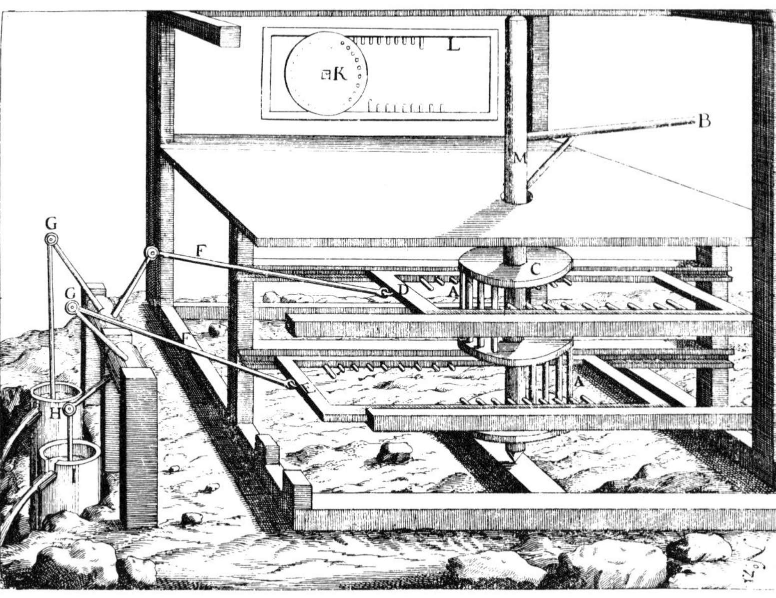

is set up on a studded drum from which the tone elements (such as carillon, metallophone, virginal, and organ) are struck.

Mechanically moving figures are controlled partly by the going train (rolling and blinking eyes), partly by the striking train (snapping jaws, raising arms). If the object as a whole travels, it is driven by a further spring mechanism. In figure clocks the motions of human figures and animals are combined with the indication of time. In automata the telling of time is dispensed with.

The various gear trains performing the different functions can be positioned in several configurations. Individual gear trains, in turn, can be arranged linearly, between narrow metal strips, or two-dimensionally, between parallel plates.

The going train needs a simple frame (see Fig. 79 and Catalog No. 1). If a striking train is added, this simple frame is no longer stable, so

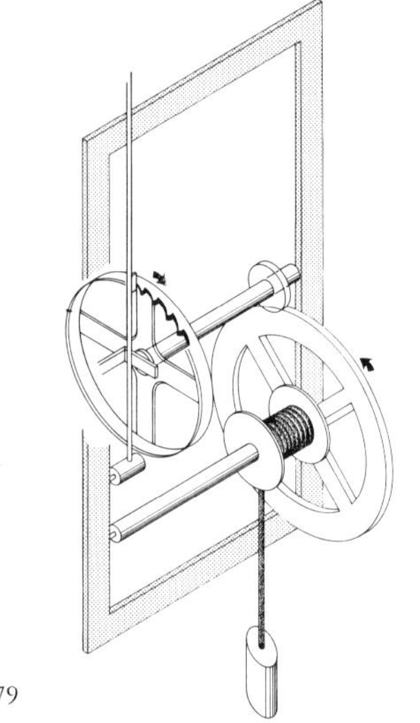

79

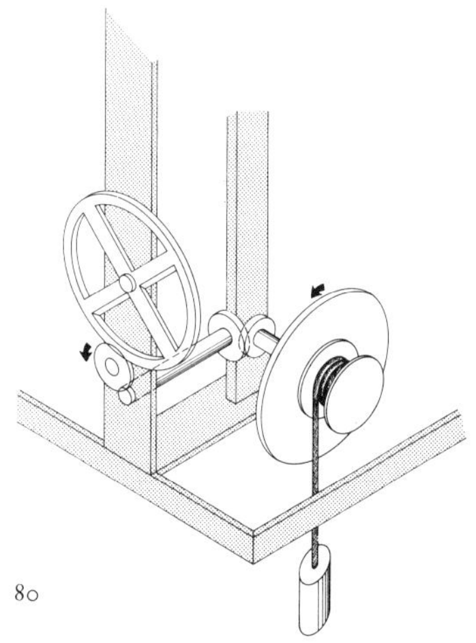

80

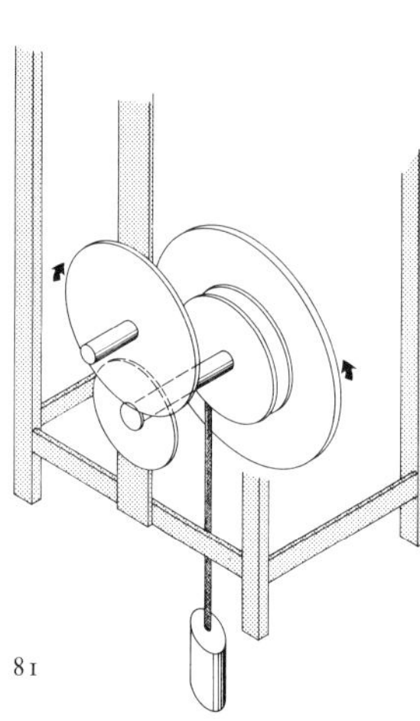

81

77, 78. Automatic adjustment for the indications of the duration of daylight and darkness of the clock shown in Catalog No. 41. The application of this technique to pump works is shown in Fig. 78. Georg Andreas Böckler, *Theatrum machinarum novum* (Nuremberg, 1661).

79. Layout of a going train in a simple frame.

80. Clockwork arranged coaxially, side by side, in a "birdcage."

81. Clockwork arranged coaxially, one behind the other, in a "birdcage."

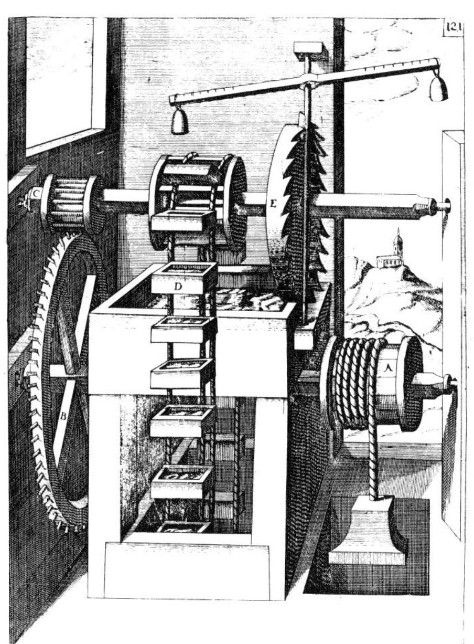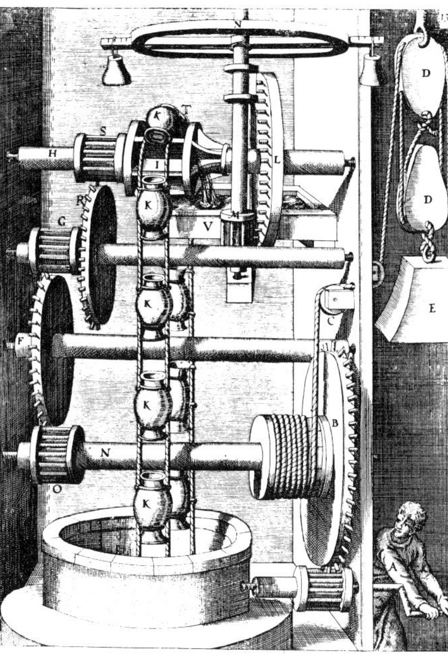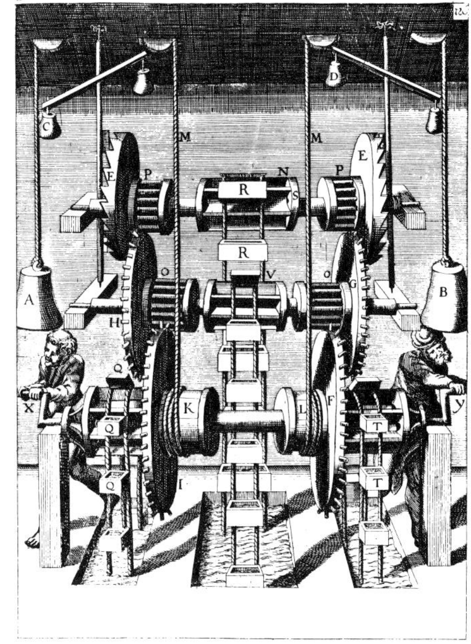

82–84. Water-pump works, which have borrowed techniques of clockwork construction. Georg Andreas Böckler, *Theatrum machinarum novum* (Nuremberg, 1961).

both trains are placed within a more solid frame which is called a "birdcage" (see Figs. 80 and 81, Catalog No. 2). The traditional way of mounting the going and striking trains is either coaxially side by side (Fig. 80) or coaxially one behind the other (Fig. 81 and Catalog No. 3). Spring-driven clocks with going, hour-striking, and quarter-striking trains have these three drives arranged in a cruciform fashion (as shown in Fig. 76 and Catalog No. 23); in this way they can be wound and disassembled more easily. Other constructions are exceptional. The arrangement of the wheel mechanism between plates, introduced generally in the second half of the 16th century, achieved a decided saving of space and made portable clocks possible. (See Figs. 85–89, Catalog Nos. 9, 21, 39–46.)

Clock design embodied the highest technological sophistication of the 16th and 17th centuries. Just as every highly developed technology has an enriching effect upon other contemporaneous technologies, so various components of clocks were applied in other technical domains, for example in the controlling of fountains and in pumping and water-raising apparatus (see Figs. 77 and 78, 82–84).

1 Wall clock

Nuremberg (?), 15th century
Darmstadt, Hessisches Landesmuseum

Frame and wheels: iron; bell: bronze
Height: 49 cm (19¼ in.)

Going movement with alarm. Alarm work released from the barrel wheel. The duration of the alarm is governed by the extent to which the cam rotates. Upon release, the lever which arrests the alarm verge wheel is raised and rests on the cam, allowing the alarm work to function while this cam rotates. The sequence of the hours on the dial follows that of the style of the Great Clock in Nuremberg: the first daylight hour begins at sunrise, the first nighttime hour at sunset. The two sections, one for daytime hours, the other for nighttime, replace each other as the change in seasons requires. This wall clock can presumably be assigned to Nuremberg in view of the Great Clock indications. For a similar piece see Maurice II, Fig. 35.

Unpublished.

2 Bracket clock

South Germany, end of 15th century
Munich, Bayerisches Nationalmuseum

Frame and wheels: iron
Bell: bronze
Height: 42 cm (16½ in.)

Within the frame the going, quarter-striking, and hour-striking trains stand one behind the other. The motion work is no longer original, the dial is missing, and the verge has been replaced. The shape of the wheel spokes is noteworthy. The Gothic style, shown in the delicate, sharp articulation of the pillars and bell strapping, places the date at the end of the 15th century.

Literature: Maurice II, Fig. 61.

3 Bracket clock

Germany (?), beginning of the 16th century
Munich, Bayerisches Nationalmuseum
(Permanent loan from the Mainfränkisches
Museum, Würzburg)

Wheels and frame: iron; height: 44 cm (17¼ in.)

Going train and striking train sounding the hours from 1 to 12; phases of the moon. The dial has been altered to conform to that of the Great Clock—that is, to indicate the hours from 1 to 16. The motion work has therefore been converted. Going and striking trains are both set at a 90° angle to the dial. The balance wheel has notched spokes on which regulating weights can be hung. The band of tracery running along the lower edge of the frame is noteworthy: three very delicate iron strips, each pierced in a different pattern, are superposed one upon another. The way the bars are secured with wedges departs completely from the usual method of fastening by means of pins. It is uncertain whether this wedging has its origin in this particular type of clock or in a particular place of origin.

Literature: Maurice II, Fig. 55.

4 Small tower clock

Switzerland (?), 1604
Basel, Historisches Museum
Frame and wheels: iron
Height: 49 cm (19¼ in.)

Going and hour-striking trains in round frame. Release of the striking train is from the great wheel of the going train. The bell was originally suspended at some distance from the movement. Dating in accordance with the museum's inventory.

Literature: Maurice II, Fig. 50.

5 Tower clock

Albrecht Ott. Hennenbach (Central
Franconia), 1569
United States, private collection

Frame and wheels: iron, with traces of protective
paint
Height: 61 cm (24 in.)

Going and hour-striking trains. Escapement is a reconstruction. Signed on the
frame: ALBRECHT OT ZV HENEBACH HAT DIE VR AVS/GEMACHT IM MERZEN
IM 1569 IAR. After the signature the following mark is repeated three times:
a cross over a star, both in a shield.

Unpublished.

6 Gearing for astronomical motion work

Laurentius Liechti. Winterthur, 1529
Winterthur, Lindengut Museum

Iron

Diameter of the main wheel: 33.8 cm (13¼ in.)

(Frame and crank are new)

This gearing belongs to an existing tower clock movement which was installed on August 21, 1529, on the Käfigtor, the original west gate of the first city wall of Winterthur. The Käfigtor was torn down in 1870. The tower clock movement drove, in addition to the astronomical dial, an hour dial of its own indicating the hours I–XII. The gearing is driven by the going train and drives the sun and moon hands, the zodiac, and the dragon hand. The lantern pinion with its 19 pins makes three rotations in 24 hours. There are the following times of revolution per 24 hours for the astronomical elements: *Sun:* one revolution in 24 hours, zero error. *Moon:* one revolution less than the sun in 29.5 days; since the synodic lunar month is 29.53058 days, the error is 0.03058 day. *Zodiac:* one revolution more than the sun, in 364.892 days; since the tropic year is 365.24241 days, the error is 0.35 day. *Dragon:* one revolution more than the zodiac in 18.45 years; since the length of the dragon year is 18.6 years, the error of the dragon hand is 0.15 year.

On the frame of the tower clock is the signature L * L 1529. There are records of the Liechti family in Winterthur from 1477 onward. In 1514 Laurentius Liechti made the clock for the Frauenkirche in Munich. In 1533 he was chosen to the grand council at Winterthur, and he died before July 1, 1554. His two sons, Laurentius and Erhard, were not yet adult at that time. Up to the middle of the 19th century twelve successive generations of Liechtis made clocks at Winterthur.

Literature: Adolf Schenk, "Die astronomische Uhr des Zeitglockenturms in Winterthur," *78. Neujahrsblatt der Hilfsgemeinschaft Winterthur 1941.* Adolf Schenk, *Die Uhrmacher von Winterthur und ihre Werke* (Winterthur, 1970), p. 28. Maurice II, Fig. 63.

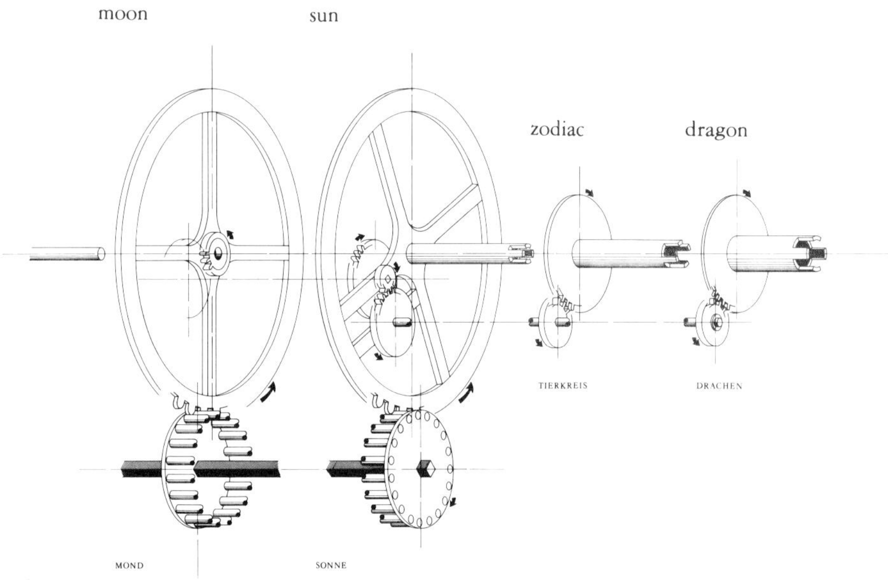

7 Bracket clock

South Germany, end of the 15th century;
modified 1537
Private collection

Frame and wheels: iron; dial: fire-gilt brass
Height: 48 cm (19 in.)

The trains for going, quarter striking, and hour striking are installed one behind the other. Frame, crosspieces, and bridges are keyed and pinned. Originally the clock had an hour dial; the astronomical motion work and the corresponding dial were added in 1537 (date on dial). At the beginning of the 18th century an indication for the days of the week was added, and on the lower dial, a minute hand. At the same time the clock—as is visible in old photographs—was modernized with a Clement anchor escapement and a pendulum. It has now been restored to a verge escapement with foliot. The outermost ring on the dial shows the Whole Clock (1–24, beginning at midnight), the innermost ring has the hours for the Italian Clock and the Bohemian Clock (1–24, beginning respectively at sunset and at sunrise). The rete is simplified to show only the zodiac; above it is the sun (hour) hand and the moon hand on which a lunar sphere rotates to show the phases.

Literature: Ernst von Bassermann-Jordan, "Astrologie, Uhrmacher, und Uhren," *Deutsche Uhrmacher-Zeitung.* 1930, 54 (No. 48): 794–798. Maurice II, Fig. 60.

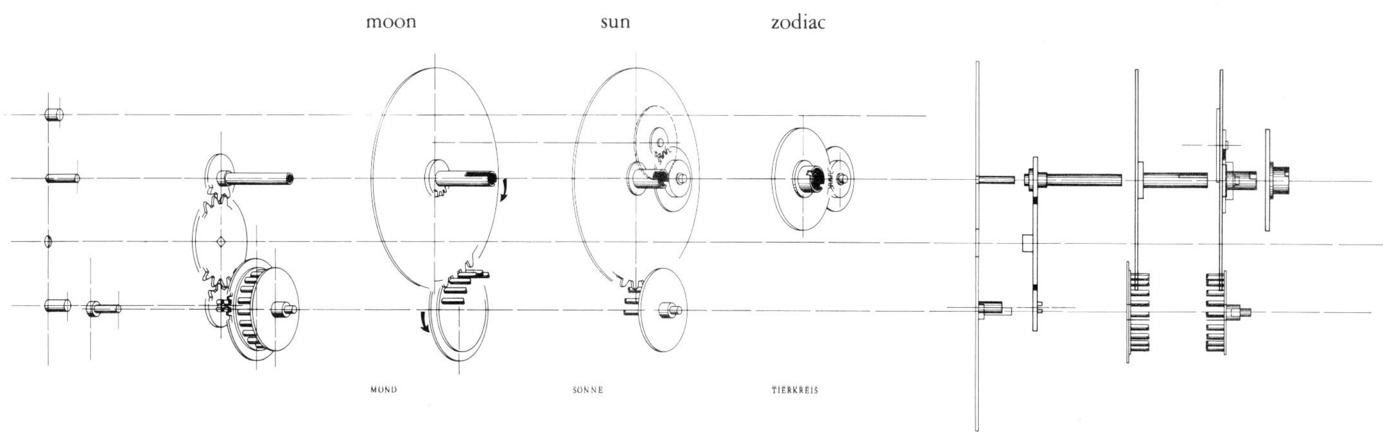

8 Bracket clock with carillon

Lower Rhine (?), 1583
Bamberg, Textor collection

Movement: iron; dial painted in colors
Height: 57 cm (22½ in.)

Going and hour-striking trains. Count wheel toothed on it inner edge and cut with a double set of 1–12 notches for notation of the hours. Carillon with six bells. The going train was subsequently given a pendulum escapement, but the original escapement has now been reconstructed. Dial with phases of the moon, I–XII ring, and a small dial on which the quarter hours are indicated in doubled series. The carillon drum is of wood (pins replaced); the hour bell is integrated into the tune. The carillon is released hourly by the motion work. After playing, the carillon in turn releases the striking train, with the jack bobbing its head and thereby striking the hours. The wedging of the bearing posts is as in Catalog No. 3.

Stylistically the costume of the jack is like figures engraved by Hendrick Goltzius (1558–1617) illustrating the dress of the 1570–1580 period. In this respect the dating 1583 on the dial is presumably correct, although the date may also originate from a reconditioning of the dial. A similar clock, restored and signed by the Liège clockmaker Jean Knaeps in 1696, was auctioned by Sotheby Parke Bernet on November 13, 1979 (No. 30). Since the chime structure and the jack of the clock exhibited here are similar to those of the former, we assume that it originated in the lower Rhine.

Unpublished.

9 Table clock

Jakob Zech. Prague, 1525
London, The Society of Antiquaries

Movement: iron; case: brass
Height: 12.5 cm (5 in.)
Diameter: 24.2 cm (9½ in.)

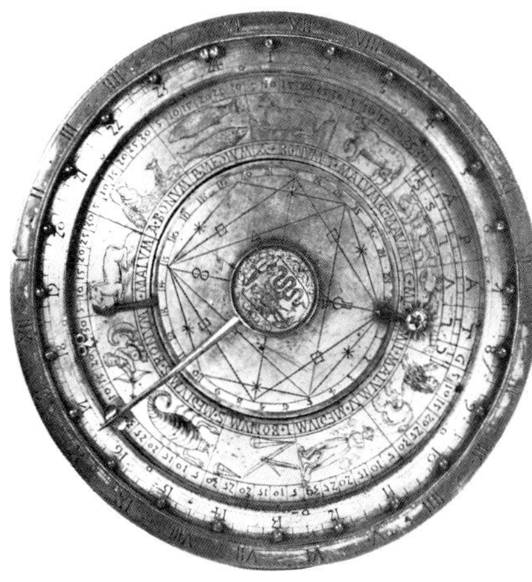

Going train with fusee and gut line, a single hourly stroke on a bell. Dial indicates I–XII hours twice repeated, adjustable ring with touch marks for 1–24 hour indication. Hour hand and two other hands show the position of the sun and the moon in the zodiac; diagram of aspects is also featured. Next to the symbols for the zodiac the astrological influence is specified. The movement runs for approximately 30 hours. The spring housing is wound with a caliper-like key—a type of winding already employed in the clock of Philip the Good, about 1430 (cf. Maurice II, Figs. 77b and c). The arms of the foliot have threaded ends on which the screw-on regulating weights can be turned inward or outward for adjustment. Signed on the spring drum: DAMAN + ZALT + 1 + 5 + 2 + 5 + IAR + DA + MAHCHT/MICH + JACOB + ZECH + ZV + PRAG + IST + BAR (i.e., *wahr* = true.) The mark of Zech is struck into the bottom of the case: in a shield the initials IZ, above these a foliot. On the case the arms of the Polish king Sigismund I, his wife, Queen Bona Sforza (of the Visconti family), and the principality of Lithuania. The same arms are repeated on the hand.

Jakob Zech had taken care of the clock on the Old Town city hall in Prague from 1515 onward. In 1534 he received a payment of 44 ducats for five clocks that he had made for Queen Anna. He died in 1540. His son-in-law Hans Steinmeissel was also a clockmaker.

Literature: W. H. Smyth, "Description of an Astrological Clock," *Archaeologia.* London, 1848, *33:* 8–25. Alfred Holinski, "Jacob Zech and a Royal Fusee Clock," *The Connoisseur.* 1963, *152:* 183–187. Maurice II, Fig. 475.

10 Automaton figure of a monk

South Germany or Spain, c. 1560
Washington, D.C., National Museum of
History and Technology

Figure: head of poplar wood; head and limbs
rendered naturalistically; modern habit
Movement: iron; height: 39 cm (15⅜ in.)

Movement has a wooden fusee. The monk is programmed to move in a square approximately 2 feet wide. Although his feet step out from beneath his robe, he is actually rolling on wheels. As he proceeds he strikes his right arm against his chest and moves the left arm up and down. He is constantly turning his head, nodding, moving his mouth, and rolling his eyes. The mechanism is pinned for the most part; only the pillars and one cam are secured by means of screws. An automaton figure of a girl playing a lute, similar to the monk in its mechanism and program, was auctioned at the Galerie d'Horlogerie Ancienne in its sale No. VI, April 9, 1978, as Lot No. 83.

The ascription to Spain is based upon the realistic conception and modelling of the monk's head and upon the datum that Emperor Charles V's clockmaker, Juanelo Turriano, had also made comparable mechanisms for automaton figures (Ambrosio de Morales, *Las antigüedades de las ciudades de España,* Alcalá, 1575, pp. 91 ff.). Yet similar moving figures were also made by the Nuremberg clockmakers. Jakob Bulmann (master in 1497) made "figures of men and women that walked around and struck their measure upon lutes and kettle-drums" (Neudorfer, 1547; cf. Maurice I, p. 58). In Emperor Rudolf II's collection of art treasures there were a number of spring-driven clothed dolls that beat upon drums and walked about (cf. Nos. 2193 and 2194 in the "Kunstkammerinventar Kaiser Rudolf II, 1607–1711," by Rotraud Bauer and Herbert Haupt, in *Jahrbuch der Kunsthistorischen Sammlungen in Wien,* 1976, 72: 113).

Literature: Uto Auktionen, 2, Zurich, Sept. 29–Oct. 3, 1975, No. 351.

11 Music work

Augsburg, c. 1670
Munich, Bayerisches Nationalmuseum

Movement: iron
Frame: beech, fir, oak
Hammers: brass
Height: 15.4 cm (6 in.)

The drum is of linden wood; a paper has been glued over the original incised lines and new pins were inserted in the 18th century. The fly has been enlarged subsequent to original manufacture. The sound-producing bars rest upon straw. Originally this mechanism, which was built into a clock, also turned two merry-go-rounds. The clockwork mechanism, still in the museum, is signed by Johann Georg Engelschalck in Friedberg. For a similar piece see Maurice II, Fig. 604b.

Literature: Ernst von Bassermann-Jordan, *Die Geschichte der Räderuhr unter besonderer Berücksichtigung der Uhren des Bayerischen Nationalmuseums* (Frankfurt, 1905), No. 48.

171

II The Guild: Quantity Production and Craft Conservatism

Around the middle of the 16th century, in the most important cities of southern Germany, special guilds established themselves which exclusively represented the clockmakers' craft. In other parts of Europe this development took place later. The 16th-century centers were Vienna, Innsbruck, Munich, Strasbourg, Nuremburg, and, far surpassing all others in excellence and productivity, Augsburg. After the middle of the 17th century German clockmaking lost its premier position, for a number of reasons.

All the guilds, including that of the clockmakers, had the immediate aim of ensuring the livelihood of their members. And many of the prescriptions of the guilds can be understood only from the standpoint of the guilds' very static and conservative ways of thinking, expressed, for example, in the conviction that increased production would dissipate demand. From this viewpoint it followed that the number of craft shops was held constant and the admission of new members was limited. The acquisition of the title of master, which alone entitled one to run an independent workshop, was subject to a number of complicated requirements, the most important of which was the making of a masterpiece.

As prescribed by the clockmakers' guild of Augsburg (see Ch. 7), the masterpiece had to incorporate the following indications: the hours according to three different methods of counting (the Small Clock, the Whole Clock, and the Great Clock), the movement of the sun and the moon in the zodiac and in the sphere of the fixed stars, hour striking adjustable for 1–12 and 1–24 hours, quarter striking, and an alarm. The shape of the clock cases might vary, but usually they were turret shaped

85–89. Masterpieces of Augsburg clockmakers: views of the movements of monstrance clocks after the clockfaces have been removed.

85. Nikolaus Rugendas. Augsburg, 1616. See Catalog No. 44.

86. Thomas Starck. Augsburg, 1620. London, British Museum.

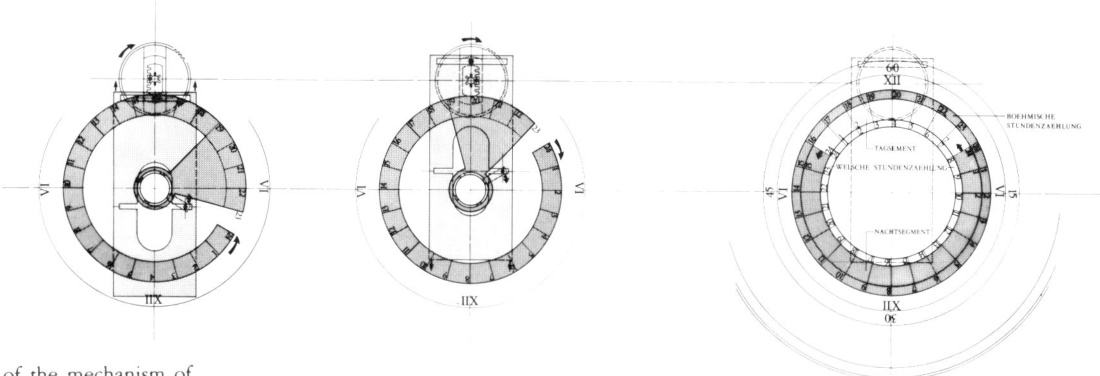

90. Schematic diagram of the mechanism of the automatic adjustment for the indications of the duration of daylight and darkness.

(as in Catalog No. 25) or in the shape of a mirror, or monstrance (Catalog No. 46). The astonishing stereotyping of these masterpieces, which continued for almost a hundred years, is accounted for by the retention of the same masterpiece rules and by the habit of repeating movement designs over and over from the same drawings. This is demonstrated, for example, by the surprising similarity, or indeed standardization, of the arrangement of the movements in the five monstrance clocks illustrated here (Figs. 85–89, 90) which were made between 1616 and 1669.

The complex aggregate of mechanical parts and case components which made up a clock required many different skills. Division of labor became not only necessary, it was also warranted by the quantities produced. The clockmaker passed out commissions to numerous subcontractors (see Catalog No. 45). For example, the Augsburg clockmakers ordered their bells from Lyon, while the clockmakers of Ulm and Strasbourg bought their ebony cases in Augsburg (see Catalog Nos. 54, 29). Elements of mechanisms and of cases, especially bronze statuettes, were in many instances manufactured as stock items and sold at fairs and through dealers (see, e.g., Fig. 93). The lion clocks (Catalog Nos. 84, 85), the Madonna figures (Catalog Nos. 62, 63), and the groups with Diana on the stag (Catalog Nos. 102, 103) all testify to this early form of mass production.

87. Nikolaus Planckh. Augsburg, 1631. Vienna, Kunsthistorisches Museum.

88. Caspar Langenbucher. Augsburg, 1649. (Slide for the control of indications of daylight and darkness has been removed.) See Catalog No. 45.

89. Johann Martin. Augsburg, 1669. (Slide for the control of the indications for daylight and darkness has been removed.) See Catalog No. 46.

12 Pedestal clock

Augsburg (?), c. 1590
Winterthur, K. Kellenberger clock
collection

Movement: iron plates, brass plates and wheels
Case: fire-gilt bronze, copper, and brass
Pedestal: ebony
Height: 102 cm (40½ in.)

Going, quarter-striking, and hour-striking trains; alarm. Original balance-wheel escapement, formerly replaced by a pendulum regulator but now restored. The three weights run within the pedestal. The upper bell and some of the drawer fronts in the base of the pedestal are strictly ornamental.

Literature: Ernst von Bassermann-Jordan and Hans von Bertele, *Uhren* (Braunschweig, 1961), p. 33. H. Alan Lloyd, *The Collector's Dictionary of Clocks* (South Brunswick/New York, 1964), p. 140. Maurice II, Fig. 121.

174

13 Tall case clock

Jørgen Eckler. Copenhagen, 1588
Kremsmünster, Abbey

Movement: iron
Case: fire-gilt bronze and brass; black-stained
wood
Height: 199 cm (6 ft 6⅜ in.)

The three weight-driven trains are arranged in a crosswise formation. They govern the following indications. *Front:* astrolabe, dragon hand, sun and moon hands, minute hand. On the two small dials below, at left, depictions of the planetary gods (the seven gods that rule the planets and the days of the week); at right, setting dial for the alarm. Hour striking and quarter striking with respective checking dials on either side. *Back:* calendar dial, engraved on each side for six months, and in the middle a circular scale of hours for the lengths of the day at Jerusalem, Nuremberg, and Stockholm. Escapement is not the original.

The signature "I.E. 1588" appears on the front between the small dials. For a long time southern Germany was taken to be the place of origin, or more precisely Nuremberg because of the length of the day in effect there. In the Mayer Amschel Rothschild collection at Mentmore there was a small table clock (Sotheby auction of May 18–23, 1977, catalog, Vol. I, lot No. 31), which is signed with the same initials and the date 1584, and bears the arms of King Frederick II of Denmark (1555–1588). Jørgen Eckler was salaried as clockmaker at the Danish court in 1585 (Bering Liisberg, *Urmagere og ure i Danmark,* Copenhagen, 1908, p. 158). The initials on the two clocks are identical; thus the Kremsmünster clock can also be attributed to Eckler. The wooden case for the weights presumably was made later, around 1625, because the vertical framing of the box overhangs the ornamental cast foot of the clock. (I am grateful to Mr. B. Hutchinson of the British Museum for providing the reference.)

Literature: S. Fellöcker, *Geschichte der Sternwarte der Benediktinerabtei Kremsmünster* (Linz, 1864). Ernest L. Edwardes, *Weight-Driven Chamber Clocks* (Altrincham, 1965), pp. 93–95, Figs. 20, 21. Maurice II, Fig. 176. *Die Kunstdenkmäler des Benediktinerstiftes Kremsmünster* (Vienna, 1977), Pt. II, p. 239, No. 47.

14 Bracket clock with weight drive

South Germany, c. 1570–1580
Private collection

Movement: iron
Case: iron painted in colors
Dial: fire-gilt brass
Height: 28 cm (11 in.)

Going, quarter-striking, and hour-striking trains and alarm. Dial with the hours I–XII twice repeated, hour hand, sun and moon hands indicating position of these bodies in the zodiac. Inside the chapter ring are segments to indicate the lengths of day and night. Lower three dials: left, the Small Clock; middle, quarter hours; right, setting dial for adjustment of the lengths of day and night. The latter mechanism is missing, but otherwise the clock is in its original condition. Both iron side doors open on hinges. Inside they are painted with an ornamental design. On the outside they depict three astronomers on one side, two on the other, debating over a celestial globe; beneath each door are two small panels showing allegorical figures of four planets. The remaining three planets appear on arched panels surrounding the bells. Two of these panels are missing.

Unpublished.

176

15 Table clock

Innsbruck (?), 1570
Cologne, private collection

Movement: iron
Case: fire-gilt brass
Height: 16 cm (6¼ in.)

Going and hour-striking trains, both with fusee and gut line. Alarm missing.
On the front the arms of the Saxon family of Günther with the initials AG and
the year 1570. On the two sides the allegorical figures of Fortune and Astrology
are engraved. On the back is a proverb: "Prima que dedit/Vitam hora carpit."
The architectonic structure of this clock is similar to that of a clock by the
Innsbruck maker Nikolaus Lanz (see Maurice II, Fig. 92), for which reason this
clock can also be assigned to Innsbruck.

Unpublished.

16 Table clock

Hans Gruber. Nuremberg, 1573
Baltimore, Walters Art Gallery

Movement: iron; spring drum of brass
Case: fire-gilt bronze and brass
Height: 23.5 cm (9¼ in.)

Going and hour-striking trains. Front has hour dial (I–XII twice repeated) and
quarter-hour dial; back side has checking dial. One side panel is engraved with
"Caesar Carolus" on the outside and an allegorical figure of Astronomy inside.
The other side panel is engraved with "Rex David" outside and a horizontal
sundial with the date 1573 inside. Escapement not original. The lion's paws are
set on balls in order to get enough height for the pendulum regulator that was
added later. On the base plate two different marks of Hans Gruber (see Maurice II, Figs. 106c, 109e).

Numerous clocks can be assigned to Hans Gruber after 1550 and until his
death in 1597. One of the best-known Nuremberg clockmakers, he also built
artillery instruments. His two sons, Hans the younger and Michael, also
worked this craft. Gruber's marks varied: he used at least five different punches.
For another work of this master see Catalog No. 24.

Literature: Maurice II, Fig. 110.

17 Table clock

Strasbourg, 1573
Furtwangen, Historische Uhrensammlung
der Fachhochschule

Movement: iron
Case: fire-gilt bronze and brass
Height: 18.5 cm (7¼ in.)

18 Table clock

Carl Gutbub. Strasbourg, last quarter
of the 16th century
Stuttgart, Württembergisches
Landesmuseum

Movement: iron, brass
Case: fire-gilt bronze and brass
Height: 27 cm (10⅝ in.)

Going train with fusee and chain, hour-striking and quarter-striking trains, alarm. Indications: I–XII hours twice repeated with touch marks, adjustable ring with 1–24 hours, alarm-setting dial, age and phases of the moon, diagram of planetary aspects. Back: checking dial for hour striking. On both sides unidentified coats of arms and the date 1573. Strasbourg hallmark on the alarm mechanism. This style of bell strapping is found on a number of Strasbourg clocks.

Literature: Maurice II, Fig. 98.

Going train without force equalization, quarter-striking train and alarm in the main case, hour-striking train in the base. Front: hour and minute hands, hour indication I–XII; alarm setting dial in center; lower dial for the days of the week, personified by the planetary gods. Back: top, outer dial with I–XII hour count twice repeated; inner dial, the zodiac with hands for the sun and the moon. Middle, a dial with the age and phases of the moon. Bottom, the checking dial for the striking of the quarter hours. On both sides of the case are engravings of warriors from classical antiquity. Alarm mechanism marked with the Strasbourg hallmark and the initials CG. Another clock identical in construction and varying only slightly in its indications is in a private collection in the Netherlands.

Because of the Strasbourg hallmark and the positioning of the hour-striking train in the base, typical for Strasbourg, the initials can be assigned to the Strasbourg clockmaker Carl Gutbub. Carl and his brother Hans, also a clockmaker, were born in Weinberg near Buchsweiler in Lower Alsace. Carl received Strasbourg citizenship in 1571. His daughter married the clockmaker Joachim Liechti, of Winterthur. Hans Gutbub acquired Strasbourg citizenship in 1587 through his marriage to Suzanne Habrecht, daughter of the clockmaker Isaac Habrecht the elder, who had made the clock in the Strasbourg cathedral (see Fig. 6).

Literature: Joseph Fremersdorf, "Seltene astronomische Strassburger Türmchenuhr mit Uhrmacher-Stadtpunze und Signatur CG = Carl Gutbub," *Freunde alter Uhren,* 1970–71, *10:* 8–26. Volker Himmelein and John H. Leopold, *Prunkuhren des 16. Jahrhunderts. Sammlung Joseph Fremersdorf* (Stuttgart, 1974), p. 79. Maurice II, Fig. 99.

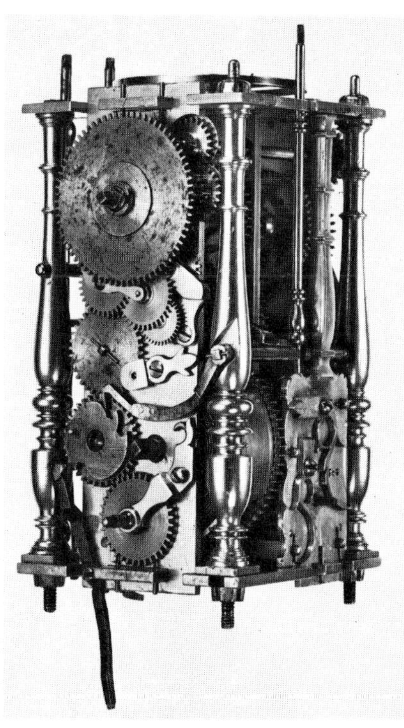

19 Table clock

Nikolaus Schmidt. Augsburg, c. 1600
United States, private collection

Movement: brass plates and wheels
Case: fire-gilt bronze and brass; base plate of iron
Original carrying case in stamped leather
Height of clock: 32 cm (12½ in.)
Height of carrying case: 35.4 cm (14 in.)

Going and hour-striking trains, alarm. The base plate for the movement is iron
and bears an etched ornament. The insides of both side plates are stamped NS,
the mark of the Augsburg clockmaker Nikolaus Schmidt.

Nikolaus Schmidt the elder was born around 1550 at Wiltz in Luxembourg
and became a master at Augsburg in 1576. In the same year he married
Katharina, the daughter of the clockmaker Hans Fronmiller, and after her
death in 1579 he married Susanne Gloninger. In 1586 he became foreman
(*Vorgeher*) of the smiths. He had three sons. Georg, born 1580, who became
a master in 1608, married Maria Schüttering (widow of Hans Fronmiller) in
that year, and died in 1630. Nikolaus the younger was born in 1582 and
became a master clockmaker in 1620. Carl (Carol) was probably born in 1586;
he became a master clockmaker in 1614. Hans Ulrich Schmidt, a grandson of
Nikolaus Schmidt the elder, became a master clockmaker in 1648.

Literature: Maurice II, Fig. 132.

180

20 Table clock

South Germany, c. 1600
Munich, Bayerisches Nationalmuseum

Movement: brass plates and wheels
Case: fire-gilt bronze and copper
Height: 44 cm (17¼ in.)

Going movement with fusee and chain; chain and escapement are no longer original. The quarter-striking and hour-striking trains are arranged side by side behind the going train. The hour-striking train is upside down. On the front is an hour chapter ring and beneath this a small quarter-hour dial. The back bears both checking dials for the striking mechanisms. Alarm missing. On the sides are crude engravings of the allegorical figures Justice and Faith. Below Faith is the head-and-shoulders figure of a man with a quadrant and the letters MB (letters probably later, not the initials of the clockmaker). Case re-gilt; crucifix and lion's feet not original.

Literature: Ernst von Bassermann-Jordan, *Die Geschichte der Räderuhr unter besonderer Berücksichtigung der Uhren des Bayerischen Nationalmuseums* (Frankfurt, 1905), p. 74, No. 28. Maurice II, Fig. 128.

21 Table clock

Andreas Yllmer. Innsbruck, 1559
Innsbruck, Tiroler Landesmuseum
Ferdinandeum

Movement: iron
Spring drum: brass
Case: fire-gilt bronze and brass
Height: 35.5 cm (14 in.)

The entire mechanism of the clock is built up between two solid plates; the going train is arranged upside down. Going, hour-striking, and quarter-striking trains all have fusees. The fact that the wheels for the differential-epicyclical drive are crossed out is noteworthy. Indications on the front: astrolabe, tympanum, engraved on both sides for 45° and 48° latitude. Rete missing. Sun, moon, and dragon hands. Dial at upper left gives the position of the sun in the zodiac; at upper right is the Roman indication, dominical letters, solar cycle, lunar cycle, Golden Number, epacts. Below, the checking dials for the hour and the quarter striking. Back: calendar dial covering six months, days of the month, saint for the day, initial letters for the day, minutes, I–XII hours twice repeated, segments for the lengths of day and night, adjustable 1–24 hour ring. The smaller dials indicate (from upper left) the planets ruling the days of the week, alarm-setting dial, the planets ruling the hours, regulatory scale. Also on the back the signature ANDREAS ILMAR VRMACHER ZVO ISBRVCK 1559. In the globe on top, age and phases of the moon; on the outside of the globe, the arms of Hesse and the inscription of the former owner: PHILIP D. G. LANDGRAVIVS HASSIAE COMES IN CATZENELNBOGEN. On the sides engravings of Fortuna, a crucifixion, Patience, and Justice, Resurrection, Faith, possibly executed by the Innsbruck goldsmith Hans Pfaundler (see Leopold Pfaundler, *Chronik der Familie Pfaundler von 1486–1915,* Salzburg, 1915, p. 10). Andreas Yllmer worked on a number of clocks for the Innsbruck court of the archdukes between 1558 and 1585. He was married to the daughter of a goldsmith named Pfaundler. He died in 1586 or 1587.

Literature: David Schönherr, "Andrä Yllmer," *Repertorium für Kunstwissenschaft,* 1876, *1:* 408. Ernst Attlmayr, "Der Hofuhrmacher Andrä Yllmer," *Beiträge zur Technikgeschichte Tirols,* 1973, *5:* 42 ff. Harro Kühnelt, "Zwei astronomische Uhren von Andreas Yllmer," *Tiroler Heimat,* 1967, *21:* 83–86. Maurice II, Fig. 162. Erich Egg, "Innsbrucker Kleinuhrmacher," *Das Fenster: Tiroler Kunstzeitschrift,* 1976–77, *19:* 2002.

22 Table clock

Augsburg, c. 1570–1580
Dresden, DDR, Grünes Gewölbe

Movement: iron
Case: fire-gilt bronze, painted with cold-applied enamel
Height: 29 cm (11⅜ in.)

Going, quarter-striking, and hour-striking trains; alarm. For the arrangement of the trains and the layout of the indications see Catalog No. 26. Noteworthy in the movement is the epicyclic train with fish-bladder-shaped crossings and the similarly shaped pierced cover plate. The layout of the movement and the indications correspond to the Augsburg masterpiece requirements (see Ch. 7). Through these masterpieces there arose a standard style, even for the case, and as a result the same case elements can be seen in a number of clocks of this type. Not all of these clocks were masterpieces. Complex clocks of this form were repeated a number of times in a single workshop (e.g., Catalog No. 23 and the reference to a *Stuckuhr* in Catalog No. 26).

Literature: Jean Louis Sponsel, *Das Grüne Gewölbe zu Dresden* (Leipzig, 1928), Vol. II, p. 190. Erwin Neumann, "Die Tischuhr des Jeremias Metzger von 1564 und ihre nächsten Verwandten. Bemerkungen zur Formengeschichte einer Gruppe süddeutscher Stutzuhren der Renaissance," *Jahrbuch der Kunsthistorischen Sammlungen in Wien,* 1961, *57:* 116. Maurice II, Fig. 165.

23 Table clock

Caspar Bohemus. Vienna, 1568 (Jeremias
Metzger. Augsburg)
New York, Metropolitan Museum of Art

Movement: iron
Case: fire-gilt bronze and brass
Height: 31.2 cm (12¼ in.)

Going, quarter-striking, and hour-striking trains, positioned crosswise; all three
with fusees and gut lines. Front with astrolabe, tympanum for 51° latitude, sun
and moon hands (missing). Above the hour indication of I–XII twice repeated,
an adjustable circular ring for 1–24 hours. Below this dial, a small dial for the
days of the week. On the central upper dial on the back, the following time
subdivisions: outermost the minutes, then I–XII hours twice repeated, segments for the lengths of day and night (missing), alarm-setting dial, hour and
minute hands. On the two small dials to left and right, indication of the dominical letters and adjustment for regulation. Below and to the right, it was possible
to set the position of the sun in the zodiac and through this the lengths of day
and night. Below this, the calendar dial: three disks engraved on both sides with
two months on each side, to be exchanged over the course of the year. On the
two sides of the clock, checking dials for the striking. On the outer plates of
both striking trains are two marks: CB in a horizontal cinched oval; a star and
a half-moon side by side, and repeated in reverse, on a shield. On the left side
of the case the etched signature ME FECIT CHAS/PARUS BOHEMUS/IN VI-
AENNA AUS/TRIA ANNO 1568. The statuette on the top is not original.

It is likely that the movement and case were made by the Augsburg clockmaker Jeremias Metzger (cf. Catalog No. 26) and that Bohemus merely added
his signature. Metzger was using this type of movement and case in his clocks
as early as 1563 and 1564, also again in 1570. The reliefs on the base are after
graphic prototypes of Heinrich Aldegrever (in the other Metzger clocks they
are after copper engravings of Sebald Beham). The bear hunt on the top of the
case is after copper engravings of Virgil Solis (see Ilse O'Dell-Franke, *Kupfer-
stiche und Radierungen aus der Werkstatt des Virgil Solis,* Wiesbaden, 1977, Plt.
12, No. g9).

The name Bohemus is written Behamb, Behaimb, Behaim in archive material; payments from the Vienna Court were delivered to him in the years 1573,
1579, 1582, and 1584. He was probably related to the Vienna clockmaker
Moritz Behaim, one of whose clocks is now in the Württembergisches Landesmuseum, Stuttgart. (Erwin Neumann, *Der königliche Uhrmacher Moritz Behaim
und seine Tischuhr von 1559,* Vienna, 1967.)

Literature: Maurice II, Fig. 159.

24 Table clock

Hans Gruber. Nuremberg, 1583
Stuttgart, Württembergisches
Landesmuseum

Movement: iron
Case: fire-gilt bronze and copper, silver
Height: 37 cm (14½ in.)

Going, quarter-striking, and hour-striking trains, arranged one behind another, each with fusee and gut line. Decoration on top not original. Indications: dial with I–XII hours twice repeated and also 1–24; a hand for each of the indications. Inside the hour rings, signs of the zodiac and two hands which indicate the positions of the sun and the moon in the zodiac. Diagram of planetary aspects. On the two small dials below, the Small Clock is at the left and the minutes on the right. The small levers are used to tone down or silence the striking and quarter striking. On the back, the checking dial for the hour striking. On the right and left sides, horizontal sundials for a latitude of 57°; on the right side also a conversion scale for telling time by moonlight. Here also the signature NORIMBERGAE FACTUM A° MDLXXXIII PER JOHAN GRUBERUM and a later engraved inscription by an owner, USUI FUI CASIMIRO DUCI DE GRIMALDI. Both side pieces have embossed reliefs with biblical scenes on their outer sides. The two scenes, from Joshua 10:1–19 and Isaiah 38:1–8, have to do with the sun's being halted by God. These two passages were the Church's main biblical arguments for its rejection of the Copernican theory. The scenes repeat motifs which Wenzel Jamnitzer had engraved on his measuring disk in the observatory at Paris (Inventory No. IA 16.5). The scenes are accompanied or elucidated by the following inscriptions:

> NON OPUS HUMANUM EST CUM LUNA SISTERE SOLEM
> NATURAEQUE ALTAS EVARIARE VICES
> SOLA HAEC IN CAELI DOMINUM VIRTUS CADIT OQUI
> SIDERA MUNDI INHIBES JOSUA QUANTUS ERAS!

(It is no work of man to bring sun and moon to a halt and to alter the exalted course of nature. This power pertains only to the Lord of Heaven. How great wast thou, Joshua, that thou hast stayed the world's constellations in their courses!)

> EZECHIAE FATUM TER QUINOS EXIT IN ANNOS
> PRAETER NATURAM VERTITUR IPSE POLUS
> OMNIA POSSE DEO FACILE EST VIS QUANTA EA DENOS
> QUOD RECLINATUR MOBILIS UMBRA GRADUS!

(The fate of Hezekiah is played out in thrice five years. The heavens themselves, contrary to nature, move backward. It is easy for God to accomplish everything. See how great is His power, that he forces backward the moving shadow of a stage in time!)

The inscription running around the entablature reads:

> LAETA VOLUBILIBUS RAPIDE LUX AVOLAT HORIS
> MAESTA TRAHIT TARDAM SORTE GRAVANTE MORAM
> QUOD TE CUMQUE MANET TEMPUS SIC TRANSIGE LETHI
> UT VIVAS AURES CUM FERIT HORA MEMOR.

(Joyous life runs rapidly away with the fleeting hours, and the sad times press hitherward under the pressure of fate; however much time may remain to thee, spend it in such fashion that thou thinkest upon death thy life long, each time thou hearest the hour strike.)

Himmelein (see below) has associated this clock with a commission from King Frederick II of Denmark, the king having ordered a table clock from Gruber in 1583. For Hans Gruber see Catalog No. 16.

Literature: Volker Himmelein and John H. Leopold, *Prunkuhren des 16. Jahrhunderts. Sammlung Joseph Fremersdorf* (Stuttgart, 1974), p. 59 and amplifications and amendments, Catalog No. 9. Maurice II, Fig. 113. Giuseppe Brusa, *L'arte dell'orologeria in Europa* (Milan, 1978), Figs. IV and V.

25 Table clock

Paulus Schuster. Nuremberg, c. 1587
Dresden, DDR, Staatlicher Mathematisch-
Physikalischer Salon

Movement: iron, brass
Case: fire-gilt bronze and brass, enamelled silver
Base: ebony
Height: 80 cm (31½ in.)

Going, quarter-striking, and hour-striking trains; alarm. A separate mechanism moves the heads of the Neptune figures seated upon the hippocamps and another moves the automaton figures in the tower—two couples and a lutenist. Indications, front: astrolabe with two tympana, engraved on both sides, for 44°, 48°, 52°, and 56°. Dragon, sun, and moon hands, diagram of aspects, age and phases of the moon. Beneath, on the small dial at the left the minutes, on the right dial the Small Clock and the alarm-setting dial. On the two dials in the upper corners the clock can be regulated, or the striking can be converted from 12 to 24 hours. On the back an annual calendar; in its center, segments for the lengths of day and night. The two lower dials indicate the days of the week and serve to set the lengths of the days and nights. The dominical letters or the Golden Number can be read off the two dials in the upper corners. On the sides are checking dials for the striking and quarter striking. In the base is a drawer with a writing table that can be set at an angle. The frieze of plaquettes on the base is after copper engravings of Sebald Beham (Adam Bartsch, *Le peintre graveur,* nouv. ed. réimpr. 1–22 in 4 vols., Würzburg, 1920–1922 [Hildesheim, 1970], Figs. 96, 101, 99).

According to the accession journal of the Elector's Art Collection at Dresden the clock was acquired by Electress Sophia from a Nuremberg clockmaker whose masterpiece it had been; she then presented it to her husband Elector Christian I (1586–1591). It entered the collection in 1590. It has always been ascribed to the Nuremberg clockmaker Paulus Schuster; he became a master in 1587 and died in 1637.

Literature: Max Engelmann, "Das Meisterstück des Nürnberger Uhrmachers Paulus Schuster," *Mitteilungen aus den sächsischen Kunstsammlungen,* 1911, 2: 31. Maurice II, Fig. 180. Helmut Grötzsch, Jürgen Karpinski, *Dresden, Mathematisch-Physikalischer Salon* (Leipzig, 1978), Figs. 37–41.

26 Table clock

Jeremias Metzger. Augsburg, 1573
Zurich, Museum der Zeitmessung Beyer

Movement: brass plates, iron wheels
Case: fire-gilt bronze and brass
Height: 38 cm (15 in.)

Going, quarter-striking, and hour-striking trains arranged in a cruciform pattern, each with fusee and line. Front: I–XII hour indication twice repeated, astrolabe, tympanum for latitude of 48°, dragon, sun, and moon hands, diagram of aspects, and phases of the moon. On small dials at the four corners one can set or read off (from right to left): striking adjustment for either 12 or 24 hours, regulation dial, alarm-setting dial, and the days of the week with the ruling planets. Within this latter dial are astrological scales with the captions HORA DIEI and HORA NOC. Back: annual calendar, hour scale I–XII twice repeated plus minute scale, hour and minute hands, segments for length of day and night. The two small dials below indicate the ruling planets and the position of the sun in the zodiac. The lengths of day and night—that is, the hours of daylight and darkness—can be set on this last dial if the clock is to be used at another latitude. On the sides are checking dials for quarter striking and the striking of 12 or 24 hours. Signature: "Jeremias Metzger von Augsburg."

Jeremias Metzger, maker of small clocks, presumably Protestant, was born at Augsburg around 1525–1530. In 1555 he received smiths' eligibility through his father, and it was probably at the beginning of the same year that he was first married. In 1564 he was an inspection master. In the following year he was accused of permitting a helper to have a so-called *Stuckuhr* ("piece-clock"), a once-acceptable custom now judged unfair competition and prohibited because of a shortage of journeyman labor. (Formerly after every five clocks a journeyman made for the master he was allowed to keep or sell the sixth one, the *Stuckuhr,* for himself. Metzger was fined 2 gulden, even though his journeyman had made sixteen clocks within two years for his master before being allowed to make one for himself.) In 1573 Metzger received permission to modify his house in the Schmiedgasse. In 1573–1574 Hans Schlottheim worked for him as journeyman. In 1578 he delivered six clocks worth over 600 gulden for the Turkish honorarium. His second marriage was to Sabina Betz in 1581; the third, to Rosina Leinauer, a cutler's daughter, in 1582. He is last mentioned in 1597.

Sources: STAA, SZB, fol. 86b. STAA, S I, 1564, May 22, 1565, fol. 325, S II, 1574. STAA, Bauamtsprotokolle 13, fol. 89r. HKA, Reichsakten, 192/I, fol. 52. STAA, HAP, Feb. 19, 1581, Dec. 2, 1582.
Literature: Maurice II, Fig. 182.

27 Table clock

Jakob von Kress (?). Augsburg, c. 1600
Stuttgart, Württembergisches
Landesmuseum

Movement: iron framing, brass plates and wheels
Case: fire-gilt bronze and brass
Height: 37 cm (14½ in.)

Going, quarter-striking, and hour-striking trains arranged crosswise (see Fig. 76), each train having a fusee. Alarm escapement no long original. The indications only on the front: calendar dial for six months (after half a year it had to be turned over); hour and minute hands, hour indication I–XII twice repeated; sun and moon hands which show, on the inner part of the dial, the position of these two bodies in the zodiac; moon hand with phases, and diagram of aspects. At the lower left a dial for the days of the week with the ruling planets, at the right an alarm-setting dial. Between these two dials the initials I.V.K. The back has a regulator dial.

Since the clock can be classified stylistically as being from Augsburg, and since Jakob von Kress worked at Augsburg around the turn of the century, the initials have been interpreted as his. For other works of this master see Maurice II, Fig. 135.

Jakob von Kress, born between 1553 and 1562, purchased his smiths' eligibility at Augsburg in 1599. In 1607/1608 he was a guild foreman (*Vorgeher*). He is last mentioned in the muster roll of 1619.

Sources: STAA, Musterregister 1610–1619. STAA, SZB, fols. 139a, 153.
Literature: Volker Himmelein, *Uhren des 16. und 17. Jahrhunderts, Württembergishes Landesmuseum* (Stuttgart, 1973), No. 24. Maurice II, Fig. 181.

28 Table clock

Augsburg, second quarter of the 17th
century
Boston, Museum of Fine Arts

Movement: brass plates and wheels
Case: fire-gilt bronze and brass
Dials: enamelled silver
Height: 54 cm (21¼ in.)

In its movement and indications this clock is similar to Catalog No. 26. The tympanum is set for a latitude of 48°. The two dials in the upper part of the calendar, which are additions here, indicate the Golden Number and the Roman indication. A similar piece presumably from the same workshop is in the Science Museum, London (cf. Maurice II, Fig. 190).

Unpublished.

29 Table clock

Heinrich Gebhard. Strasbourg,
c. 1630–1640
Stuttgart, Württembergisches
Landesmuseum

Movement: brass plates and wheels
Case: fire-gilt bronze and copper, silver appliqués
Base: ebony, by the Augsburger cabinetmaker
David Krieger (?)
Height: 54 cm (21¼ in.)

Going, quarter-striking, and hour-striking trains, all in parallel, arranged between the plates which stand in the same plane with the front. All winding from the back. Alarm. The going train is upside down. The escapement is no longer the original. Indications on the astrolabe side: chapter ring I–XII twice repeated, astrolabe, sun and moon hands. The three small dials are checking dials for the hour striking, quarter striking (outer sides), and days of the week (middle). On the opposite side: hour dial, hour and minute hands, outside these a calendar for the whole year; inside the age and phases of the moon; below, an alarm-setting dial. The two engraved scenes and the verses on the front relate to Psalms 143 and 119. On the back, engraving of the allegorical figures Astronomy and Geometry. The sides are glazed, with the signature: "Heinrich Gebhard von Strassburg" and the city hallmark of Strasbourg. The ebony base is stamped with the Augsburg pine cone, the mark EBEN (indicating real ebony), and the master's initial DK in a horizontal rectangle, presumably the mark of the Augsburg cabinetmaker David Krieger.

Heinrich Gebhard, the son of a tailor, was baptized at Strasbourg on December 19, 1602. On November 29, 1627, he married Elisabeth Haut. He occupied various honorary offices within the guild from 1631 to 1661. He died before 1662.

Literature: Volker Himmelein, *Uhren des 16. und 17. Jahrhunderts, Württembergisches Landesmuseum* (Stuttgart, 1973), No. 36. Maurice II, Fig. 201. Giuseppe Brusa, *L'arte dell'orologeria in Europa* (Milan, 1978), Figs. 263, 264.

30 Table clock

David Buschmann. Augsburg, 1652
United States, private collection

Movement: brass plates and wheels
Case: fire-gilt bronze and brass
Dials: enamelled silver
Height: 58 cm (22¾ in.)

In its movement and indications the clock is similar to Catalog No. 26. On the outer plate of the striking train is the signature "Davidt Buschman Aug." with the pine cone, the Augsburg hallmark. The year 1625 appears on the small armillary sphere; this should presumably be 1652, since Buschmann was not born until 1626. The acanthus ornament over the bell also indicates a later origin; but the shell ornament on the base on the other hand shows that certain decorative elements were repeated over many years in the workshop. The armillary sphere is similar to that on the year clock of Hans Buschmann (Catalog No. 57) made in 1651/1652; it was also made larger as a separate instrument by David's brother Johann (see Fig. 28).

David Buschmann, small-clock maker, Protestant, was born at Augsburg in 1626. He learned clockmaking in the workshop of his father Hans (= Johann), and as his son he received with the smiths' eligibility the right to open his own workshop. He nevertheless frequently worked with this father and his brother Johann. David married Barbara Schäffler on July 15, 1657, and after her death Maria Wappler, widow of Sedelmair, on June 13, 1689. He paid taxes from 1658 to 1700. In 1668 he collaborated on the Turkish honorarium. David Buschmann represented the fifth generation of the Buschmann family of clockmakers. Among his five children, however, none is known to have been a clockmaker. He died April 6, 1701, in Augsburg.

Literature: Maximiliam Bobinger, *Kunstuhrmacher in Alt-Augsburg* (Augsburg, 1969), p. 107 (though on p. 96 he is erroneously entered as a son of Matthäus Buschmann). Maurice II, Fig. 192.

31 Table clock

Peter Anton Schegs. Nuremberg,
third quarter of the 17th century
Wuppertal, Historisches Uhrenmuseum

Movement: brass plates and wheels
Case: fire-gilt bronze and brass
Base: black-stained wood, gilt
Height: 47 cm (18½ in.)

In its movement and indications the clock is similar to Catatog No. 26. On the front with the calendar dial is a small additional dial by which one can turn on or off a striking bell for any given minute. The escapement is no longer the original. On the outer plate of the quarter-striking train is the signature "Peter Antoni Schegs in Nürnberg."

Peter Anton Schegs died on March 16, 1700. His father, Abraham, also a clockmaker, obtained burgher status in 1656 and died in 1695. Schegs' clock shows how long a design, which was initiated at the middle of the 16th century, was retained in Nuremberg.

Literature: Jürgen Abeler, *5000 Jahre Zeitmessung* (Wuppertal, 1968), p. 19. Maurice II, Fig. 200.

32 Table clock

Georg Kostenbader. Strasbourg, 1583
Zurich, Museum der Zeitmessung Beyer

Movement: iron plates and wheels
Case: gilt bronze, brass, and copper
Height: 35 cm (13¾ in.)

Going, quarter-striking, and hour-striking trains arranged crosswise (see Fig. 76); the two striking trains have exposed springs. The going train has a later fusee and a barrel beneath that was added at a later date. Alarm. Indications on the front: astrolabe, sun and moon hands (an image of the sun moves around the ecliptic between the tropics, the moon hand with rotating ball for the lunar phases). Left side: calendar dial for twelve months, indication of the zodiac, number of hours of daylight and darkness. Beneath, an inscription in which the customer, the clockmaker, and the engraver are named: "Beernhart Fritschman liess machen und merck Georg Costenbader dies Uhrwerk und durch Hans Meister gestochen fürwahr als man zählt 1583 Iar." On the right side: a panel engraved on both faces with a calendar for a quarter of the year, the saints' days, and dominical letters. On the back: dial for the days of the week.

There is another signed clock of Georg Kostenbader, dated 1588, in the Casteel Museum at Gaasbeek in Brabant. In the literature there appears only a Lorenz Costenbader, who in 1585 is mentioned in the guild tribunal of the smiths at Strasbourg (cf. Maurice II, Fig. 220).

Literature: Maurice II, Fig. 229.

33 Table clock

Augsburg, 1600
Munich, Bayerisches Nationalmuseum

Movement: brass plates and wheels, iron wheels
for the epicyclic train
Case: fire-gilt bronze and brass; silver
Dials: silver, partly enamelled
Base: black-stained wood
Height: 52 cm (20½ in.)

Going, quarter-striking, and hour-striking trains, all with fusee and chain.
Unlike the preceding clocks, which have trains arranged crosswise, this clock
has the two striking trains set side by side behind the going train. The escapement is no longer the original. The drawer in the base has been reconstructed.
Indications on the astrolabe side: astrolabe, tympanum for latitude of 50°,
dragon, sun, and moon hands, diagram of aspects, age and phases of the moon.
The rete is signed AS (cf. Fig. 24). On the dial beneath it the ruling planets and
days of the week are indicated. On the opposite side: minute and hour rings,
minute and hour hands, segments for length of day and night; beneath, a setting
dial for the length of day and night. To either side are two calendar disks each
covering half a year; they show in addition to the calendar saints, in tabular
form, the dominical letters, the Golden Number, the epacts, and the dates of
Easter for the years 1600–1657 and 1658–1687. Beneath each calendar disk,
two smaller dials, the checking dial for the two striking trains, the regulator
dial, and the alarm-setting dial. The clock is made to rotate on its base. The
decorative architectonic elements, modelled with uncommon sharpness, are
impressive. The first date named in the calendar suggests the year in which the
clock was made. A comparable clock marked with the Augsburg pine cone
hallmark is located in the Württembergisches Landesmuseum, Stuttgart.

Literature: Maurice II, Fig. 237.

34 Table clock

Augsburg, end of the 16th century
United States, private collection

Movement: brass plates, iron wheels
Case: fire-gilt bronze and brass
Dials: enamelled silver
Height: 39 cm (15¼ in.)

In its indications the clock is similar to Catalog No. 33. Tympanum for 50° latitude. Rete and moon hand missing. Base is a reconstruction. Quarter- and hour-striking trains are arranged side by side and stand behind the going train; the quarter-striking train upside down. Altered at a later date to pendulum escapement. On the dial with the indications for the lengths of day and night there is a minute hand in addition to an hour hand. A similar piece is in the Victoria and Albert Museum, London.

Literature: Maurice II, Fig. 235.

198

35 Table clock

Augsburg, beginning of the 17th century
Wuppertal, Historisches Uhrenmuseum

Movement: brass plates and wheels,
epicyclic train of iron
Case: fire-gilt bronze and brass
Dial: enamelled silver
Height: 43 cm (17 in.)

Going, quarter-striking, and hour-striking trains, all with fusee and gut line; alarm. The trains are arranged crosswise (see Fig. 76). Indications on the astrolabe side: astrolabe, tympanum for 48° latitude, dragon, sun, and moon hands. On the opposite side: calendar disk, engraved on each side for six months. Minute ring, hour indication of I–XII twice repeated, hour and minute hands, segments for lengths of day and night. The two lower dials are for the ruling planets, days of the week, and a setting dial for the lengths of day and night. At the sides are checking dials for the striking trains. Hour striking adjustment for either 12 or 24. Alarm-setting dial.

Unpublished

199

36 Table clock

Augsburg, second quarter of the 17th
century
New York, Metropolitan Museum of Art

Movement: brass plates and wheels
Case: fire-gilt bronze, copper, and brass
Height: 63.5 cm (25 in.)

In its movement and indications this clock is similar to Catalog No. 26. Escapement is no longer the original. On the embossed base the allegorical figures of the four seasons. Directly comparable to the clock of the Augsburg clockmaker Nikolaus Rugendas the younger in the Kunsthistorisches Museum, Vienna (cf. Maurice II, Fig. 241).

Unpublished.

37 Table clock

Richard Ledertz. Strasbourg, c. 1640
Amsterdam, Rijksmuseum

Movement: brass plates and wheels
Case: fire-gilt bronze and copper
Height: 39.5 cm (15½ in.)

Going, quarter-striking, and hour-striking trains; alarm. Both striking trains are arranged behind the going train. Escapement is no longer the original. Indications: minute ring, hours from I to XII, hour and minute hands, alarm-setting dial. Beneath, on the dial at left, the age and phases of the moon, at the right the reigning planets and the days of the week. On the back, checking dials for the two striking trains; sides glazed. On the alarm mechanism the signature "Reichard Ledertz in Strasburg." The assignment of a date around 1640 arises from a similar piece dated 1641 in a private American collection.

Literature: Maurice II, Fig. 653.

38 Table clock

Heidelberg (?), before 1540
Munich, Bayerisches Nationalmuseum

Movement: iron; alarm dial and fusee of brass
Case: engraved copper, partly fire-gilt, silvered
(silvering restored)
Height: 11 cm (4⅜ in.)
Diameter: 15.6 cm (6⅛ in.)

Going train between solid plates; alarm. Dial with two hands, one for an indication of I–XII hours, the other for 1–24 and I–VI hours, the latter four times repeated. Duration of movement, eight days. Backplate etched with tendril ornament. Verge cock no longer original. Hunting scenes engraved on the case walls, with gilt hunters and animals; silvering has worn away to the last trace and then has been restored. On the bottom of the case the arms of the Bavarian Palatinate, between the helmet crests the motto of Ottheinrich: MIT DER ZEIT. At the edge the signature OTT. HEINRICK. VON GOTTES. GNADEN. PFALZGRAF BEI DEN REIN. HERTTZOG. IN. OBEREN UND NIDEREN BAI. Ottheinrich (1502–1559) employed Jörg Leberer in 1539 as court clockmaker; possibly he was the maker of this clock. His diploma of appointment has been published by Joseph Baader (in *Anzeiger für Kunde der deutschen Vorzeit,* 1875, *22:*cols. 379–380). Leberer had to take care of the court clock and the "little clock that we have in our chamber." For this he received annually a salary, 20 gulden retainer fee, and two suits of court clothing. Clocks that he made were paid for separately. In October 1550 Leberer became a burgher of Regensburg: "Jörg Leberer, clockmaker, of Neuburg was accepted as a burgher upon his departure. Rendered oath on the Monday after St. Gall's day." (Regensburg, Stadtarchiv, Bürgerbuch. Communication from Dr. Wolfgang Pfeiffer, Regensburg.)

Literature: Ernst von Bassermann-Jordan, *Die Geschichte der Räderuhr unter besonderer Berücksichtigung der Uhren des Bayerischen Nationalmuseums* (Frankfurt, 1905), p. 69, No. 11. Maurice II, Fig. 493. Giuseppe Brusa, *L'arte dell'orologeria in Europa* (Milan, 1978), Fig. 71.

39 Table clock

Jakob Marquart. Augsburg, c. 1560
Chicago, Adler Planetarium

Movement: iron
Case: brass, etched and fire-gilt
Height: 7.5 cm (3. in.)
Diameter: 16 cm (6¼ in.)

Going and hour-striking trains; going train with fusee. Movement incomplete. On the bottom of the case a vertical sundial (a so-called *organum Ptolemaei*) and the signature IACOB MARQUART.

The clockmaker Jakob Marquart, presumably Protestant, came from an old Augsburg family which—according to his own statement—had made clocks for two hundred years. He was supposedly born about 1524 as the son of the clockmaker Benedikt Marquart. He spent his apprenticeship in his father's workshop, subsequently worked twenty-five years in Italy and France for "Princes and Lords," and from 1560 was back in Augsburg. In 1567 he married Katharina, widow of Budani; their sons were Anton and Jakob. In 1567 he also acquired his father's smiths' eligibility, a prerequisite for opening an independent clockmaker's establishment, which became fully valid only upon passing the examination for the title of master. But in 1569 Jakob Marquart had still not made a masterpiece and thus was in trouble with the guild (see Ch. 7). With imperial support he secured a special temporary permit. His first work for the court was a round bracket clock, according to bills that Marquart sent on July 16, 1568. At the beginning of the year 1569 another clock was paid for by the provincial deputy Georg Ilsung on behalf of the Emperor, and a new commission was given to Jakob Marquart for a particularly ingenious clock mechanism "which perhaps no one, or at best very few, of those here could make." For the two years that were needed, Marquart was to be allowed to work undisturbed by guild prescriptions and helped by one journeyman. Shortly after expiration of this term Marquart left Augsburg, and there is evidence that he was in Florence in 1573. He died before June 28, 1575.

Sources: STAA, U I, fol. 137. STAA, HAP 1567, fol. 157v. STAA, Pflegschaftsbuch, July 28, 1573, June 18, 1575. STAA, SZB, fol. 100. HKA, HZAB No. 23 (1568), fol. 89v. STAA, RPR 1569, fol. 133. STAA, Steuerbücher, 1560–1572.
Literature: Max Engelmann, *Sammlung Mensing, altwissenschaftliche Instrumente* (Amsterdam, 1924), No. 381. Maximilian Bobinger, *Alt-Augsburger Kompassmacher* (Augsburg, 1966), pp. 95 f. Maurice II, Fig. 470.

40 Table clock

South Germany, c. 1560–1570
Munich, Bayerisches Nationalmuseum

Movement: iron, partly blued
Dials: brass
Case: fire-gilt bronze
Height: 7.4 cm (3 in.)
Diameter: 22.8 cm (9 in.)

Going, quarter-striking, and hour-striking trains; alarm. Indications: hour and minute hands (missing); rete simplified to show only the zodiac; sun (hour) and moon hands give the positions of these two bodies in the zodiac; age and phases of the moon; diagram of aspects. The annual calendar associated with the zodiac has been repositioned to conform to the Gregorian calendar. On the tympanum, lines for the temporal hours and the hours according to the Italian Clock. Escapement no longer original. The backplate, decorated with blued ornament and birds, is noteworthy. The frieze with its representation of the Orpheus legend is found on a total of ten clocks; nine are published in Coole and Neumann (see below). Concerning the tenth Orpheus clock see Sotheby, Zurich, May 6, 1977, No. 111.

The mechanisms of these table clocks are constructed in very different ways. There is evidence that the movements, as far as they are original, were made in four different workshops. The relief of the casing for all these clocks originated in the workshop or within the circle of Hans Kels. Although basically identical, the cases are distinguishable as to the fineness of the engraving of their reliefs. There are also differences in that part of the frieze which was added to the relief in order to make it long enough to encircle the movements, which vary in their diameters. It is still a completely open question whether the clocks were made at one time in different shops or at different times as well.

Literature: Philip G. Coole and Erwin Neumann, *The Orpheus Clocks* (London, 1972), pp. 51 ff. Maurice II, Fig. 524. Giuseppe Brusa, *L'arte dell'orologeria in Europa* (Milan, 1978), Figs. 117–119.

41 Table clock

Augsburg, c. 1560–1570
United States, private collection

Movement: iron
Case: fire-gilt brass
Height: 8.5 cm (3⅜ in.)
Diameter: 16.5 cm (6½ in.)

Going, quarter-striking, and hour-striking trains; going train with fusee. Escapement no longer original. On the concentric rings the following indications are given by the seven hands: I–XII hours on the outermost scale, the hours divided into quarter sections; touch marks, I–XII hours twice repeated; the hand with an image of the sun (marked "sol") completes a rotation once in 24 hours. The two small hands with images of the sun indicate the rising and setting of the sun in one case, the Nuremberg hours in the other. The two small hands with half-moons indicate the rising and setting of the moon and the number of hours of daylight and darkness (see also Fig. 77). Position of sun and moon in the zodiac; age and phases of the moon; diagram of aspects. On the backplate are two correction dials by which the rising and setting of the sun and the moon can be set within the month and can be corrected between different degrees of latitude. These function by mean of eccentrics. The indication for rising and setting of the moon is unique. On the base plate, a vertical sundial (a so-called *organum Ptolemaei*) and the Augsburg hallmark, plus a master's mark not yet identified.

Literature: Edward S. Jones and Peter Guggenheim, "A Sixteenth Century Table Clock," *Antiquarian Horology,* 1970, 6: 433. Clare Vincent, *Northern European Clocks in New York Collections,* Metropolitan Museum of Art, Jan. 4–March 28, 1972, p. 17, No. 13. Maurice II, Fig. 519.

42 Monstrance clock

Jeremias Metzger. Augsburg, 1564
London, Victoria and Albert Museum

Movement: iron
Case: fire-gilt bronze and brass; silver
Height: 37.6 cm (14¾ in.)

Going, quarter-striking, and hour-striking trains; alarm. Going train with fusee. Escapement and some other parts of the movement no longer original. Alarm bell in the drum at the top is supported by dolphins; striking bell inside the case, which has a pierced frieze. This style of case has frequently appeared with minor variations; it is first represented in a portrait, dated 1556, of Duchess Anna of Bavaria (Vienna, Kunsthistorisches Museum; cf. Maurice II, Figs. 539, 540). The clock is signed on the base: IEREMIAS · METZGER · VHRMACHER · 1564 · IN · AVGSPVRG.

For Metzger see Catalog No. 26.

Literature: Maurice II, Fig. 541.

43 Monstrance clock

Paulus Braun. Augsburg, c. 1600
Winterthur, K. Kellenberger clock
collection

Movement: brass plates and wheels; all wheels of
the motion work of iron
Case: fire-gilt bronze, copper, and brass; silver
Base: ebony; height: 66 cm (26 in.)

Going, quarter-striking, and hour-striking trains, all with fusee and gut line. In
the top is a separate hour-striking train with exposed spring, which repeats the
strike of hours after each of the first three quarters. The hour striking is adjustable for 12 or 24 hours, that of the striking train over 33 hours. Indications:
the motion work can be set from the minute hand on one side. Moon, dragon,
and sun hands and rete can be set individually, as can the ruling planet for the
day. On the front an annual calendar, an astrolabe for 48° latitude, sun, moon,
and dragon hands. On the sides: top right, I–XII hours; bottom right, ruling
planet for the day; top left, position of sun in the zodiac and lengths of daylight
and darkness (alteration of these two segments for another latitude was provided for, but was never carried out); lower left, alarm-setting dial. On the case
the Augsburg pine cone stamped twice. On the back a mark PB in a shield, to
be identified as that of Paulus Braun.

Paul Braun (probably identical to Paulus Bauer) was mentioned as senior
journeyman of the clockmakers in Augsburg in 1596, indicating that by then
he must have been there a long time. In that year he rebelled against the
practice of engraving the locksmith arms ahead of the clockmaker arms on the
silver beaker which on each New Year's Day was given to the common hostel
keeper of the journeymen in the smiths' guild. Braun, who is described as
having been "particularly brash," convoked secret gatherings of the clockmaker journeymen. In 1600 he married the daughter of the public official
Esaias Busch, through whom Braun came to possess smiths' eligibility and the
right to operate a workshop of his own in Augsburg. In 1610 he was registered
in the muster roll of the city as a clockmaker forty-six years old; by 1615 he
was no longer registered.

Sources: STAA, Musterregister 1610. STAA, S V, fols. 167 ff. STAA, SZB, fol. 139b.
Literature: Rudolf Lepke, Auktion 2000, *Kunstwerke aus den Beständen Leningrader Museen
und Schlösser,* Berlin, Nov. 6, 1928, No. 124, Plt. 50. Maurice II, Fig. 580.

44 Monstrance clock

Nikolaus Rugendas. Augsburg, 1616
United States, private collection

Movement: brass plates and wheels
Case: fire-gilt bronze, copper, and brass
Dials: enamelled silver
Height: 72 cm (28¼ in.)

Going, quarter-striking, and hour-striking trains; alarm. For the plan of the mechanism and the indications see pages 172–173 and Fig. 85. Tympanum set for 48° latitude. On the alarm mechanism the marriage arms of the first owner, Wratislaus I, Count von Fürstenberg (1586–1631) and his wife, Anna von Croy. Marked with the Augsburg pine cone on the case. Backplate signed "Nicolaus Rugendas." This clock is Rugendas' masterpiece.

Nikolaus Rugendas the elder, clock- and compass-maker, Protestant, was born in 1585 at Melsungen in Hessen. From 1608 he lived in Augsburg; he became an independent master in 1616, and in the same year he married Sara Schmidt. Two of their nine children, Hans Jakob and Nikolaus the younger (born 1619), were also clockmakers. The daughter Katharina married the small-clock maker Wilhelm Pepfenhauser in 1647. In the muster roll of Augsburg for 1619 there is an entry to the effect that a journeyman from Schlander was working in Rugendas' shop. In 1638/1639 he held the office of "sworn inspection-master." He died May 20, 1658.

Sources: STAA, Musterregister 1619. STAA, SZB, fols. 167 f. Bayerisches Staatsarchiv Neuburg a.d. Donau, Reichsstadt Augsburg Lit., 559a. STAA, HAP 1668, fol. 97. EMA, Taufbuch Barfüsser, June 15, 1619. EMA, Trauungsbuch Barfüsser, Sept. 16, 1667. STAA, U II, Nov. 29, 1639.
Literature: Franz Schestag, *Katalog der Kunstsammlungen des Freiherrn Anselm von Rothschild* (Vienna, 1866), p. 23, No. 161. Maurice II, Fig. 551.

45 Monstrance clock

Caspar Langenbucher. Augsburg, 1649
Braunschweig, Herzog Anton
Ulrich-Museum

Movement: brass plates and wheels
Case: fire-gilt bronze, copper, and brass
Dials: enamelled silver
Base: ebony
Height: 97 cm (3 ft 2⅛ in.)

Going, quarter-striking, and hour-striking trains; alarm. Going movement with fusee and chain. For the indications and plan of the movement see pages 172–173 and Fig. 88. Tympanum for 48° 15′ latitude. The bell for the alarm and for striking the quarters is in the ornamental crown at the top, that for striking the hours in the pierced metal base. The case is stamped twice with the Augsburg pine cone, and the backplate is signed "Caspar Langenbucher Augspurg." The clock is Langenbucher's masterpiece.

Caspar Langenbucher, Protestant, purchased smiths' eligibility in 1649, married in the same year, for a second time in 1658, and for a third time in 1667. He died in 1677/1678. (For a further discussion of the family see Ch. 12.)

Langenbucher sold this clock in 1650 to Duke August the Younger of Braunschweig. According to his instructions for setting up the clock, the following craftsmen collaborated in building the case: an artist-cabinetmaker for the octagonal ebony foot, a turner for the round metal parts of the case, a goldsmith for the embossed work and the final chasing, another worker in gold for the enamelled silver dials and the silver rings of the calendar dial, an engraver for the flower ornamentation on the bridges and on the housing for the alarm, a script-embosser and a bookbinder for the leather travelling case (which no longer remains).

Literature: August Fink, "Die Uhren Herzog August des Jüngeren," *Kunsthefte des Herzog Anton Ulrich-Museums,* No. 8 (Braunschweig, 1965), p. 11. Maurice II, Fig. 554. Volker Himmelein, "Die Uhren," in *Sammler, Fürst, Gelehrter, Herzog August zu Braunschweig und Luneburg 1579–1666* (Woltenbüttel, 1979), p. 165, No. 350.

46 Monstrance clock

Johann Martin. Augsburg, 1669
Munich, Bayerisches Nationalmuseum

Movement: brass plates and wheels; epicyclical
drive of iron
Case: fire-gilt bronze, copper, and brass
Dials: enamelled silver
Height: 70 cm (27½ in.)

Going, quarter-striking, and hour-striking trains; alarm. For the layout of the
mechanism and the indications see Fig. 89. Tympanum for 48° latitude. Backplate signed "Johann Martin, Augspurg." The clock is the masterpiece of this
clockmaker. The case has a shape similar to that of Catalog No. 45.

Johann Martin was born on February 13, 1642, in Frankfurt am Main; he
moved to Augsburg, acquired smiths' eligibility there, and married Maria
Barbara, daughter of the small-clock maker Elias Weckherlin, on May 19,
1669. Martin died on December 12, 1721, in Augsburg.

Literature: Ernst von Bassermann-Jordan, *Die Geschichte der Räderuhr unter besonderer
Berücksichtigung der Uhren des Bayerischen Nationalmuseums* (Frankfurt, 1905), p. 48, No.
49. Maurice II, Fig. 557.

47 Table clock

Johann Sayller. Ulm, 1617
Ulm, Museum der Stadt Ulm

Movement: brass plates and wheels
Case: ebony, oak, and fir; fire-gilt and silvered
brass
Height: 52 cm (20½ in.)

Going, quarter-striking, and hour-striking trains; escapement no longer original. Hour and minute hands. Signed on the dial "Johann Sayler Ulm A° 1617."

Johann Sayller was born on June 7, 1597, in Angelsberg in Lower Bavaria,
came to Ulm in 1617, became a burgher, was master of the guild in 1646, and
died on September 16, 1668. In 1626 he made as his masterpiece the rolling
ball clock, Catalog No. 54, which found its way to the council chamber of the
magistrate and was presented to King Friedrich I of Württemberg in 1812. In
1649 he made on commission from the magistrate three clocks which were
presented to the Swedish Count Douglas on his departure for home in 1650.
Sayller's son Christopher, also a clockmaker, died in Ulm in 1688. (Albrecht
Weyermann, *Neue historische, biographische und artistische Nachrichten von Ge-
lehrten und Künstlern,* Vol. II, Ulm, 1829, p. 248.)

Literature: Maurice II, Fig. 659.

III The Court: Source of Support and of Challenge

91. From the description of the Kunstkammer in Vienna. Edward Brown, *Sonderbare Reisen.* 1677 (Nuremberg, 1686).

92. Clocks in the Kunstkammer. Jan Brueghel the elder, *The Sense of Hearing* (detail), Madrid, Prado.

For rulers, collecting has always been an inevitable necessity; collections were just as much part of their residences as was the state treasure. In the course of the 16th century this timeless activity took on a new dimension, manifested in the princely *Kunstkammer* (cabinet of art). Here, collecting became subjected to a proto-scientific discipline: artifacts and natural curiosities were preserved and displayed in a systematic fashion (see Figs. 91–93).

The clock has always been presented as a gift of state. There is good documentation for the clocks that Theodoric had made for German princes and the ones Charlemagne and Emperor Frederick II Hohenstaufen received as gifts. More recent examples are the presents of clocks delivered to the Chinese court by Jesuits (see Ch. 4) and the clocks which the Vienna court gave to the Sublime Porte in Constantinople as honoraria up to the beginning of the 17th century (Ch. 5).

To have clocks on hand for presentation purposes, or in order to collect singular devices themselves, many rulers called clockmakers to their courts. Emperor Charles V and all the other Hapsburgs, the Electors of Saxony, Landgrave Wilhelm IV of Hesse, Duke Augustus the Younger of Braunschweig, the Dukes of Schleswig—all were enthusiastic supporters of this craft, patrons of clockmakers individually employed at their courts, and also frequently participated themselves in the clockmaking.

The most ingenious mechanician of the time was Jost Bürgi, maker of uncommonly complicated and precise clocks (see Ch. 8). His achievements could only have developed to their height at the courts of Kassel and Prague, to which he had been invited; they would not have flowered in the egalitarian atmosphere of the imperial cities with their republican administrations. In contrast to the ever-constant types of clock mechanisms which the craftsmen produced in the cities controlled by the guilds, uniquely original designs evolved at the courts. The search for new approaches and solutions concentrated on technically critical points in the clock and on achieving an exact replication of the epicyclical movements of the planets.

New systems for constant power sources were invented (the remontoire, Catalog Nos. 52, 55; the ball drive in Catalog No. 60), and precise escapements (the vane governor, Catalog No. 53; the cross-beat escapement, Catalog No. 56; the rolling-ball clock, Catalog No. 54). Immensely complex clockwork mechanisms were made in order to simulate the elliptical motions of the planets. Clocks became analog computers for the positions of the planets (see Ch. 6). Much ingenuity and artistic dexterity were also devoted to producing surprising external effects and to the costly ornamentation of the mechanisms. Techni-

212

93. The Wenzel Hall in the castle of Prague (detail). Clocks are displayed in the cabinet in the right side and back. Egidius Sadeler, Prague, 1607.

cal and functional interests were frequently sacrificed in the search for the unusual and sensational. Clocks took the shape of vases, mirrors, and columns (Catalog Nos. 50, 51, 59).

Yet a well-regulated mechanism and reliable, predetermined operation were always regarded as the preponderant virtues of the mechanical clock. On that account the clock became the symbol for a state of order, a symbol frequently transferred to the State itself. "A Prince and Ruler is the nation's clock; everyone guides himself by the Prince's example in his good works, just as he guides himself by the clock in his business affairs," wrote Christoph Lehmann in 1630 in his collection of proverbial sayings. This literary reference forms part of a long tradition, a counterpart of which is the custom of introducing clocks into portraits of dignitaries (as shown in Figs. 1–3 in Ch. 1). It was an authoritarian order that the clock embodied.

213

48 Table clock

Austria, 1545
Vienna, Kunsthistorisches Museum
Movement: iron
Case: painted iron, bronze
Height: 47.6 cm (18¾ in.)

Going and hour-striking trains. Numerous indications are controlled not by gears but by pulley and line transmission. Indications on the front: central dial with diagram of aspects, hour hand, position of the sun and moon in the zodiac, calendar. The Latin abbreviations IN, MA, BO relate to the cupping days; that is, they indicate the times when the planets do not affect the letting of blood (*indifferens*), when they are unfavorable (*malus*), or propitious (*bonus*). At lower left, indication of the days of the week through the reigning planets; at lower right, the rising and setting of the moon. Indications on the back: in the middle, the phases of the moon; at upper left, the dominical letters; at right, the days of the month from 1 to 30; at lower left, 1–24 hours; at right, indication of the planets ruling the hours. Arms of Austria, Castile, Hungary, and Bohemia. On each of the two side doors an astronomer is depicted.

On the balance there are screw-adjustable weights for regulation. To set the hands, the verge arbor is pressed out of its bearings by means of a lever. Force equalization in the going train—which runs about 24 hours—is effected by means of a spring which, like that of the wheellock of a gun, is pulled taut by means of a chain (see the figures at the left). In the striking train there is a device similar to the Geneva stopwork (to prevent overwinding the spring).

Literature: Maurice II, Fig. 88.

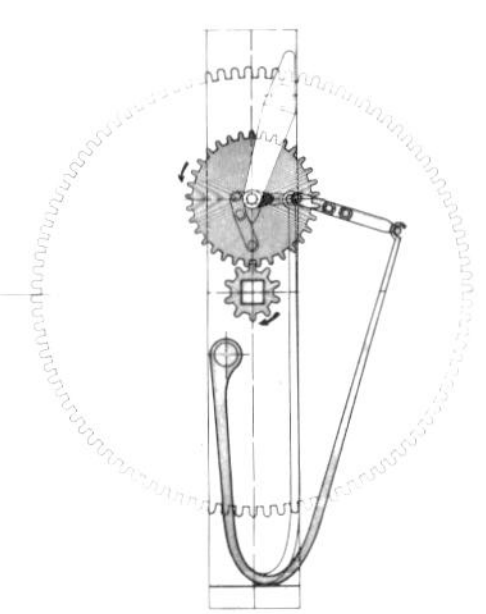

49 Table clock

Steffen Brenner. Copenhagen, 1558
Lübeck, St. Annen-Museum

Movement: iron
Case: fire-gilt bronze and brass; silver
Height: 21 cm (8¼ in.)

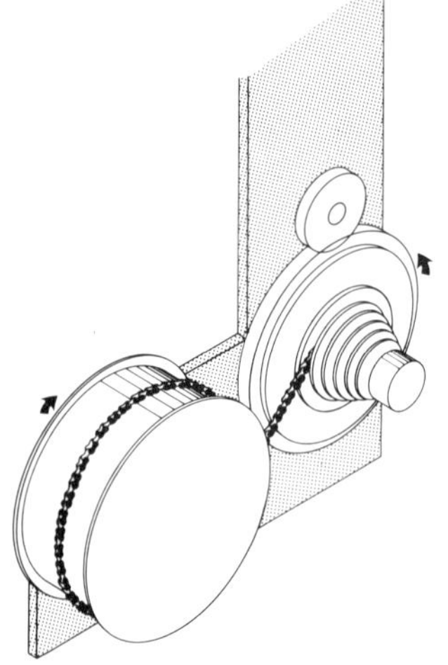

Going, quarter-striking, and hour-striking trains; alarm. All trains with fusee and chain. The gearing of the spring drum to the fusee is angled in singular fashion at 90° to the fusee, for which reason the bars are L-shaped (see the figures below). Steffen Brenner was thereby able to make a cubical case with five dials of equal size. The escapement and the upper calendar dial are no longer original. Indications on the five dials: (1) Dial with astrolable, dragon hand with rotating moon image, sun hand. On the tympanum there are engraved lines for the twelve astrological houses and for the temporal hours, 1–12 twice repeated. (2) The dial for age of the moon, phases of the moon, diagram of aspects, and duration of moonlight in terms of hours and minutes; also the position of the sun in the zodiac. (3) The dial to indicate the length of day and night, with the correction dial in the center. (4) The dial for the days of the week, with the planets governing the hours of the day engraved on each segment; in the middle the alarm-setting dial. (5) On the top is the calendar disk on the outside with the saints' days; in the middle a dial for the years 1885–1911, which has been revised to follow the original scheme and which bears engraving of the dominical letters, the number of weeks and days between Christmas and Shrove Tuesday, the dates of Easter, the circuit of the sun, the Golden Number, and a restored minute dial. On the bottom of the case the date 1558, among the digits the initials SB, below this a curved shield bearing a lily in a lozenge. A number of repair signatures on the movement, including "F. A. Rebers, Lübeck, 1847, 1848."

In 1550 Steffen Brenner received an annual stipend as a "clock-hand maker" at the court of Christian III at Copenhagen. In 1554 he became court clockmaker to the Danish King Friedrich II. Probably he had emigrated to Denmark, but from what country is not known. In some of his numerous surviving clocks he used ready-made parts that were manufactured primarily in Nuremberg. In the extremely clear architectonic articulation of this clock case, Brenner is almost thirty years ahead of the stylistic evolution of south German clock cases.

Literature: Vilhelm Slomann, "Gamle Ure," *Tilskueren,* 1931, p. 182. Maurice II, Fig. 222.

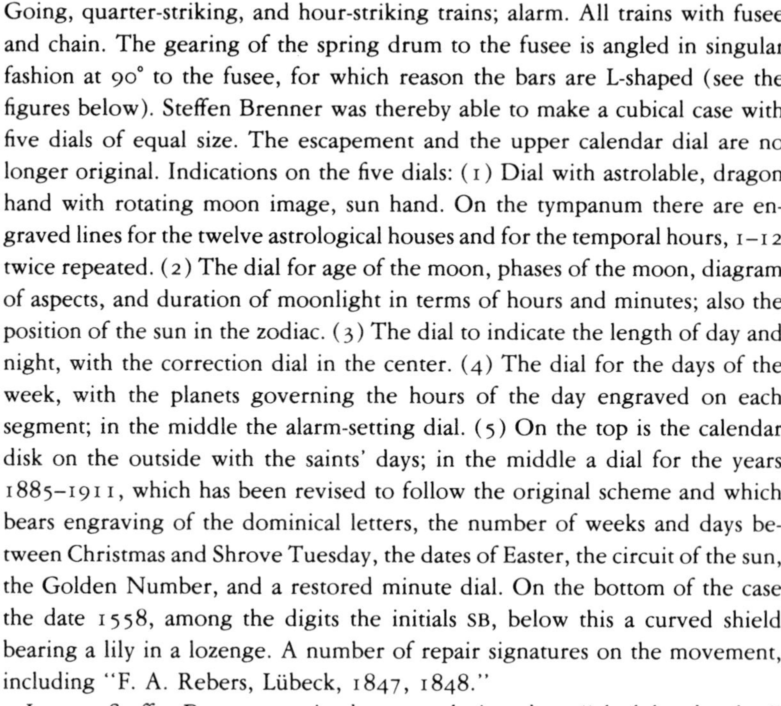

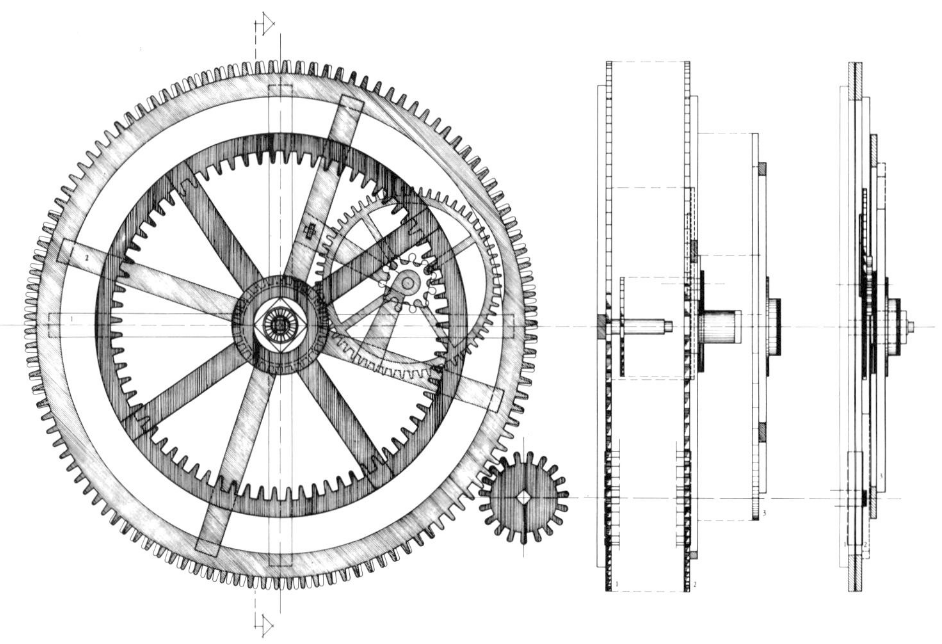

50 Table clock in the form of a vase

David Fronmiller. Augsburg, c. 1600
Baltimore, Walters Art Gallery

Movement: brass plates and wheels
Case: fire-gilt bronze and brass; plaques of silver
Height: 22.5 cm (8⅞ in.)

Going and hour-striking trains; alarm. Original verge escapement with balance spring added. Hour and minute hands. On the body of the vase four plaques with representations of the seasons. In the foot of the base three small plaques with embossed flowers. Opposite the hour dial, the checking dial for the striking. The movement is stamped with the letters DF; above appears the Augsburg pine cone on a shield. The initials can be interpreted as those of the Augsburg clockmaker David Fronmiller.

David Fronmiller, a large-clock maker, Protestant, was born around 1546 the son of the Augsburg clockmaker and burgomaster Hans Fronmiller and his wife Katharina, née Laminit. In 1572 he became independent upon receiving his father's smiths' eligibility, and in the same year he married Sara Koch, daughter of the master butcher Hans Koch. His children were Christoph, Carl, and Rosina, who married the Augsburg clockmaker Daniel Scheurer in 1608. David Fronmiller was called to honorary positions by his guild on numerous occasions, and in 1586 and 1601/1602 he was foreman of the guild. In 1610 he was named in the muster rolls as a clockmaker sixty-four years old. He paid taxes at Augsburg from 1575 until his death at the turn of the year 1618/1619.

Literature: Friedrich Streng, "Augsburger Meister der Schmiedgasse um 1600," *Blätter des Bayerischen Landesvereins für Familienkunde,* 1963, 26 (No. 1): 247–287. Maurice II, Fig. 363.

51 Wall clock

Augsburg or Munich, c. 1600
Munich, Bayerisches Nationalmuseum

Movement: brass plates, iron and brass wheels
Case: ebony, fire-gilt bronze and brass, reliefs in silver
Height: 128 cm (4 ft 2⅜ in.)

Going, quarter-striking, and hour-striking trains; hour and minute hands. All trains arranged between solid plates. The hours and quarters that have struck show within the small openings on the inner part of the dial. Escapement no longer original. The ebony case is edged with volute and strapwork ornamentation with large stylized flowers. At the top the figure of Actaeon with a dog, flanked by two other dogs (the right one is missing), to either side a female half-figure, two cupids below. Inlaid silver reliefs: above, Diana resting, in the background Actaeon; below, Perseus and Andromeda. On the panel in the middle, scroll-like foliage ornamentation. There are two doors and in the second there is a mirror. Within, pierced scrollwork ornaments with silver relief: resting Diana, spied upon by a satyr. The bronze sculpture is stylistically comparable to works of Hans Krumper. The clock comes from Schloss Neuburg on the Danube.

Literature: Ernst von Bassermann-Jordan, *Die Geschichte der Räderuhr unter besonderer Berücksichtigung der Uhren des Bayerischen Nationalmuseums* (Frankfurt, 1905), p. 69, No. 12. Maurice I, Plt. III. Giuseppe Brusa, *L'arte dell'orologeria in Europa* (Milan, 1978), Figs. 225, 226.

52 Experimental clock

Jost Bürgi. Kassel, c. 1585–1590
Kassel, Staatliche Kunstsammlungen,
Astronomisch-Physikalisches Kabinett

Movement: iron
Case: ebony, glass doors
Height: 36.5 cm (14⅜ in.)

Going train with weight-driven remontoire, the weight being drawn up daily
for three months (see Fig. 56). Dial with hours I–XII twice repeated, hour
hand and annual calendar. On the back, two sundials and two dials showing the
days of the week and the number of times the remontoire has functioned. The
piece has a verge escapement. The mechanism for the remontoire is the same
as in the Vienna planetary clock, Catalog No. 53. The case can be revolved on
its base, presemably to orient the sundials properly.

Bürgi made three so-called experimental clocks in which he experimented
with escapements and trains. As early as 1567–1587 Tycho Brahe mentions
brass wheels for clocks, but since the movement of this clock is made entirely
of iron, and since it has a conventional verge escapement with wheel balance
(the other two experimental clocks have cross-beat escapements), this is presumably the first of the three experimental clocks, constructed around 1585–
1590.

On Jost Bürgi see Chapter 8.

Literature: Hans von Bertele, "Präzisionszeitmessung in der Vor-Huygensschen Peri-
ode," *Blätter für Technikgeschichte,* 1954, 6: 166 ff. Hans von Bertele, "Precision Timekeeping in the pre-Huygens Era," *Horological Journal,* Dec. 1953, pp. 2–24. Hans von
Bertele, "Jost Bürgis Beitrag zur Formenentwicklung der Uhren," *Jahrbuch der Kunsthistorischen Sammlungen in Wien,* 1955, 15: 169 f. Maurice II, Fig. 635. Giuseppe Brusa,
L'arte dell'orologeria in Europa (Milan, 1978), Figs. 163–166. Ludolf von Mackensen (with
contriburtions from Hans von Bertle and John H. Leopold), *Die erste Sternwarte Europas
mit ihren Instrumenten und Uhren. 400 Jahre Jost Bürgi in Kassel* (Munich, 1979), p. 135,
Cat. No. 19.

53 Planetary clock

Jost Bürgi (?). Prague, c. 1604
Vienna, Kunsthistorisches Museum

Movement: iron plates with brass bushings; brass
and iron wheels
Case: fire-gilt bronze and brass, painted parchment
Height: 39.3 cm (15½ in.)

Going movement with weight-actuated remontoire, vaned fly as escapement.
The weight is drawn up hourly by a mainspring working through a wooden
fusee. Release of the remontoire takes place from the motion work via a
twelve-pointed star wheel. In this process the motion work is moved forward
at an interval of one hour through a connecting rod. For the mechanism see
Figs. 51–54.

Indications: On the front there are two dials one above the other, an engraved glass plate in front of them. On the lower dial the zodiac is set in a
recess, over which rotate the sun, moon, and dragon hands. Above this, on the
same plane as the dial, the annual Gregorian calendar is engraved, with the
most important saints' days (including that of Rudolphus, in honor of Bürgi's
employer Emperor Rudolf II); only every second day is marked off. This lower
dial presents a geocentric planetarium.. The hour hand itself is missing; the
hour numerals I–XII are etched upon the glass plate in front. On the upper dial
hands bearing the symbols for the planets circle about the sun, thus presenting
a heliocentric planetarium. Lines are drawn on the glass from a point that lies
between the planets Venus and Mars; formerly a small image of the earth was
inserted at this point. This lattice of lines radiating outward from the earth
makes it possible to read the phase angles of the planets. On the back there is
a calendar disk with the dominical letters, the dates for Easter, the Golden
Number, and the cycle of the sun for the years 1600–1640. Hans von Bertele
first ascribed the clock to Jost Bürgi, subsequently to Christoph Margraf. Un-
questionable evidence of Bürgi's hand is the originality of the remontoire, the
highly structured nature of the design, and the precise workmanship, all of
which were observed in the examination documented here for the first time
(see Ch. 8). As the case was made in the shops of the Prague Court, the clock
must have been made after Bürgi arrived in Prague, that is, after 1604.

For Jost Bürgi see Chapter 8.

Literature: *Kunsthistorisches Museum, Wien, Katalog der Sammlung für Plastik und Kunst-
gewerbe,* II, Vienna, 1966, p. 91, No. 342 (with additional references there). Maurice II,
Fig. 639 (with ascription to Margraf). Henry C. King, *Geared to the Stars* (Toronto, 1978),
Fig. 5.21.

54 Rolling-ball clock

Johann Sayller. Ulm, 1626
Stuttgart, Württembergisches
Landesmuseum

Movement: brass plates and wheels
Case: ebony, fire-gilt brass
Height: 30 cm (11¾ in.)

Going, quarter-striking, and hour-striking trains. Dial with hour and minute hands, date, age and phases of the moon. The clock has a rolling-ball escapement, an idea developed in Prague, in which a ball runs on an inclined sheet of glass along wires stretched in a zigzag pattern. The running time of the ball is always the same, and thus the motion work can be regularly advanced each time the ball has run its course. At the same time a lever raises a second ball onto the course and transports the first ball back to the start of the course. There is a mirror in the lid of the case which reflects the movement of the ball, making it appear to move up and down. The case is stamped with EBEN and the Augsburg pine cone. Signed on the dial: "Johann Sayller Ulm." According to Weyermann, Sayller produced this clock as his masterpiece (see Catalog No. 47).

Literature: Hans von Bertele and Erwin Neumann, "Der Kaiserliche Kammeruhrmacher Christoph Margraf und die Erfindung der Kugellaufuhr," *Jahrbuch der Kunsthistorischen Sammlungen in Wien.* 1963, 59: 56,74 f. Volker Himmelein, *Uhren des 16. und 17. Jahrhunderts* (Stuttgart, 1973), No. 40. Maurice II, Fig. 646.

224

55 Table clock

Johann Sayller. Ulm, c. 1630
Stuttgart, Württembergisches
Landesmuseum

Movement: brass plates and wheels
Case: silver, partly fire-gilt
Height: 57 cm (22½ in.)

Going and hour-striking trains. The going train is wound from the fusee wheel of the striking train when the clock strikes. The spring for the striking train, located in the base of the clock, is of unusually large proportions. This spring, connected through a gear train on the right side of the fusee (see figure at left) to the small spring of the going train, serves to keep the going-train spring always fully wound. Presumably the spring of the going train has a friction clutch to prevent the spring from overwinding and breaking.

Indications: on the front, hour and minute hands; on the back, the date (by month and day), and the position of the sun in the zodiac, separate dial for the days of the week and checking dial for the hour striking. The clock is signed on the regulating dial: "Johann Sayller fecit Ulm." The case of solid silver is by the Ulm goldsmith Hans Jerg Merckle (born 1588, master 1613, died 1634); mark on the dial on the back (Rosenberg[3] 4775). A similar clock, but with a simpler case, is in the same museum.

For Sayller see Catalog No. 47.

Literature: Volker Himmelein, *Uhren des 16. und 17. Jahrhunderts* (Stuttgart, 1973), Figs. 42–45. Maurice II, Fig. 655. Giuseppe Brusa, *L'arte dell'orologeria in Europa* (Milan, 1978), Figs. 243, 244.

56 Wall clock

Schleswig, 1642
Copenhagen, Nationalmuseet

Movement: iron
Case: fire-gilt brass
Height: 58 cm (22¾ in.)

Going, quarter-striking, and hour-striking trains, all weight driven. Hour hand and (missing) minute hand; originally also a calendar ring and phases of the moon. Peculiarities of this clock, which are also emphasized by an inscription, are the regulation of the running of all three trains through one cross-beat escapement each, and the clockwork mechanism itself, which is reduced to a single wheel in the case of each train. Rings that function as count wheels are attached to the wheels for the quarter- and hour-striking trains. Each quarter-hour is struck by a hammer of its own; corresponding to the relevant hour, there are pegs on the wheel that is at the same time a pin wheel and an escape wheel. The escape wheel of the going train has 360 teeth, and the oscillation period of the cross-beat is about 5 seconds.

226

Figures of Copernicus and Tycho Brahe are engraved upon the sides of the case; they stand pointing to their world systems. The engravings are explained by text within cartouches. The one for Tycho reads: "Hypotyposis Systematis Mundani a Tychone Inventa" and "Quid si sic?" The one for Copernicus reads "Hypotyposis Systematis Mundani Copernica" and "Sic movetur mundus." Below the dial is a comment concerning the novelty of the clock's construction and the date: "Nova Inventio Horologii primi Mobilis et Luminarium aemuli Anno Salutis 1642."

The clock, which comes from the ducal art collections at Gottorf Castle, has certainly been influenced by Bürgi's designs. Duke Friedrich III (1616–1659) promoted the making of timepieces just as his brother Friedrich Johann, Prince-Bishop of Lübeck, did (see Catalog No. 120).

Literature: *Gemessene Zeit. Uhren in der Kulturgeschichte Schleswig-Holsteins. Ausstellung Schloss Gottorf 1975.* pp. 157–159. Maurice II, Fig. 640. Giuseppe Brusa, *L'arte dell'orologeria in Europa* (Milan, 1978), Figs. 271–274.

57 Year timepiece

Hans Buschmann. Augsburg, 1651/1652
Braunschweig, Herzog Anton
Ulrich-Museum

Movement: brass plates and wheels
Case: ebony, fire-gilt bronze and brass
Dials: silver; miniature painting on ivory
Height: 64.3 cm (25¼ in.)

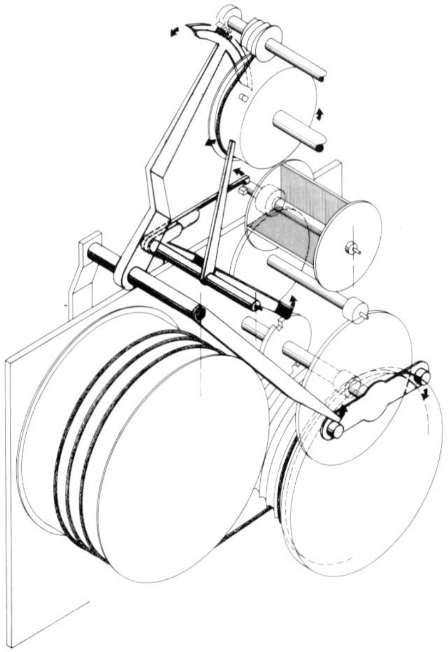

Going train wound every 24 hours by a remontoire, which is supposed to give the clock a running time of one year. Hour dial; below it a quarter-hour dial and a calendar. On the back, indication of the days of the week by a disk with the reigning planets engraved on it, after engravings of Virgil Solis. Beneath this, a dial that adds up the number of day-by-day windings. The crowning sphere is in the form of a small armillary sundial; below this is a lunar sphere showing the phases of the moon. Escapement no longer original.

Hans Buschmann first proposed this clock to Duke August the Younger of Braunschweig and Lüneburg in a drawing (see Fig. 31) which specified all its properties and indications. The archival material published by Himmelein has been confirmed through the restoration of the clock in the workshops of the Bayerisches Nationalmuseum: the clock did not work because the power of the spring in the base was insufficient, and it was necessary to add another diagonally positioned spring drum. Probably Caspar Langenbucher added this improvement when he attempted to repair the clock in 1657. Langenbucher complained about its shortcomings, among them the fact that the crown wheel had been spoiled by improper filing. He could not promise that this "sleepy year clock" would run for a year, because many things about it had been tinkered with. After two years he sent it back to the Duke unrepaired. His claims for the time he had spent working on it were rejected. It seems that Wolf Griessbeck, the court clockmaker at Wolfenbüttel, finally got the clock to run.

Buschmann's design conception is simple and convincing: the independent going train is would up daily by a powerful remontoire in the base. The construction drawing elucidates the mechanism. The difficulty lies, however, in the remontoire spring, which broke even as the clock was being transported, and in the heavy friction which arises from the fact that the ratchet, levers, and wheels that engage the ratchet are all made of the same material. Hans Buschmann made similar clocks for the Grand Duke Ferdinand II of Tuscany, for Archduke Karl, and in 1656 for the Emperor Ferdinand III. The price of each of these clocks was 400 gulden. The Augsburg pine cone is struck on the going train along with Buschmann's mark HB in a shield. The ebony base is stamped with the pine cone and EBEN.

Hans Buschmann, small-clock maker, Protestant, was born after 1591, son of the clockmaker Caspar Buschmann III at Augsburg. He learned the trade in his father's shop, presumably spending his journeyman period in Prague. On June 28, 1620, he received his smiths' eligibility from his father and set up a workshop on the Mauerberg. In the same year he married Rosina Scheiggert; after her death in 1658 he married Elisabeth Stürzlin, widow of Böglin. There were eight children of the first marriage. His sons Hans (Johann) and David were also clockmakers and worked in their father's shop. Hans Buschmann, who belonged to the fourth of seven generations of Buschmanns active in Augsburg's clockmaker's trade, was the most outstanding representative of the family. He paid taxes between 1621 and 1662. In 1638/1939 he was a sworn inspection-master. He died on December 15, 1662.

Sources: HKA, HZAB No. 87 (1641), fol. 154v. STAA, U III, fol. 27v.
Literature: Maximilian Bobinger, *Kunstuhrmacher in Alt-Augsburg* (Augsburg, 1969), pp. 83–89. August Fink, *Die Uhren Herzog August d. J.* (Braunschweig, 1953), pp. 7–10. Maurice II, Fig. 666. Giuseppe Brusa, *L'arte dell'orologeria in Europa* (Milan, 1978), Fig. 278, 279. Volker Himmelein, "Die Uhren," in *Sammler, Fürst, Gelehrter, Herzog August zu Braunschweig und Lüneburg, 1579–1666* (Wolfenbüttel, 1979), pp. 166 f.

58 Table clock

David Buschmann. Augsburg, c. 1660
United States, private collection

Movement: brass plates and wheels
Case: fire-gilt bronze and brass, silver
Height: 16.7 cm (6½ in.)

Going and hour-striking trains; alarm. Escapement and hands no longer original. For the case of the clock David Buschmann used a construction element already available from his pillar clock (see Catalog No. 59). Backplate signed "D. Buschman." For David Buschmann see Catalog No. 30.

Literature: Maurice II, Fig. 600.

59 Table clock in the form of a pillar

David Buschmann. Augsburg, c. 1660
Vienna, Kunsthistorisches Museum

Movement: brass plates and wheels
Case: fire-gilt bronze, silver, wooden core with tortoiseshell applied
Height: 75 cm (29½ in.)

At the lower part of the pillar a rotating drum indicates the annual calendar and the position of the sun in the zodiac. In the spiral convolutions of the pillar an indicator (no longer present) screws its way upward to mark the hours. Probably the indicator had to be returned manually. The quarter-hours are indicated on the upper side of the capital. Signed: "Davidt Buschman Aug."

A similar clock is in the same collection in Vienna; Buschmann adapted the central part of the case to a clock of his own (see Catalog No. 58). For David Buschmann see Catalog No. 30.

Literature: Kunsthistorisches Museum Wien, Katalog der Sammlung für Plastik und Kunstgewerbe, II, Vienna, 1966, No. 324. Maurice I, Plt. VII.

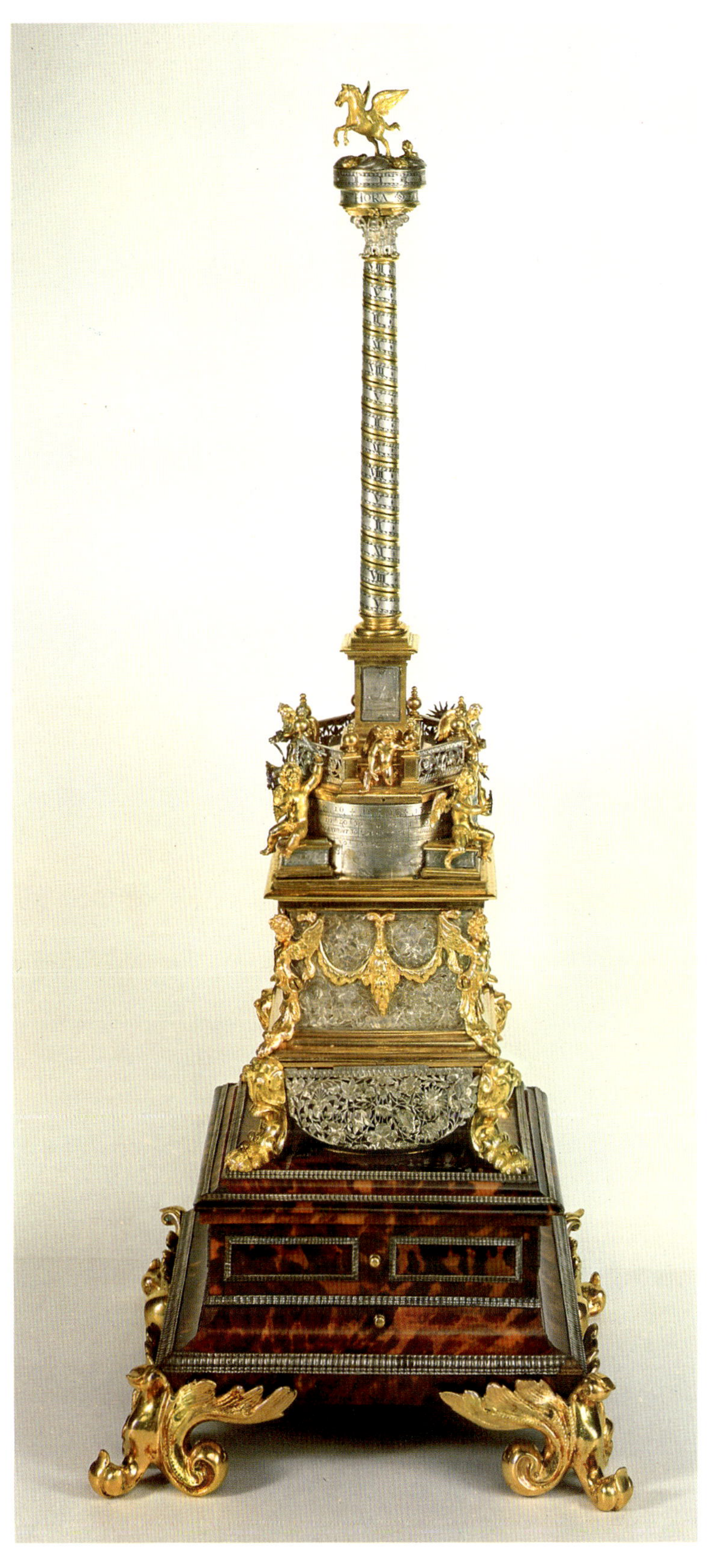

60 Table clock

Nikolaus Radeloff. Schleswig, 1654
Copenhagen, Nationalmuseet

Movement: brass plates
Case: ebony, fire-gilt brass, glass
Height: 93.6 cm (3 ft ¾ in.)

Going and hour-striking trains. Cross-beat escapement; hour and minute hands. The going train is given a constant power from stone balls that descend within a cage. The gravitational force of the balls causes the cage to rotate, and this drives the train. To wind the clock, the balls are restored to place in a magazine. Dial with coat of arms and initials: "V(on) G(ottes) G(naden) F(riedrich) E(rbe) z(u) N(orwegen) H(erzog) z(u) S(chleswig) H(olstein)." Below this the year 1654. Above, an engraving of a naval battle; at the back, Saint Peter praying.

Two other clocks by Radeloff on this design principle are at the Rosenborg Castle, Copenhagen (Inv. No. 1256) and in a private collection in Vienna (see Maurice II, Fig. 658).

Nikolaus Radeloff was invited by the polyhistor Adam Olearius, librarian to the dukes of Holstein-Gottorf (1559–1671), to come there from Saxony. From 1647 onward Radeloff lived in Schleswig, and there is record of him there as late as 1666.

Literature: Gemessene Zeit. Uhren in der Kulturgeschichte Schleswig-Holsteins. Ausstellung Schloss Gottorf 1975. p. 96. Maurice II, Fig. 657. Giuseppe Brusa, *L'arte dell'orologeria in Europa* (Milan, 1978), Figs. 280, 281.

94, 95. Manually propelled gala carriages from the triumphal procession of Emperor Maximilian I. Hans Burgkmair, Augsburg, 1512–1526.

IV Artificial Life: Automata and Figure Clocks

In the form of legends, the desire to simulate life reaches far back into history: mechanically functioning figures of men and beasts are mentioned by Homer (8th century B.C.), Archytas of Tarentum (4th century B.C.), Hero of Alexandria (1st/2nd century A.D.), and Philostratos (3rd century A.D.). In mythology we are familiar with the artificial creations of Daedalus, Pygmalion, and Prometheus. In the Middle Ages there is the legend of a mechanical figure of a woman reputedly made by Albertus Magnus and smashed to bits as an invention of the devil by his disciple Thomas Aquinas. In this same tradition are the artificial fly and the eagle built by Regiomontanus in 1436–1437 which flew during a visit of the Emperor in Nuremberg. All of these simulations of the act of creation, although brought forth by rational and mechanical means, were sensed as being magic, but a nondemonic and hence acceptable magic.

Since classical antiquity, parks, theaters, festivals, and other celebrations have provided occasions for creating and displaying lifelike effects produced by machinery. Part of this tradition are the carriages in the triumphal procession depicted in Fig. 97. In miniaturized form, such automated moving carriages and other automata were also built as table ornaments (Fig. 95) and as clever components of drinking games (see Catalog Nos. 102–104). Part of the automata—the Madonna figures and the flagellation and crucifixion scenes—served the purpose of heightening religious feeling (Catalog Nos. 62–69).

It was precisely the mechanical clock that could provide the illusion of lifelike animation. The earliest example still preserved in the West is the monumental cock of 1352–1354 (Fig. 5) which crowned the clock of the cathedral at Strasbourg and which crowed, spread its feathers and wings, and turned about. The realistic quality of such automata became enhanced through artificially produced music or animal

96. Festive table setting on the occasion of the wedding of Johann Wilhelm of Jülich, Cleve, and Berg, Düsseldorf, 1587. Dietrich Graminaeus, *Fürstliche Hochzeit* (Cologne, 1587).

noises. Here, too, the acoustic repertoire was mechanically produced and was pre-programmed.

The figure clocks and the automata of this exhibit convey an idea of the numbers and variety in which these mechanisms were produced. But for larger figures and for more lavish groups we now have only descriptions. Only one example (shown in Catalog No. 100) remains of the once common stuffed animals that were seemingly brought back to life by a clockwork mechanism.

Whereas clockwork creatures were first built as magical attempts to imitate creation, at the end of our period clocks and automata were used as means to explain life mechanically. The revolutionary physiology of Descartes (1596–1650) derived from his experience with mechanisms. The automaton analogy lies at the foundation of it; the animals are nothing more than soulless, programmed machines. During the Enlightenment, Julian La Mettrie carried the Cartesian idea to its ultimate conclusion: his famous book bears the title *L'homme machine* (1748).

97. Procession in Antwerp, Edward Brown, *Sonderbare Reisen*. 1677 (Nuremberg, 1686).

61 Figure clock—Adam and Eve

Nuremberg (?), c. 1580
Brussels, Musées Royaux d'Art
et d'Histoire

Movement: brass plates and wheels
Case: fire-gilt bronze and copper
Height: 30 cm (11¾ in.)

Hour-striking clock with stackfreed on going train. Escapement no longer original. The hours are indicated on a ball which crowns the tree, while the serpent's tongue serves as the fixed indicator. The convex base with the rich acanthus foliage is also found in a monstrance clock atop a gaming cabinet (Maurice II, Fig. 549). This cabinet with plaques of the Nuremberg goldsmith Elias Lencker was sold by the Nuremberg merchant Sebald Schwarz to Dresden (Monika Bachtler, "Die Nürnberger Goldschmiedefamilie Lencker," *Anzeiger des Germanischen Nationalmuseums,* 1978, p. 116, No. 6). Because of the close similarity between the two bases, our figure clock can be assigned to Nuremberg.

Literature: Anne-Marie Berryer and Liliane Dresse de Lébioles, *La mesure du temps à travers les ages aux Musées Royaux d'Art et d'Histoire* (Brussels, 1961), Plt. VIIIb. Maurice II, Fig. 390.

62 Figure clock—Madonna

Nikolaus Rugendas the elder.
Augsburg, c. 1635
United States, private collection

Movement: brass plates and wheels
Case: fire-gilt bronze and brass
Base: ebony
Height: 31.5 cm (12⅜ in.)

Going train with fusee and chain, hour-striking train. The hours are indicated by the rotating crown of the Madonna. The base is stamped with the Augsburg pine cone and EBEN. Backplate signed A, beneath which is N·R· (= Augustanus Nikolaus Rugendas). The Madonna figure with the same case exists in a number of other examples which are signed by the Augsburg clockmakers Nikolaus Schmidt the younger (Maurice II, Fig. 395) and Jeremias Pfaff, dated 1643 (*The Connoisseur,* 1971, *176:* 94), and by the Strasbourg clockmaker Isaak Habrecht III, Catalog No. 63. On Nikolaus Rugendas the elder see Catalog No. 44.

Literature: Maximilian Bobinger, *Alt-Augsburger Kompassmacher* (Augsburg, 1966), pp. 156 f. Clare Vincent, *Northern European Clocks in New York Collections,* Metropolitan Museum of Art, Jan. 4–March 28, 1972, p. 18, No. 20. Maurice II, Fig. 394.

63 Figure clock—Madonna

Isaak Habrecht III. Strasbourg, c. 1635
Munich, Bayerisches Nationalmuseum

Movement: brass plates and wheels
Case: fire-gilt bronze and brass
Base: ebony
Height: 32 cm (12⅝ in.)

The figure and the base are ready-made parts from Augsburg (see Catalog No. 62), into which the Strasbourg clockmaker has built his movement. Signed on the backplate: "Isaak Habrecht A Strassbourg." Base stamped EBEN and the Augsburg pine cone.

Isaak Habrecht III, grandson of the builder of the Strasbourg Cathedral clock, was born in 1611 and died in 1686.

Literature: Maurice II, Fig. 393.

63

62

64

61

64 Figure clock – Madonna

Munich, c. 1640
Silver figure by the
goldsmith Stephan Hötzer
Munich, Bayerisches Nationalmuseum

Movement: brass plates and wheels
Case: silver, partly gilt
Base: pearwood stained black
Height: 30.5 cm (12 in.)

Going train with fusee and gut line, hour-striking train. Madonna with Child on a crescent moon. Indication of hours on the rotating crown. Originally the Madonna raised her arm when the clock struck. Movement in the wooden base. Struck with the mark of the Munich goldsmith Stephan Hötzer (Rosenberg[3] 3440).

Literature: Maurice II, Fig. 400. Giuseppe Brusa, *L'arte dell' orologeria in Europa* (Milan, 1978), Fig. 275.

237

65 Figure clock– flagellation of Christ

South Germany, c. 1625–1630
United States, private collection

Movement: brass plates and wheels
Case: fire-gilt bronze and brass
Base: ebony
Height: 34.5 cm (13½ in.)

Going train with stackfreed; quarter-striking and hour-striking trains; alarm. Christ flanked by two executioner's henchmen, at the pillar of martyrdom; upon this a ball with an hour ring. The bronze figures of the flagellation group are modelled after the famous and frequently imitated group which is ascribed to Algardi and Duquesnoy (Jennifer Montagu, "A Flagellation Group: Algardi or du Quesnoy," *Bulletin des Musées Royaux d'Art et d'Histoire*, 1966/1967, 38/39: 153).

Literature: Maurice II, Fig. 403.

66 Figure clock– Christ at the flagellation pillar

Zacharias Erb. Vienna (?), c. 1630
United States, private collection

Movement: brass plates and wheels
Case: fire-gilt bronze and brass
Base: ebony
Height: 32 cm (12½ in.)

Going train with fusee and chain; quarter-striking and hour-striking trains. On the pillar a ball with hour ring, also indication of the quarter hours on the surface of the base. Signed "Zacharias Erb." The bronze Christ standing to the side of the pillar perhaps is derived from the school of Adrian de Vries; compare especially the resurrected Christ on the gravestone of Prince Ernst von Schaumburg-Lippe in Stadthagen (Olof Larsson, *Adrian de Vries*, Vienna/Munich, 1967, Figs. 153, 155; a reference received from Dr. Peter Volk of Munich).

The Erb family of clockmakers worked in Vienna for a number of generations (Maria Habacher, "Mathematische Instrumentenmacher, Mechaniker, Optiker und Uhrmacher im Dienste des Kaiserhofes in Wien (1630–1750)," *Blätter für Technikgeschichte,* 1960, *22:* 5–80, wherein there is no mention of a clockmaker with the given name Zacharias).

Literature: Sotheby, London, Oct. 19, 1964, *The Percy Webster Collection,* No. 196. Maurice II, Fig. 404.

67 Figure clock—flagellation of Christ

Nikolaus Schmidt the younger
Augsburg, c. 1630
United States, private collection

Movement: brass plates and wheels
Case: fire-gilt bronze and brass, partly painted in colors
Base: ebony
Height: 28 cm (11 in.)

Going train with fusee and gut line; hour-striking train; alarm. The hours are indicated on the dial, the quarter hours on the pillar. When the hours strike, the henchmen lash Christ. On the backplate the signature "Nicolaus Schmidt der Jünger." The tall base is stamped with EBEN and the Augsburg pine cone. On Nikolaus Schmidt see Catalog No. 19.

Literature: F. J. Britten, *Old Clocks and Watches and Their Makers* (5th ed., London, 1922), Fig. 118. Maurice II, Fig. 405.

239

68 Figure clock—crucifixion

Isaak Ebert. Steyr, c. 1640–1650
Munich, Bayerisches Nationalmuseum

Movement: brass plates and wheels
Case: fire-gilt bronze, copper, and brass
Height: 38 cm (15 in.)

Going and hour-striking trains. Numerous repairs and restorations. The horary ball rotates upon the upper tip of the cross, and the fixed pointer shows the time. On the backplate the signature: IS.EB.STEYR. Another crucifixion by the same master is located in the Schlossmuseum at Linz (see *Edles Silber—Kostbare Uhren,* Linz, 1972, p. 49, No. 7).

Literature: Ernst von Bassermann-Jordan, *Die Geschichte der Räderuhr unter besonderer Berücksichtigung der Uhren des Bayerischen Nationalmuseums* (Frankfurt, 1905), No. 40. Maurice II, Fig. 411.

240

69 Figure clock–Calvary

South Germany, c. 1580–1590
Wuppertal, Historisches Uhrenmuseum

Movement: iron
Case: fire-gilt bronze and copper
Height: 62 cm (2 ft ⅜ in.)

Going, quarter-striking, and hour-striking trains, all with fusee and chain; alarm. On a very tall base borne by Death and demons, the clock case with half-figures at the corners (like Catalog No. 55), surrounded by a revolving procession of figures including John, the two Marys, and Roman soldiers. On the bell housing a wooden cross with a bronze Christ. The hour-striking train and a train for the procession lie within the base as separate units. In their structure and arrangement the trains resemble those of the Turkish boat in Catalog No. 70.

Literature: Maurice II, Fig. 406. Giuseppe Brusa, *L'arte dell orologeria in Europa* (Milan, 1978), Figs. VIII, IX.

241

70 Figure clock–Turkish ship

Augsburg, c. 1580–1590
Innsbruck-Ambras
Kunsthistorisches Museum
Sammlungen Schloss Ambras

Movement: iron
Case: fire-gilt bronze and copper
Height: 45 cm (17¾ in.)

Going, quarter-striking, and hour-striking trains. All trains with fusee and gut line. Hour dials and checking dial for hour striking on the sides of the case, which stands in the middle of the boat in place of a cabin. The Turk's eyes are connected to the going train. When the hours strike, he raises his arm (presumably he once held a sword) and at the same time the two oarsmen paddle. On the striking of the quarter hour, the ape sitting on the bow moves and the forward oarsman turns his head. In its structure the clock movement resembles that of No. 69; the embossed work is like that on ship No. 71.

Literature: Maurice II, Fig. 271.

242

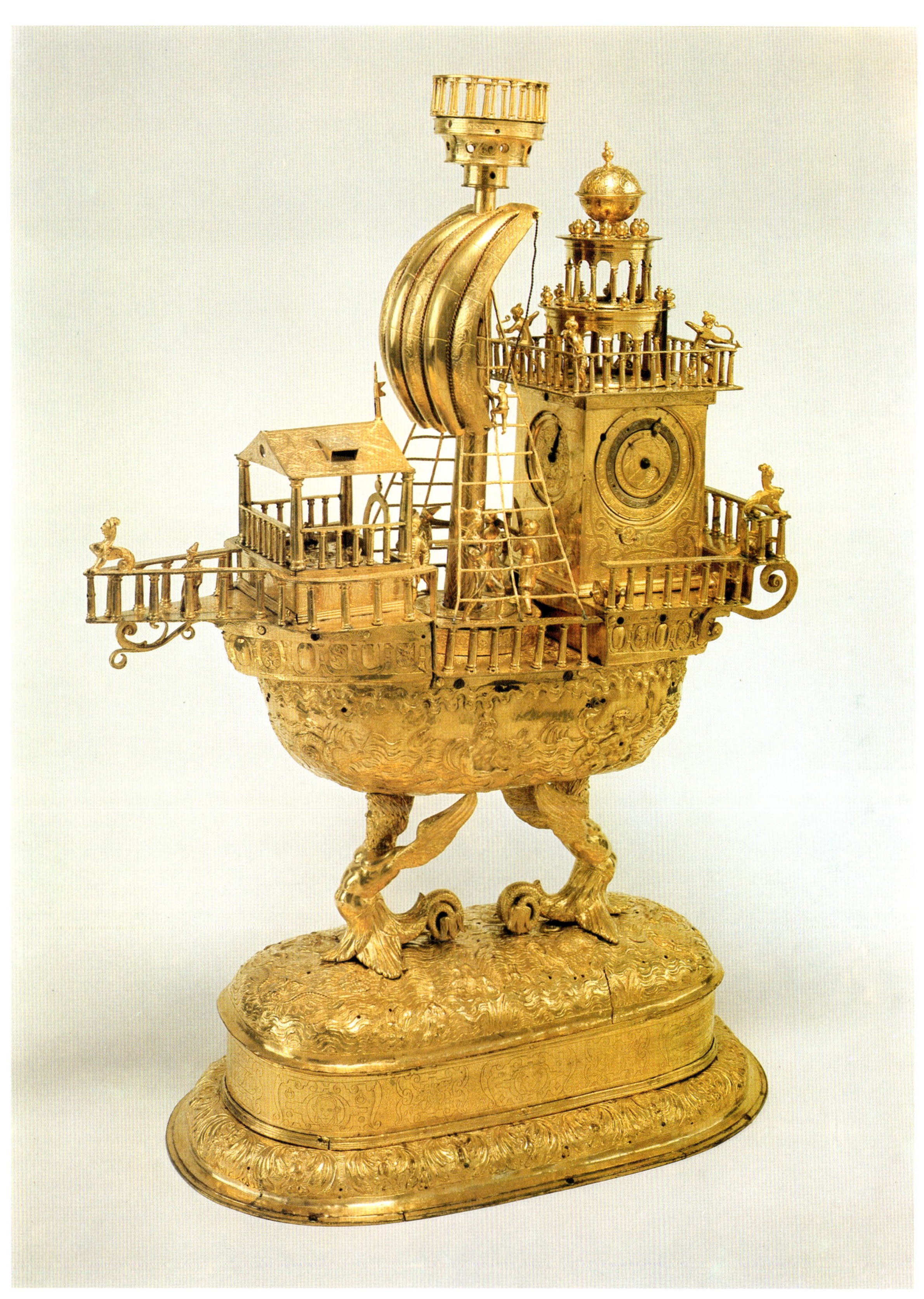

71 Clock in the form of a ship

South Germany, end of the 16th century
United States, private collection

Movement: iron plates
Case: fire-gilt bronze and copper
Height: 71 cm (28 in.)

Going, hour-striking, and quarter-striking trains; alarm. Going train upside down. The hours and the age and phases of the moon are indicated on the stern. When the clock strikes, a separate (restored) train in the ship's hold moves a guard of Turks around the ship's mast. Originally there were other mechanically operated motions in the mast top. The embossing work probably comes from the same workshop as the figure clock representing a Turkish ship (No. 70). Two other vessels with abundant automata have survived (see Maurice II, Figs. 268, 270). Both certainly were created within the school of the Augsburg clockmaker Hans Schlottheim: descriptions of this type of complicated automata have come down from him. The embossing work on this ship can, however, also be compared to that of a small ship on wheels which was made by the Nuremberg goldsmith Esaias zur Linden (see Charles Oman, *Medieval Silver Nefs,* London, 1963, Fig. 7). The Middle Ages were already familiar with ships as table ornaments.

Literature: F. J. Britten, *Old Clocks and Watches and Their Makers* (5th ed., London, 1922), Fig. 95. Maurice II, Fig. 269.

244

72 Figure clock— mounted Turkish Pasha

South Germany, end of the 16th century
Moscow, Kremlin Museum

Case: fire-gilt bronze and copper
Height: 47.5 cm (18¾ in.)

Going and hour-striking trains; alarm (?). The eyes of the horse are connected with the movement; the other automatic movements are not known. There are four similar clocks: (1) in the Staatlich Mathematisch-Physikalisches Salon in Dresden (Maurice II, Fig. 273); (2) in the Adler Planetarium in Chicago (Inv. No. M 384); (3) illustration, Plate III of Samuel Guye and Henri Michel, *Uhren und Messinstrumente des 15. bis 19. Jahrhunderts* (Zurich, 1971); (4) *Rothschild Collection at Mentmore,* Sotheby, May 20, 1977, No. 1047. The model for the bronze statue of the mounted Turk may have been a horseman on the woodcut shown above by Pieter Coecke van Aelst, *Des moeurs et fachons de faire du Turcz* (Antwerp (?), 1553).

Literature: Opis Moskovskoi Oruzheinoi Palaty (Description of the Armory in the Kremlin) (Moscow, 1884–1893), Vol. I, Plt. 113. Maurice II, Fig. 274.

73 Figure clock–horseman

Nikolaus Schmidt. Augsburg, end of the
16th century
Rockford, Illinois, The Time Museum

Movement: brass plates and wheels
Case: fire-gilt bronze and copper
Height: 27.9 cm (11 in.)

Going and hour-striking trains. Hour dial and checking dial for striking are on
the upper side of the base. When the hours strike, the horseman turns his head.
Backplate marked NS within a shield (= Nikolaus Schmidt). Three other
examples of the horseman figure are known: (1) illustration in Ludwig Hovesi,
"Austellungen von Fächern und Uhren, Palais des Ungarischen Ministers,
Wien 1903," *Kunst und Kunsthandwerk,* 1903, 6: 196 ff.; (2) photo in archives
of the Bayerisches Nationalmuseum (present location unknown); (3) illustration in *Catalogue des objets d'art et de haute curiosité . . . ,* Collection Spitzer (Paris,
1893), p. 171, No. 2689, which is perhaps identical to (1).

On Nikolaus Schmidt the elder see Catalog No. 19.

Unpublished.

74 Figure clock—eagle

Augsburg, c. 1630
New York, Metropolitan Museum of Art

Movement: brass plates, iron wheels
Case: fire-gilt bronze and brass
Base: ebony
Height: 33 cm (13 in.)

Going, quarter-striking, and hour-striking trains. The dial is engraved with an outer indication of I–XII hours and an inner indication of 13–24 hours. This continuation of the count, however, is shifted on the dial by six hours (i.e., to begin at 7 instead of 1). The intent of this is obscure, unless it is a mistake on the part of the engraver. On the case is the quarter-hour dial. Going train with fusee and chain; escapement no longer original. Upon the striking of the hour the scepter moves; on the striking of the quarters the eagle moves its beak and rolls its eyes. The mechanism is marked on the top plate with the Augsburg pine cone within a shield; on the base is stamped EBEN and the mark of the cabinetmaker MS in ligature within a shield. A similar clock is located in the Staatlich Mathematisch-Physikalischer Salon in Dresden (Maurice II, Fig. 308).

Literature: Metropolitan Museum of Art Bulletin, 1929, 24:100.

247

75 Figure clock—fowl

Johann Ott Hallaicher. Augsburg,
middle of the 17th century
Munich, private collection

Movement: brass plates and wheels
Case: fire-gilt bronze and brass; pearwood stained
black; capitals of ivory stained black
Height: 32 cm (12½ in.)

Going, quarter-striking, and hour-striking trains. Hour chapter ring restored.
The striking train controls the movements of the bird's wings and beak. Signature on the backplate: "Johann Ott Halleicher." The wooden case and the trains have been repaired several times and alterations have been made. For that reason it is not clear to what extent Hallaicher—who is known to have made other figure clocks (see Catalog No. 78)—devised the present form of the clock.

Johann Ott Hallaicher, small-clock maker, Protestant, was born on November 23, 1612, son of the Augsburg pastor M. Tobias Hallaicher and his wife Katharine née Kayser. He acquired smiths' eligibility as a clockmaker in 1636 through his marriage to Regina, daughter of the cutler Matthäus Kistler. Their son Matthäus, who also became a clockmaker, was born in 1641. Johann was married for a second time in 1648 to Judith Braun, and for a third time in 1686 to Hanna Hermann. In the years 1638–1640 he exported clocks to Venice via his journeyman Hans Ferdinand Mehrer. In 1666 he was a witness to the marriage of the Augsburg clockmaker Johann Sommer VI; in 1673 he was foreman of the smiths' guild. In 1699 he is mentioned for the last time as witness to the marriage of his son Matthäus.

Sources: STAA, SZB, fol. 195a, EMA, Barfüsser I, p. 32, No. 243, Oct. 24, 1641. STAA, SZB, fol. 220a (1672). EMA, Barfüsser I, p. 117, No. 25. EMA, Hl. Geist I, p. 36, No. 3. STAA, U II, fol. 267. STAA, HAP, 1666, p. 425. STAA, U II, fol. 237. STAA, HAP, 1699, p. 150 (Oct. 11).

Unpublished.

76 Figure clock—cock

South Germany, c. 1580–1590
New York, private collection

Movement: iron
Case: fire-gilt bronze and copper; comb and tail
feathers painted
Dial: silver
Height: 38 cm (15 in.)

Going and hour-striking trains. Quarter-hour dial and regulating dial on base. The eyes of the cock are connected to the going train, the beak and wings to the striking train.

Literature: Clare Vincent, *Northern European Clocks in New York Collections,* Metropolitan Museum of Art, Jan. 4–March 28, 1972, p. 19, No. 26. Maurice II, Fig. 300.

77 Figure clock–griffin

South Germany, c. 1580–1590
London, Victoria and Albert Museum

Case: fire-gilt bronze and copper
Height: 33 cm (13 in.)

Going and hour-striking trains. Hour dial no longer original. On the base is a small dial for the quarter hours and a checking dial for the striking. When the hour strikes, the griffin moves its beak and wings. This bronze figure has been modelled after the copper engraving of a griffin by Martin Schongauer (Max Lehrs, *Martin Schongauer,* Berlin, 1914, p. 93).

Literature: Maurice II, Fig. 302.

78 Figure clock–griffin

Johann Ott Hallaicher. Augsburg, c. 1640
Zurich, Museum der Zeitmessung Beyer

Movement: brass plates and wheels
Case: fire-gilt bronze and copper
Base: ebony
Height: 42 cm (16½ in.)

Going, quarter-striking, and hour-striking trains. The eyes of the griffin are connected to the going train, the beak and wings to the striking train. Signed on the backplate "Hans Ott Halaicher." For this master see Catalog No. 75.

Literature: Maurice II, Fig. 307.

79 Figure clock–parrot

South Germany, c. 1600
Munich, Bayerisches Nationalmuseum

Movement: iron
Case: fire-gilt copper
Mountings: silver
Base: pearwood stained black
Height: 40.1 cm (15¾ in.)

Going train located in bird's body, striking train in the base. On the hour, the going train, acting by a mechanical linkage that goes through the bird's left foot, releases a mechanism in the base (brass spring, wooden fusee) that drives a bellows. Thereupon the hours are whistled, the parrot moves its bill, wings, and eyes, and balls fall out of a magazine at the tail-end through a trapdoor.

Literature: Maurice II, Fig. 310

251

80 Figure clock–unicorn

Augsburg (?), c. 1640
Schwäbisch Gmünd, Städtisches Museum

Movement: brass plates, iron wheels
Case: fire-gilt bronze and brass
Base: ebony, mountings of silver
Height: 39.5 cm (15½ in.)

Going and hour-striking trains. Going train with fusee and chain. The eyes are connected to the going train, and with every striking of the hour the unicorn raises his foreleg. Dial plate engraved with small animals and a landscape. The unicorn is modeled after the ancient bronze horses (Greek, 4th–3rd century B.C.) now on the facade of Saint Mark's Cathedral, Venice. Another example of this clock, but with an embossed copper base and a different motion upon the striking of the hour, is found in the Grassy collection, Madrid (Maurice II, Fig. 295).

Literature: Uto Auktionen, Zurich, Nov. 2, 1976, No. 225.

252

81 Figure clock–dromedary

South Germany, c. 1600
Munich, Bayerisches Nationalmuseum

Movement: brass plates, iron wheels
Case: fire-gilt bronze and copper
Dials: enamelled silver
Original base replaced after the middle
of the 18th century
Height: 26.6 cm (10½ in.)

Going and hour-striking trains, the dial being on the lower part of the saddle-cloth. The eyes of the dromedary are connected with the going train. A separate mechanism in the base makes the entire unit roll forward while the driver swings his mace.

Literature: Ernst von Bassermann-Jordan, *Die Geschichte der Räderuhr unter besonderer Berücksichtigung der Uhren des Bayerischen Nationalmuseums* (Frankfurt, 1905), p. 86, No. 55. Maurice II, Fig. 298.

253

82 Figure clock—reclining lion

Augsburg, c. 1630
Furtwangen, Fachhochschule,
Historische Uhrensammlung

Movement: brass plates and wheels
Case: fire-gilt bronze
Base: ebony
Height: 25 cm (9⅞ in.)

Going, quarter-striking, and hour-striking trains; alarm. The eyes are connected with the going train, and when the hour strikes, the lion snaps his jaws. Three similar clocks are known: Maurice II, Figs. 325, 326, and *The Connoisseur*. 1971, *176:* 101.

Literature:　Maurice II, Fig. 324.

83 Figure clock—lion of Saint Mark

Augsburg, c. 1630
Rockford, Illinois, The Time Museum

Movement: brass plates and wheels
Case: fire-gilt bronze
Base: ebony
Height: 25.3 cm (10 in.)

Going train with fusee and chain; hour-striking train. The eyes of the lion are connected with the going train. When the hour strikes, the beast snaps his jaws. Escapement original; only the balance spring has been replaced. Base stamped with EBEN and the Augsburg pine cone. On the cover plate a regulating slot with the indications T(arde) and P(resto). The clock was probably intended for the Italian market. A similar clock is in a private collection.

Literature:　Maurice II, Fig. 327.

84 Figure clock – lion rampant

Philipp Trump. Crailsheim, c. 1630–1640
Munich, Bayerisches Nationalmuseum

Movement: brass plates and wheels
Case: fire-gilt bronze and brass; silver
Base: ebony
Height: 40 cm (15¾ in.)

Going, quarter-striking, and hour-striking trains. Going train with fusee and chain. The Whole Clock (1–24 hours) is indicated on the rotating ball, the Small Clock (1–12 hours) on the dial on the base. The second dial shows the quarter hours; the third is the checking dial for the hour strike of 1 to 6. The eyes of the lion are connected with the going train. When the hour strikes, the lion moves its jaw. The base is stamped with EBEN and the Augsburg pine cone. Signature on the backplate: "Philipp Trump in Crailsheim." Trump ordered ready-made parts, the lion, and the ebony base from Augsburg and built his own clockworks into these. For a similar clock see No. 85.

Literature: Maurice I, Plt. V.

85 Figure clock – lion rampant

Marked MD. Augsburg, beginning
of the 17th century
Cologne, private collection

Movement: upper plate iron, lower plate brass;
brass wheels
Case: fire-gilt bronze
Dial: enamelled silver
Base: pearwood stained black
Height: 34 cm (13⅜ in.)

Going train with fusee and gut line. The eyes of the lion are connected with the going train; when the hour strikes, the lion moves its jaw. Backplate marked MD in a shield. The bronze figure of the lion has frequently been repeated among Augsburg and south German clockmakers: Maurice II, Figs. 318, 319–323; Sotheby, December 9, 1977, No. 258 (signed Pfaff); Sotheby, October 27, 1978, No. 65 (signed Benedict Muller).

Three Augsburg clockmakers come under consideration in resolving the initials: (1) Michael Debetshauser (born 1585, still in the records in 1637); (2) Matthäus Degen (born 1578/1579, died after 1619); (3) Michael Dirr the elder (born between 1571 and 1573, died 1631).

Literature: Uto Auktionen, Zurich, Oct. 17, 1977, p. 60, No. 151.

86 Figure clock – standing lion

Hans Buschmann. Augsburg, middle
of the 17th century
Munich, Bayerisches Nationalmuseum

Movement: brass plates and wheels
Case: fire-gilt bronze and brass; silver
Dials: enamelled silver
Base: ebony
Height: 60 cm (23⅜ in.)

Going, quarter-striking, hour-striking trains (*grande sonnerie* with count wheel);
alarm. All three movements have fusees. Indications: hours, age and phases of
moon, date, day of the week, repeat striking work for the hours. Signature on
the rear dial: "Hans Buschmann Augspurg." Separate drive mechanism in the
lion's body to control the movement of the eyes and the snapping of the jaw
every quarter hour. The hourglass on the shield is turned over every quarter
hour. The plinth with the dial plates can rotate on the base. The plinth is
stamped with EBEN and the Augsburg pine cone.

For Hans Buschmann see Catalog No. 57.

Literature: Maurice II, Fig. 315.

87 Figure clock—seated lion

South Germany, c. 1630–1640
Chicago, Adler Planetarium

Movement: no longer original
Case: fire-gilt bronze
Original base missing
Height: 22.9 cm (9 in.)

Of this figure clock only the bronze lion remains. There is a complete example in a private collection in France (cf. Maurice II, Fig. 317). Originally the going train controlled the eyes, the striking train the lion's jaw.

Literature: Max Engelmann, *Sammlung Mensing. Altwissenschaftliche Instrumente* (Amsterdam, 1924), No. 387.

88 Figure clock—standing lion

South Germany, c. 1630
Newark, Newark Museum

Movement: no longer original
Case: fire-gilt bronze and copper
Height: 36.5 cm (14⅜ in.)

Original movement replaced with a Parisian pendulum movement. The lion originally moved its eyes and jaw. Two similar lion figures are known, one in the Adler Planetarium, Chicago (Inv. No. M 386), and another, with different chasing, in a London private collection (*The Connoisseur.* 1961, *147:* 253).

Literature: "European Clocks, 1550–1830," *The Museum.* Newark, N.S., Winter 1968, No. 3. Maurice II, Fig. 329.

258

89 Figure clock – striding lion

South Germany, first quarter
of the 17th century
Stockholm, Kungl. Husgerådskammaren

Case: fire-gilt bronze and copper
Enamelled dial not original
Height: 46 cm (18⅛ in.)

Going train which controls the movement of the eyes. Hour-striking train,
connected to the lion's jaw and to the head and arm of its keeper. This clock
was acquired in 1654 by Friedrich III, Duke of Schleswig-Holstein-Gottorf
(1616–1659).

Literature: Sten Lundberg, *Ur och Urverk* (Stockholm, 1951), pp. 64, 65. Ernst Schlee,
*Gemessene Zeit. Uhren in der Kulturgeschichte Schleswig-Holsteins. Ausstellung Schloss Gottorf,
1975.* p. 87. Maurice II, Fig. 316.

90 Figure clock—standing lion

Philipp Miller the elder or the younger
Augsburg, c. 1580–1590
United States, private collection

Movement: iron
Case: fire-gilt bronze and copper
Height: 34 cm (13⅜ in.)

Going and hour-striking trains, both with fusee and gut line; alarm. The eyes of the lion are connected with the going train, and on striking the hour the lion moves its jaw and paw. The two small Apostle figures serve as keys. Below the dial the arms of the first owner, the Abbott Gangenrieder of the Tierhaupten Monastery. Beneath this the master's mark FM, in ligature in a shield. The initials can be interpreted as being those of the Augsburg clockmaker Philipp Miller or of his son bearing the same name.

Philipp Miller the elder, clockmaker, Protestant, acquired smiths' eligibility presumably about 1550 from his father, an Augsburg locksmith of the same name. In 1550 he married; from 1560 onward he was often named as guardian and frequently as witness at marriages. He was four times inspection master between 1569 and 1581. He owned a house in the Schmiedgasse. He died between October 7, 1590, and September 28, 1599.

Philipp Miller the younger, large- and small-clock maker, Protestant, was born in Augsburg around 1550–1551. He received smiths' eligibility from his father in 1587 and became an independent master through this process. On January 11, 1587, he married Maria Beck. He is named in the Augsburg muster rolls of 1610 and 1615 (age sixty and sixty-four) but is no longer in the roll of 1619.

Sources: STAA, HAP, Apr. 24, 1569, fol. 269r. STAA, Suppl. Sept. 28, 1599. STAA, Musterregister of 1610, 1615, 1619. STAA, Pflegschaftsbücher, and HAP, from 1560. STAA, SI–SIII, 1569, 1570, 1571, 1581.
Literature: Clare Vincent, *Northern European Clocks in New York Collections.* Metropolitan Museum of Art, Jan. 4–Mar. 28, 1972, p. 18, No. 25. Maurice II, Fig. 301.

91 Figure clock—elephant

Nikolaus Schmidt. Augsburg, c. 1580

Munich, private collection

Movement: iron, fusee of brass
Case: fire-gilt bronze and copper
Dial: enamelled silver
Height: 43 cm (17 in.)

Going and hour-striking trains both within the palanquin. Opposite the dial is the checking dial for the striking. The going train is upside down so that the balance wheel is in a position to move the eyes of the elephant directly (at present the clock has a pendulum regulator). The elephant is embossed; the limbs are soldered on. The saddle-cloth, the sides of the base, and the animal's harness are decorated with the letter T and stars. A T thrice repeated was the arms of the Bologna families of the Tartagnia and the Terzi. Yet neither explanation is convincing, because of additional figures in the arms. Below the checking dial, the master's mark NS in a shield.

For Nikolaus Schmidt see Catalog No. 19.

Literature: *Catalogue des objects d'art, Collection Spitzer,* Paris, 1883, No. 2695. Ernst von Bassermann-Jordan, *Die Kunstsammlungen Wilhelm von Miller* (Munich, 1906), No. 239. Maurice II, Fig. 290.

92 Figure clock—elephant

Augsburg, c. 1600
Munich, private collection

Case: fire-gilt bronze and copper
Dial and arms: silver
Height: 39.5 cm (15½ in.)

The original mechanism is no longer present and has been replaced with a Parisian pendulum movement. The base has been reconstructed. On both sides of the caparison are the arms of a member of the Tanari family of Bologna. This figure of an elephant has been repeated on a number of occasions; see Maurice II, Figs. 286, 287; Sotheby, October 27, 1978, No. 66 (with the mark of David Haisermann, Augsburg), and *Antiquarian Horology.* 1978, *10:* 827.

Literature: Maurice II, Fig. 292.

93 Figure clock—elephant

Augsburg, beginning of the 17th century
Silver figure by the goldsmith
Georg Jungmair
Munich, Bayerisches Nationalmuseum

Movement: brass plates and wheels
Case: fire-gilt silver and brass
Base restored
Height: 31 cm without base (12¼ in.)

Going and hour-striking trains in the elephant's howdah. Going train is upside down; it moves the elephant's eyes, rolling them from side to side; when the hour strikes, the mahout pulls on the chain. Escapement no longer original. The goldsmith's mark was read for a long time as being that of Christoph Lencker (Rosenberg[3], 412 i); but research carried out by Helmut Seling makes it apparent that only the goldsmith Georg Jungmair could have made the silver figure. Jungmair was born about 1563, became a master before 1624, and died in 1634.

Literature: Maurice I, Plt. IV. Helmut Seling, *Die Kunst der Augsburger Goldschmiede. 1520–1868. Geschichte. Werke. Marken* (Munich, in press), 1177a.

94 Silver figure of a dog

Elias Zorer or Zacharias Flicker
Augsburg, c. 1600
Munich, private collection

Silver
Height: 8 cm (3⅛ in.)

This figure, with a removable head, was intended to be used as a drinking vessel. The same figure was also cast in bronze: see Catalog No. 95. Similar silver and bronze dogs were used to make up figure clocks (examples of the silver dog are in Aspreys, London, *The Connoisseur,* 1977, 196, advertisement in December issue; the bronze dog, see Catalog No. 96). Master's mark (Rosenberg[3] 420) perhaps Elias Zorer (married 1585, died 1625) or Zacharias Flicker (married 1580), and Augsburg assay mark (Rosenberg[3] 131).

Literature: Maurice II, Fig. 330.

95 Bronze figure of a dog

Augsburg, c. 1600
Hamburg, private collection

Bronze
Height: 7.5 cm (3 in.)

See the preceding and following catalog numbers.

Literature: Maurice II, No. 331.

96 Figure clock–dog

Augsburg, c. 1625–1630
Wuppertal, Historisches Uhrenmuseum

Movement: brass plates and wheels
Case: fire-gilt bronze and brass
Base: ebony
Height: 21 cm (8¼ in.)

Going and hour-striking trains. The former moves the eyes of the dog, the latter its mouth. Five similar pieces are known: see Maurice II, Figs. 332,334, 335; F. J. Britten, *Old Clocks and Watches and Their Makers* (London, 1922), Fig. 112; Adler Planetarium, Chicago, Inv. No. M 395 B. This type of base was used for numerous Augsburg figure clocks; for the most part these bases are marked with the ebony stamp. Since this stamp was introduced for the first time in 1625 (see p. 124), this clock should be dated later than the individual dog figures in Catalog Nos. 94 and 95.

Literature: Maurice II, Fig. 333.

97 Figure clock—turtle

Georg Fronmiller (?). Augsburg, c. 1610
Darmstadt, Hessisches Landesmuseum

Movement: iron
Case: fire-gilt copper
Wooden polychrome figure
Height: 15 cm (6 in.)

Going, hour-striking, and automaton trains. The automaton train controls the crawling motion of the animal, the repeated forward thrust of its head, and the motion of the rider.

In 1611 Archduke Ferdinand paid Georg Fronmiller 25 gulden for "a turtle with clockwork that we have purchased" (*Jahrbuch der Kunsthistorischen Sammlungen in Wien*, 1898, *19*, Reg. 17082). Probably the specimen named there is identical with the automatically functioning turtle in the Kunsthistorisches Museum in Vienna (*Katalog der Sammlung für Plastik und Kunstgewerbe*, II, Vienna, 1966, No. 378).

Georg Fronmiller, small-clock maker, Protestant, was born in 1565, a son of the Augsburg clockmaker and burgomaster Hans Fronmiller and his wife Katharine née Laminit. He was the younger brother of David Fronmiller (see Catalog No. 50). Georg Fronmiller became an independent master in 1580, at which time he received his father's smiths' eligibility. In 1593 he married Sibylla, daughter of the city attorney Dr. Hieronymous Fröschel. They had five children, and his son Georg the younger also became a clockmaker. From 1598 he lived for long periods in Italy and France and was active at the Elector's Court at Cologne and at the Prague Court. During these years, however, he made uninterrupted tax payments at Augsburg and acquired a house in the Kautzengässchen. In 1616–1617 he was imprisoned for a considerable time for manslaughter in Augsburg, and he was subsequently exiled from the city for ten years. Emperor Matthias and Emperor Ferdinand II made numerous written requests for leniency on his behalf. From 1608 onward he was imperial clockmaker and spent the time of his exile in Vienna. He died after October 21, 1621.

Sources: HKA, HZAB No. 63 (1613/1614), No. 1385, fol. 721v.
Literature: Friedrich Streng, "Augsburger Meister der Schmiedgasse um 1600," *Blätter des Bayerischen Landesvereins für Familienkunde.* 1963, *26:* 247–287. Maurice II, Fig. 312.

98 Figure clock – Bacchus

Augsburg, beginning of the 17th century
United States, private collection

Movement: iron
Case: fire-gilt bronze and brass;
traces of cold enamel
Base: wood
Height: 35.4 cm (14 in.)

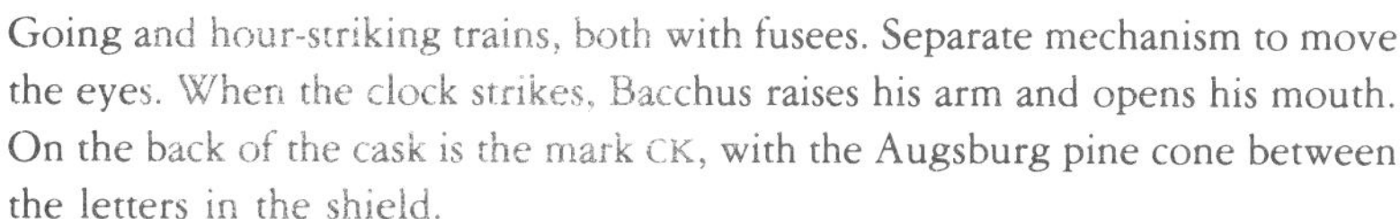

Going and hour-striking trains, both with fusees. Separate mechanism to move the eyes. When the clock strikes, Bacchus raises his arm and opens his mouth. On the back of the cask is the mark CK, with the Augsburg pine cone between the letters in the shield.

A member of the Kreitzer family may be considered for a reading of this mark. A precise ascription is difficult, since there is an abundance of variants for the writing of the name, and two masters of the same name, presumably unrelated, were active at the same time.

Conrad (2a) Kreytzerer, bachelor, of Augsburg, married Catharina Baltas on October 26, 1586, and acquired the smiths' eligibility of his deceased father Conrad (1a). It was probably of this marriage that a son Conrad (3a) was born, who is listed in the muster roll for 1615 as being twenty-eight years old (therefore born in 1587, one year after the marriage). In 1614 he acquired his father's smiths' eligibility. On May 26, 1620, his mother married Michael Debetshauser; thus Conrad (2a) must have died previously. Hans Christoph Kreutzerer was probably a younger son of 2a; he too acquired smiths' eligibility in 1622 from his deceased father Conrad, and in the same year he married Catharina Haym; thereafter on November 27, 1651, Susanne Hilsenbeck, widow of Johann Jakob Rehlin. It was evidently the second Conrad Kreyzerer of generation 2a who married Barbara Stegler on January 21, 1614, and Catharina Spix, widow of Sigmund Eckhart, on July 20, 1621. All the bearers of this name mentioned above were Protestant. In 1577 one Hans Kreutzerer acquired smiths' eligibility; in the 1619 muster roll his age was given as seventy, in 1615 as sixty, and he died on May 8, 1624. In the 1615 muster roll a Joseph Kreitzerer thirty years old is mentioned. One Conrad Kreitzer (Coenraet Kreyser) was buried on November 14, 1658, at the Anthonies Cemetery in Amsterdam, listed as a clockmaker from Augsburg (Enrico Morpurgo, *Nederlandse klokken- en horlogemakers vanaf 1300*, Amsterdam, 1970, p. 74).

Literature: E. J. Britten, *Old Clocks and Watches and Their Makers* (London, 5th ed., 1922), Fig. 113. Maurice II, Fig. 351.

99 Figure clock – Bacchus

South Germany, c. 1580–1590
United States, private collection

Movement: iron
Case: fire-gilt bronze and copper
Painted garments
Height: 32 cm (12½ in.)

Going, quarter-striking, and hour-striking trains. The going train controls the motion of the eyes, the striking train the action of drinking. There is a similar clock in a German private collection (Maurice II, Fig. 349), and another—with more ample automatic movements of the figure and with a larger base—in the Kunsthistorisches Museum, Vienna (Maurice II, Fig. 350).

Literature: Maurice II, Fig. 348.

100 Bear with clockwork

Saxony, second quarter of the 17th century
Dresden, DDR, Staatlicher Mathematisch-
Physikalischer Salon

Movement: iron
Case: mounted bearskin
Height: 97 cm (3 ft 2¼ in.)

The bear has within its body going and hour-striking trains, and alarm. The bear's eyes are connected with the going train. When the hour strikes, the bear snaps with its mouth, and when the alarm is released, it beats on a drum.

The bear was a gift of Duke Julius Heinrich of Saxony to the Elector Johann Georg I; it entered the Kunstkammer in 1655. Stuffed animals which were given a lifelike appearance by means of clockwork were common, as were mechanically functioning wax figures. But because of the impermanence of the skins used, almost all have disappeared; this bear is perhaps the only surviving specimen.

Literature: Maurice II, Fig. 346.

271

101 Figure clock—bear and trainer

South Germany, c. 1580–1590
United States, private collection

Movement: iron
Case: fire-gilt bronze and copper
Height: 33 cm (13 in.)

Going and hour-striking trains, dial on the upper surface of the base. When the clock strikes, the bear and the trainer turn their heads, and the trainer pulls on the chain. A comparable clock was made in 1582 by the Augsburg clockmaker Asmus Pierenbrunner (see Maurice II, Fig. 291). The pomegranate-like feet of the base are also to be found in a number of Augsburg clocks.

Literature: Maurice II, Fig. 296.

◁ 102 Automaton—Diana on a stag

Augsburg, c. 1620
Goldsmith work by Joachim Fries
New York, Metropolitan Museum of Art

Movement: iron, fusee of wood
Case: silver, partly fire-gilt
Height: 37.5 cm (14¾ in.)

When the head of the stag is removed (the joint is concealed by a collar), the body serves as a drinking vessel and was used in drinking games. In the base there is a mechanism which upon release causes the assembly to roll on a table for a certain distance and then stop by itself; the one before whom the automaton stops must then drain the vessel. On the neck of the stag there is a mark of the goldsmith Joachim Fries (Seling 1284e) and the Augsburg assay mark (Seling 44). Twenty similar automata are known: eight are marked by the Augsburg goldsmith Joachim Fries (d. 1620), seven by the Augsburg goldsmith Matthias Walbaum (d. 1631 or 1632), and five by the Augsburg goldsmith Jakob Miller the elder (d. 1618).

Because of the great number of these drinking vessels that are so similar, it has long been thought that they were prizes in a competition that was held on the occasion of the coronation of the Emperor Matthias in Frankfurt. In order for this number of objects to have been made quickly, the commission would have had to be put in the hands of the three Augsburg masters who would then have traded individual models back and forth among each other. But as a result of the researches of Helmut Seling it has turned out that there are differing assay marks, so that not all of the animals can have derived from the same occasion (if indeed there was any such occasion at all).

Literature: Karl A. Dietschky, "Der Werthemannsche Hirsch und seine Verwandten," *Historisches Museum Basel. Jahresbericht 1967.* pp. 29 ff. and supplement in *Jahresbericht 1968.* p. 30 (with reference there to earlier literature). Regina Löwe, *Die Augsburger Goldschmiedewerkstatt des Matthias Walbaum* (Munich, 1975), pp. 83 ff. (for the specimens marked by Walbaum). Seling 1284e.

103 Drinking vessel–Diana on a Stag

Augsburg, c. 1620
Goldsmith work by Matthias Walbaum (?)
Munich, Schatzkammer der Residenz

Case: silver, partially fire-gilt; coral
Height: 32.5 cm (12¾ in.)

As in the case of No. 102, this group too can be used as a drinking vessel. But there is no mechanism in the base, presumably as specified—along with the use of coral as antlers—by the person placing the order. The silver group is not marked, but the whole design and the decoration of the base point toward the Augsburg goldsmith Matthias Walbaum (see Catalog No. 102).

Literature: Regina Löwe, *Die Augsburger Goldschmiedewerkstatt des Matthias Walbaum* (Munich, 1975), p. 84 (earlier literature cited here).

104 Automaton–
Saint George as dragon-slayer

Augsburg, c. 1613–1615
Goldsmith work by Jakob Miller the elder
Darmstadt, Hessisches Landesmuseum

Movement: iron; fusee of wood
Case: silver, partially fire-gilt
Height: 40 cm (15¾ in.)

Like the preceding Diana automata, this automaton falls in the category of table diversions; it has a drive train as most of the others do. A similar drinking vessel is found in the Grünes Gewölbe at Dresden, but it is marked by the Augsburg goldsmith Joachim Fries. It was included in the Saxon Kunstkammer as early as the beginning of 1612 (Jean Louis Sponsel, *Das Grüne Gewölbe zu Dresden,* Leipzig, 1928, Vol. II, p. 210). The Darmstadt example bears the master's mark of the Augsburg goldsmith Jakob Miller the elder (Seling 976c) and the Augsburg assay mark for 1613–1615 (Seling 36).

Jakob Miller became a master before 1583; he died in 1618.

Literature: Bayern, Kunst und Kultur Ausstellung (Munich, 1972), Cat. No. 972 (earlier literature cited here). Seling 976c.

105 Figure clock—centaur

Augsburg, c. 1600–1610
Goldsmith work by Melchior Mair
Vienna, Kunsthistorisches Museum

Movement: iron
Case: silver, partially fire-gilt; enamel, gems
Base: stained black
Height: 39.5 cm (15½ in.)

Going and hour-striking trains in the belly of the centaur; separate drive train
in the base. Upon release of this train the group rolls along a table, the centaur
moves its eyes and shoots an arrow. Diana and her two hounds turn their heads
and one of the hounds opens its mouth.

Two similar groups are known: in the Grünes Gewölbe at Dresden, by the
same master and acquired before 1610; the second was given by Duke Maximilian of Bavaria to the Jesuits before 1616 for their mission to China (see Ch.
4). The deerhound is repeated in the groups of Diana on a stag shown in
Catalog Nos. 102 and 103.

The goldsmith Melchior Mair became a master in 1598 and died in 1613.

*Literature: Kunsthistorisches Museum Wien, Katalog der Sammlung für Plastik und Kunst-
gewerbe. II (Vienna, 1966), No. 368. Maurice II, Fig. 311. Seling 1131b.*

106 Table carriage with clock

South Germany, beginning of
the 17th century
Moscow, Kremlin Museum

Movement: numerous restorations
Case: fire-gilt bronze and copper
Height: 50 cm (19⅝ in.)

Composite of individual parts from various carriages, built up high behind and
with a seated Bacchus, the whole drawn by an elephant which bears the clock.
On top a fortress-like structure with cupola roof and double eagle. A similar
table carriage varying in detail was on loan at the Bowers Museum in Durham
(Maurice II, Fig. 279).

Literature: Opis Moskovskoi Oruzheinoi Palaty (Moscow, 1884–1893), Plt. 114. Ernst
Zinner, "Clocks in Russian Museums," *Antiquarian Horology.* 1959, 2(No. 11): 207.
Maurice II, Fig. 278.

107 Table carriage with Diana

South Germany, beginning of
the 17th century
Milan, Museo Poldi Pezzoli

Movement: brass plates and wheels
Case: fire-gilt bronze and copper; ebony
Dial and mountings: silver
Height: 30.5 cm (12 in.)

Going and hour-striking trains. Drive train for forward motion of carriage is
in the ebony box. Diana's eyes are connected with the going train; as the
carriage rolls, the panthers leap and turn their heads. Opposite the hour dial
there is a setting dial for the striking. Originally there was a volute-like stem
at the front of the carriage upon which Diana set her foot.

Three similar pieces are known: Statens Historiske Museum, Stockholm;
Yale University Art Gallery, New Haven (Maurice II, Fig. 280); and Maurice
II, Fig. 281 (location unknown).

Literature: Giuseppe Brusa, *Gli orologi, Cataloghi del Museo Poldi Pezzoli,* I (Milan, 1974),
p. 31. Maurice II, Fig. 282. Giuseppe Brusa, *L'arte dell'orologeria in Europa* (Milan, 1978),
Figs. 207–209.

108 Table carriage with Minerva

Augsburg, c. 1630
Vienna, Kunsthistorisches Museum

Movement: brass plates and wheels;
fusee of wood
Case: fire-gilt bronze and brass; ebony
Dials: silver
Height: 54 cm (21¼ in.)

Going and hour-striking trains inside Minerva's throne-like seat. Two trains are in ebony boxes, one for forward motion of the carriage, the other to operate the mechanical organ. The eyes of Minerva are connected to the going train. As the carriage rolls, the horses leap and the satyrs and the apes on the back of the carriage turn. The organ has twelve wooden and twelve tin pipes, and both rows of pipes can be coupled with a link. There are two melodies set up on the cylinder. In order for the melodies to play in sequence, the keyboard, not the barrel, is shifted. The case is stamped with EBEN and the Augsburg pine cone.

A slightly different version of this table carriage was once in the Staatlicher Mathematisch-Physikalischer Salon at Dresden (destroyed in 1945). Albert Protz, who studied the Dresden specimen (*Mechanische Musikinstrumente,* Kassel, 1939, pp. 49–53), determined that the organ mechanism came from the same workshop as the automatic organ works of the Pomeranian art cabinet (1617). Since Achilles Langenbucher probably built that organ, he may also have been the master involved with the organ in the Vienna table carriage. But since Langenbucher also worked as a goldsmith and a wax embosser, even the figures here might also come from his hand. He received burgher rights in Augsburg in 1611 and died there in 1650.

The development of automatic musical mechanisms in Augsburg was set forth in a court case involving the brothers Bidermann, which is discussed in Chapter 12.

Literature: Kunsthistorisches Museum Wien, Katalog der Sammlung für Plastik und Kunstgewerbe, II (Vienna, 1966), No. 368. Maurice II, Fig. 283.

109 Figure clock—Cupid on a carriage

Andreas Stahel (?). Augsburg, c. 1600
Karlsruhe, Badisches Landesmuseum

Movement: iron, wood
Case: partially fire-gilt bronze, brass, and silver;
lapis lazuli, ebony, glass
Height: 30 cm (11¾ in.)

Going and hour-striking trains. The going train operates the eyes of the Cupid when one pushes upon a lizard mounted on the platform of the carriage. Separate drive train in the carriage; when it is released, the carriage rolls, the goats buck and turn their necks, the hound jumps out of the carriage, and the Cupid shoots an arrow. The case is marked AS with a small cross within a shield and with the Augsburg pine cone. The initials may be read as those of the Augsburg clockmaker Andreas Stahel.

Andreas Stahel, small-clock maker, probably Protestant, was born in Augsburg in 1560/1561, a son of the clockmaker Bernhard Stahel. In 1589 he secured smiths' eligibility from his deceased father and in the same year he married Judith, daughter of the clockmaker David Haisermann. He lived in his parents' house on the Mauerberg; it was transferred to him, for payment, by his widowed mother. In 1605 he took in an apprentice, Georg Christoph Lutzenberger, who had been born illegitimate, in exchange for Lutzenberger's promise to later marry Andreas' daughter. But Lutzenberger subsequently refused—something which he himself made public. In the muster lists of the city of Augsburg Andreas Stahel is recorded in 1615 as being a fifty-four-year-old clockmaker with two journeymen (from Augsburg and Breslau). Stahel died between October 4, 1634, and March 22, 1635.

Sources: STAA, Musterregister 1610, 1615, 1619. STAA, SZB, fols. 55a, 89b, 123b. STAA, Pflegschaftsbuch 1565–1569. STAA, HAP, 1589, fol. 154a. STAA, Consens, Sept. 16, 1589, March 22, 1635. STAA, U I, fol. 470; STAA, Schuldbrief, Oct. 4, 1634. *Literature: Badisches Landesmuseum. Neuerwerbungen 1952–1965* (Karlsruhe, 1966), p. 115. Maurice II, Fig. 284.

110 Table clock with automata

Jakob Marckstain (?). Augsburg, c. 1580–1590
Newark, Newark Museum

Movement: brass plates and wheels
Case: fire-gilt bronze and copper
Height: 17 cm (6⅝ in.)

Going and hour-striking trains; alarm. Going train has fusee and gut line. The striking train drives the carousel composed of marching and riding Turks. On the cover plate of the movement are the Augsburg pine cone and two differentiated marks of a master. The initials IM vary, as do the shapes of the shields. For a reading of the initials probably only the small-clock maker Jakob Marckstain is suitable.

Jakob Marckstain, Protestant, was born at Heitersheim in Breisgau. He acquired Augsburg smiths' eligibility through his marriage to Judith, daughter of the Munich clockmaker Asmus (or Erasmus) Pierenbrunner. Pierenbrunner emigrated to Augsburg by 1544 at the latest and was active there until his death in 1574. Marckstain's marriage witnesses were the clockmakers Hans Runggel and Asmus Pierenbrunner the younger, his brother-in-law. In 1586 Marckstain married Ann Reichart of Oberhausen. In 1591 he was nominated for the office of sworn inspection master but was not elected. In 1610 he is mentioned for the last time in the Augsburg muster lists as being a clockmaker sixty-six years old.

Sources: STAA, SZB, fols. 108a, 176a. STAA, Pflegschaftsbuch, 1572–1576, fol. 271 (1575). STAA, HAP, 1576, fol. 30v, 1586, fol. 35v. STAA, S IV, S V.
Literature: Maximilian Bobinger, *Kunstuhrmacher in Alt-Augsburg* (Augsburg, 1969), p. 21. "European Clocks 1550–1830," *The Museum,* Newark, N.S. Winter 1968, No. 11. Maurice II, Fig. 614.

111 Automaton with trumpeters

Hans Schlottheim. Augsburg, 1582
Vienna, Kunsthistorisches Museum

Movement: iron
Case: ebony
Figures: cold-enamelled silver; fire-gilt brass
Height: 33.4 cm (13⅛ in.)

There is an organ with ten pipes in the ebony base. When it has been set going, the eleven little figures are also put into motion. The musicians lift and lower their trumpets; the drummer drums. The drum tones are produced through the striking of two clappers upon a membrane stretched across the underside of the automaton. On the front there are two enamelled round shields on rectangular fields, one with the arms of Duke Wilhelm V of Bavaria, the other with his initials W.H.I.B. and the date 1582.

The organ is like that in Schlottheim's crèche automaton (Fig. 9), which resembles the architecture of the trumpet automaton. This crèche automaton was given, at the latest in 1588, by Electress Sophie to her husband Christian I of Saxony. In 1589 Schlottheim delivered another "trumpet movement" to the sister of Wilhelm V of Bavaria. The similarity of the present drive, in terms

286

of singularity and complexity, to the two documented organ mechanisms can only point, within Augsburg, to Schlottheim.

Hans Schlottheim, large- and small-clock maker, Protestant, was born between 1544 and 1547 at Naumburg an der Saale, the son of a Naumburg master of the same name and his wife Anna. He lived in Augsburg presumably from 1567 onward, at the latest from 1573. In 1573–1574 he was journeyman to Jeremias Metzger for four months. In 1573, still before his master's examination, he married, which guild regulations did not permit. His wife Ursula was the widow of the master locksmith Hans Schitterer and the daughter of locksmith Thomas Geiger. She brought to the marriage the craft eligibility and a locksmith's workshop on the Schmiedberg. Through the marriage Schlottheim became incidentally related to the Fronmiller family of clockmakers. In 1576 Schlottheim became a master. In 1577 a journeyman goldsmith was allowed to work in his shop temporarily, although Schlottheim was quite able to make clock cases himself, as is indicated by the objections he aroused. From his first marriage there was a son of the same name, born in 1575, who acquired smiths' eligibility in 1591 and who is recorded at Augsburg up to 1603. After that he moved to the Naumburg area.

The elder Schlottheim married for the second time in 1606, Euphrosina Osswald. In 1579 he acquired the neighboring house of the gunsmith Hans Sumer in the Schmiedgasse and thus became a neighbor of Nikolaus Schmidt the younger, his relative by marriage. In 1586 he was the foreman of his guild. In 1589 and 1593 he spent a considerable period at Dresden, in 1587 and 1601 at Prague. Around the turn of the century Schlottheim experienced financial difficulties, since the imperial court lagged behind in paying his considerable bills. The city of Augsburg supported his demands for payment in 1599 and again in 1615. He died in 1625–1626.

Sources: STAA, HAP 1573 (Dec. 20). STAA, S III, fol. 640 (July 14, 1583). STAA, SZB, fols. 108b., 128b. STAA, Steuerbücher, 1600–1603. EMA, St. Anna, 1. Hochzeitsbuch, p. 48, No. 48. STAA, S III, fols. 175 ff. HKA, Reichsakten, 192/I, fol. 54. HKA, Gedenkbücher No. 146, fol. 144r.
Literature: Julius Schlosser, *Die Sammlung alter Musikinstrumente* (Vienna, 1920), p. 76, A. 135. Friedrich Streng, "Augsburger Meister der Schmiedgasse um 1600," *Blätter des Bayerischen Landesvereins für Familienkunde,* 1963, *26* (No. 1):247–287. Maximilian Bobinger, "Der Augsburger Uhrmacher Hans Schlottheim," *Schriften der Freunde alter Uhren,* 1971–1972, *11:* 8ff. Maurice II, Fig. 388.

112 Figure clock–procession of Bacchus

Nuremberg (?), c. 1580
Augsburg, Städtische Kunstsammlungen

Movement: iron
Case: fire-gilt bronze, brass, and copper; cold enamel
Height: 70 cm (2 ft 3½ in.)

Going, quarter-striking, and hour-striking trains. A separate drive train in the base for the procession is set going either by the hour-striking train or manually. The carousel revolves and individual figures go into motion. Simultaneously with the satyr's drumming, two hammers beat upon a drumhead that is stretched across the substructure. Indications: date, saints for the day, months, signs of the zodiac, chart of the hours, hours of the day and night in five concentric circles, adjustable chapter ring from 1 to 24, segments for the length of daylight and darkness.

Attribution to Nuremberg is based on the prominence of architectural elements and a fine articulation of ornamentation, both aspects typical of Nuremberg clocks (see Catalog No. 25). In 1628 Philipp Hainhofer saw this clock, or one similar, at Dresden.

Literature: Bruno Bushart, *Kostbarkeiten aus den Kunstsammlungen der Stadt Augsburg* (Augsburg, 1967), p. 71. Alma Helfrich-Dörner, "Grosse Prunkuhr mit Figurenwerk aus dem Maximilianmuseum in Augsburg," *Die Uhr,* 1968, *17:* 58–64. Maurice II, Fig. 383.

V The Clockwork Universe: Mechanical Models
of the Heavens

"The universe behaves no differently than a clockwork," wrote Christian Wolff in 1719; and this sentence became an axiom in his later *Cosmologia generalis.* His student Johann Christoph Gottsched popularized this principle in his book *Erste Gründe der gesamten Weltweisheit* (1733):

Insofar as the universe is a machine, it has to that extent a resemblance to a clock; and it is in a clock that we can on a small scale make plainer to one's understanding that which takes place in the universe on a large scale. The wheels of the clock represent the parts of the universe; the motions of the hands, the events and changes taking place in the universe. Just as in the clock all positions of the wheels and the hands ensue from the inward arrangement, shape, size, and linkage of all its parts in accordance with the rules of motion, so everything that takes place in the universe also produces its effect.

Furthermore, just as all future shifts of the wheels, and all positions of the hands, have their certainty and truth in a properly functioning clock, even before they take place (inasmuch as everything is determined by past and present changes, so that it is possible to tell in advance all changes that may take place as of any given moment), just so do all events in the universe have their defined truth and certainty before they take place; and he who has perfect insight into its structure can see every future thing from its past and its present state of arrangement.

Yet just as in a clock, nevertheless, there is no unconditional necessity that it shall alter, and on that account something in its parts or its motions can easily be altered through some external cause, as for example in case the clockmaker wishes to set it otherwise: so there is likewise no geometrical inevitability in the events taking place in the universe, but it is instead possible that a force that exists outside it, such as the almighty divine power, as shall be demonstrated at the proper place, shall bring about a change within it that could never have taken place through its own institution.

Further, just as in a clock something can be altered by an outward cause and, as time goes on, all future positions of the hand, or all its strikings, will take place otherwise than would have been the case, with the result that, for example, it will go faster or slower than hitherto, before the alterations in it had taken place: in just the same way must the universe, too, upon something's being altered in it by the almighty divine power, be constituted differently for all future time than the way it would otherwise have been—be the traces of such action ever so small. And finally, as in a clock the change in this way brought about through some outward cause may be cancelled in the event that the hand of such a clockmaker should be hap set everything the way it was before, so that no trace of the extraordinary movement that had occurred should remain within it: just so can one easily conceive of quite the same thing in the universe, if it should please the almighty divine power utterly to cancel the consequences of its direct intervention therein, and to set everything once more in its original state.

This passage puts in simple terms the philosophy of the Enlightenment. This is the way that the century which follows our exhibit saw the universe, explained by analogy to a clock. How did that era—which witnessed the greatest revolution in intellectual history since the days of Scholasticism—arrive at this deterministic, mechanistic philosophy? The image of the clock provided the various elements of the new philosophy with a defined form and structure. Rationality in construction and harmony in its predetermined operation were the prevailing impressions of the universe, and ones that came directly from the ubiquitous clock—mundane measurer of time on church steeples and municipal buildings, monumentally conceived in cathedrals and cloisters.

Precisely calculated clock mechanisms were even used for astronomical observations, when mechanical models of the heavens simulated the motions of the planets against the celestial backdrop of the fixed stars (see Catalog Nos. 113–120). By their precision they made it possible to project the movement of constellations at successive intervals in the past and in the future. They were analog computers for the purpose of deriving time from stellar positions, or conversely for determining stellar positions from chronological data. Familiarity with these models and the biblical certainty of a universe ordered as to mass, number, and weight (Wisdom of Solomon) rendered inevitable the chain of argument leading to the formula of the clockwork universe which Wolff coined at that time.

113 Celestial globe with clockwork

Gerhard Emmoser, Vienna, 1579
New York, Metropolitan Museum of Art

Movement: iron plates, wheels partially brass
Case: silver, partially fire-gilt
Height: 27.3 cm (10¾ in.)

Going movement with fusee and gut line. The mechanism in the globe drives the calendar in the horizontal ring via a telescoping rod (see figure at left). A small image of the sun moves along the ecliptic; speeding or slowing its movement at the perihelion or aphelion is achieved by means of an irregularly toothed wheel. (Subsequent alterations of design are dealt with in a forthcoming publication by Clare Vincent and Bruce Chandler.) Neither the identity of the outstanding goldsmith who modelled the group nor the iconological meaning of the combination of Pegasus with the celestial globe is known. (The *Penitential Psalms* of Orlando di Lasso contains an illustration of a horse with a terrestrial globe. See Lieselotte Schütz, *Hans Mielichs Illustrationen zu den Busspsalmen Orlando di Lassos,* Munich, 1966, pp. 92 ff.).

The globe was at one time in the collections of Emperor Rudolf II and of Queen Christina of Sweden. Signed on the meridian ring: GERHARD EMMOSER SAC(RAE) CAES(AREAE) M(AI)E(STAT)IS HOROLOGARIUS F(ECIT) VIENNAE A(NNO) 1579.

Gerhard Emmoser, small-clock maker, Catholic, was born at Rain on the Lech probably before 1540, and before 1559 he was sent by the Count Palatinate Ottheinrich from Heidelberg to Philipp Imser at Weil as a collaborator in working on the latter's planetary clock (see Ch. 9). In 1563 Emmoser became a burgher of Augsburg. In 1566 he received a firm appointment to the imperial court at Vienna, and he died on November 31, 1584.

Literature: Christina Queen of Sweden, Exhibition, Stockholm 1966, Cat. No. 1225. Maurice II, Fig. 251. Clare Vincent, "Renaissance Timepieces," *The Triumph of Humanism,* Exhibition, San Francisco, 1977–1978, No. 144.

114 Armillary sphere with clockwork

Josias Habrecht. Strasbourg, 1572
Copenhagen, Danske
Kongers Kronologiske Samling
Rosenborg Castle

Movement: iron
Case: fire-gilt brass
Height: 31 cm (12¼ in.)

Going and hour-striking trains, neither with force equalization. The four sides of the case show the following indications: (1) hours, (2) quarter hours in doubled count, (3) latitude from 51° to 55°, (4) days of the week with the reigning planets of the day and the setting dial for the striking. At the South Pole of the sphere is the motion work for the sun and moon hands; on its cover plate there is a notice: "If the movement has been at a halt, set it here by hand until the upper indications revolve as well" (So das Werck stil gestanden so wind das hie mit der Hand vm so gat das ober werck also mit vm), and the initials IH. In the armillary sphere an image of the sun is moved along the ecliptic. The sun and moon hands are mechanically driven, but the dragon hand must be set manually. On the dial of the sphere the age and phases of the moon are given in addition to the hours. On the meridian ring there is a compass and a sundial. Signature on the bottom of the case: IOSIAS HABRECHT 1572. Presumably acquired by Tycho Brahe and presented to King Christian IV of Denmark.

Literature: Bering Liisberg, *Urmagere og Ure i Danmark* (Copenhagen, 1928), p. 145. Théodore Ungerer, *Les Habrecht, une dynastie d'horlogers Strasbourgeois au XVIᵉ siècle* (Strasbourg, 1925), p. 10. Hans von Bertele, *Globes and Spheres* (Lausanne, 1961), p. 34. Maurice II, Fig. 250. Henry C. King, *Geared to the Stars* (Toronto, 1978), Fig. 5.14.

115 Celestial globe with clockwork

Eberhard Baldewein, Marburg 1575
London, private collection

Globe: silvered copper and bronze
Movement: iron
Height: 55.6 cm (21⅞ in.)

Going movement with fusee and (since 1694) chain, wound at the North Pole. The movement drives the hour hand, the globe, an image of the sun, and (via a telescoping arbor extending from the South Pole) the calendar on the horizon ring. At the zenith there is a small compass, and beneath it the bearing for the adjustable quadrant, which is used to determine where any given star lies relative to the horizon.

We know from documents published by von Drach that the Nuremberg goldsmith Wolff Meyer produced the stand with figures of the four seasons, that he silvered the globe, and that he painted on the images of the stars with mastic colors. The Giessen goldsmith Herman Diepel worked and engraved the sphere of the globe, and the clockwork was probably the result of a collaboration with Hans Buch (Bucher), a clockmaker of Augsburg.

Baldewein fixed the clockwork firmly to the globe so that both must rotate together. The hour hand and the sun index are linked with the globe via an epicyclic drive. The apparent irregularity of the movement of the sun—running faster in winter than in summer—was achieved by setting the teeth of the solar ring farther apart for the winter and closer together for the summer. The calendar ring has to be adjusted manually to correct for leap year. The globe can be disengaged from the clockwork and can thus be used as an instrument for demonstrations. A friction clutch permits independent setting of the globe, solar image, and hour hand, in order to synchronize the various functions of the instrument. Baldewein's design is more complicated than that of his successor Bürgi (see Catalog No. 116), but his astronomical cycles are not so precise. Whereas Bürgi indicated the positions of the stars as corrected by Wilhelm IV in the Hessian star catalog, the star positions on Baldewein's globes still followed the old tables, and Baldewein furnished his stars with their astrological characters.

On his travels Wilhelm IV always carried this globe with him in its leather case. It was probably he who determined right ascension and declination of the fixed stars and likewise those of the planets and comets if their predicted paths had been indicated on the globe. Eberhard Baldewein (c. 1525–1593) was originally a tailor, but from 1560 onward he was employed as a light-chamberlain (*Lichtkämmerer;* i.e., in charge of heating and lighting) at the court of the Landgrave Philipp the Magnanimous, Wilhelm's father. Subsequently he was architect to Wilhelm's brother Ludwig. From 1568 onward he made clocks and instruments for Wilhelm IV. The correspondence contains a number of references to his collaboration with the clockmaker Hans Buch.

Literature: C. Alhard von Drach, *Die zu Marburg im Mathematisch-Physikalischen Institut befindliche Globusuhr Wilhelms IV. von Hessen* (Marburg, 1894). J. H. Leopold and K. Pechstein, *Der kleine Himmelsglobus 1594 von Jost Bürgi* (Lucerne, 1977), pp. 20 ff. (with corrections to Drach and a summary of all authors subsequent to Drach). Ludolf von Mackensen (with contributions by Hans von Bertele and John H. Leopold), *Die erste Sternwarte Europas mit ihren Instrumenten und Uhren: 400 Jahre Jost Bürgi in Kassel* (Munich, 1979), pp. 15 ff.

116 Celestial globe with clockwork

Jost Bürgi, c. 1600
Kassel, Staatliche Kunstsammlungen,
Astronomisch-Physikalisches Kabinett

Globe: fire-gilt copper and brass
Movement: brass plates and wheels,
iron striking train
Height: 48 cm (18⅞ in.)

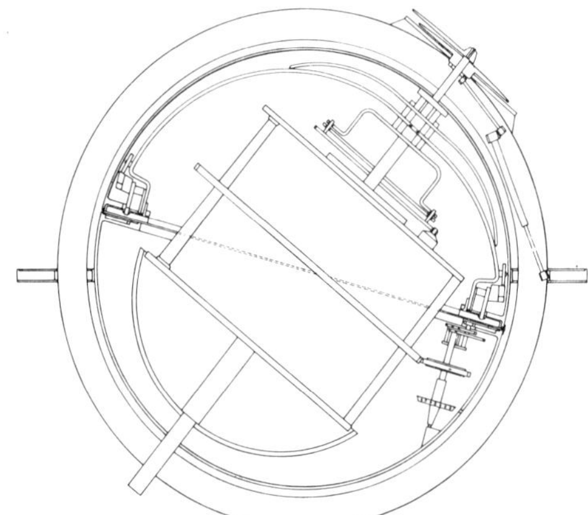

Going train with fusee and gut line; hour-striking train. A single winding square at the South Pole serves for winding both trains, one winding to the left, the other to the right. From this point one can also regulate the amplitude of the balance wheel. Dial with hour and minute hands at the North Pole. Small compass at the zenith, beneath it the altitude quadrant (see Catalog No. 115). Within the globe the fixed clockwork drives the motion work. Attached to the hour wheel is a sector which carries an image of the sun along the ecliptic inside the globe. The drive for the sector is via an epicyclic gear mounted in a frame on the inner wall of the globe. This gear meshes with a toothed ring which surrounds the middle plate and also with another toothed ring sitting on the sector below the image of the sun. The epicyclical gear receives its impulse from the turning of the sector and thus rolls along the toothed ring around the middle plate and controls the rotation of the globe. The gear is calculated so that the globe rotates 4 minutes faster per day than the solar image does; that is, the solar image moves along the ecliptic once a year. The apparent irregularity of the sun's movement—faster in winter than in summer—was achieved by Bürgi through a separate gear which gives the solar image an additional motion within the sector. The calendar on the horizon is likewise advanced by the hour wheel, via an arbor inside the meridian ring, which is constructed telescopically so that the globe may be elevated. The calendar ring in the horizon rotates once in 365¼ days. On the inner side of the horizon ring a small gear train lies concealed, which moves the calendar index ¼ day backward per year; thus the index constantly shows 365 days. But in a leap year the train advances the index by one day, thus correcting the date. A second ring, concentric with the calendar, bears the symbols for the days of the week and small tongue-like indicators. Set correctly at the beginning of the year, it always shows the day of the week appropriate for the date, and with the small indicators it shows the date of Easter and of the movable feasts that depend upon Easter. The stars, engraved by Anton Eisenhoit of Warburg (see Catalog No. 119), are positioned according to the Hessian catalog of Wilhelm IV; L. von Mackensen and J. H. Leopold have published detailed drawings of the globe.

Five globes are ascribed to Bürgi: Paris, Conservatoire Nationale des Arts et Métiers; Kassel (two examples); Vienna, Kunsthistorisches Museum (unpublished and incomplete); Dresden, Staatlicher Mathematisch-Physikalischer Salon. A sixth in Zurich, Schweitzerisches Landesmuseum, is actually signed by Bürgi and dated 1594. The pieces at Paris, Vienna, Zurich, and the second specimen at Kassel have stands with figures similar to that of the Baldewein globe (Catalog No. 115). The first globe at Kassel, included in the exhibit, has a plain stand, as does the one at Dresden. The mechanisms in the globes vary, certainly reflecting the wishes of those commissioning the pieces. For stylistic reasons the Kassel globe and its Dresden counterpart should be dated later than the Zurich one of 1594. Leopold and Pechstein, on the other hand, are inclined toward a date of 1584; they also date to the same time, c. 1585, the Bürgi-Eisenhoit sphere of Catalog No. 119. The more precise dating of the armillary sphere which has now become possible (after 1595 and before 1603) is thus

the one arising from arguments, already presented indirectly by these two authors, for the later dating around 1600. For Bürgi see Chapter 8.

Literature: Maurice II, Fig. 255. J. H. Leopold and K. Pechstein, *Der kleine Himmelsglobus 1594 von Jost Bürgi* (Lucerne, 1977), pp. 22 ff. Ludolf von Mackensen (with contributions by Hans von Bertele and John H. Leopold), *Die erste Sternwarte Europas mit ihren Instrumenten und Uhren—400 Jahre Jost Bürgi in Kassel* (Munich, 1979), pp. 74 ff., Cat. No. 17.

117 Celestial globe with clockwork

Johann Reinhold, Georg Roll
Augsburg, 1584 (?)
Leningrad, Hermitage

Globe: fire-gilt bronze and copper
Movement: iron plates
Height: 58 cm (22⅞ in.)

Going, quarter-striking, and hour-striking trains. Winding of the three trains at the South Pole by means of a winding square, which can be pushed inward in three stages to wind each train separately. Regulation by means of screwed weights which can be turned in or out along the foliot or by altering the depth of the pallets with the crown wheel. Hours and quarter hours are indicated on the dial at the North Pole. The clockwork drives the globe and a solar image and a lunar image upon two arcs so that they move around the globe. This simple method is far from being the ingenious design used by Bürgi in Catalog No. 116. Only an average value is given for the motion of the moon, the same as with clocks having a lunar indication; the complicated motion of the moon along the ecliptic is not taken into consideration. (In order to avoid such inaccuracies Bürgi left the motion of the moon out of his globes and his armillary sphere.) The clockwork operates the calendar in the horizon ring via a telescoping arbor at the South Pole. The calendar ring has to be adjusted manually for the leap-year correction. Under the celestial globe there is a small terrestrial globe; below this, four sundials are engraved upon the base plate. In its indications and design this globe is simpler than those of Bürgi. Six globes similar in many respects still survive: they are dated between 1584 and 1589, and most of them are signed jointly by Reinhold and Roll. On this globe the date has been removed from the signature area, yet since the dominical letters commence with 1584, this is presumably the date of its completion.

George Roll (1546–1592) was born in Liegnitz and worked first as a clockmaker journeyman at Friedberg, which was outside Augsburg and belonged to the Duchy of Bavaria. Roll had great aptitude for commerce and he dealt in clocks, including ones by Augsburg masters, as far afield as Rome and Naples. At Augsburg he contrived to win the authority to carry on general trade *(Kramergerechtigkeit)* in the face of the opposition of the local clockmakers. He also owned a workshop in which there were at times twenty-five craftsmen working for him. Johann Reinhold (c. 1550–1596) also came from Liegnitz and

300

was active as a clockmaker at Augsburg starting in 1567; in 1584 he became a master. Presumably Reinhold made the astronomical gears for the globes.

Literature: Maximilian Bobinger, *Kunstuhrmacher in Alt-Augsburg* (Augsburg, 1963), pp. 14 ff., 29 ff. John H. Leopold, *Die grosse astronomische Tischuhr von Johann Reinhold* (Lucerne, 1974), p. 78 (with a summary of all earlier literature). Maurice II, Fig. 259. J. H. Leopold and K. Pechstein, *Der kleine Himmelsglobus 1594 von Jost* Bürgi (Lucerne, 1977), p. 17.

118 Celestial globe with clockwork

Isaak Habrecht III. Strasbourg, 1646
London, National Maritime Museum

Movement: brass plates
Case: fire-gilt bronze and copper
Height: 49 cm (19¼ in.)

Going, quarter-striking, and hour-striking trains within the globe. On the meridian ring a 24-hour dial with minute hand (one revolution in 2 hours). An image of the sun is driven along the zodiac on the outer surface of the globe upon a ring, and an image of the moon is similarly driven by another ring. On the horizontal ring there is a calendar with both Julian and Gregorian reckoning and a representation of the zodiac; both are engraved. To determine the date, one must first determine the position of the sun in the zodiac circle on the globe and then find the same position on the zodiac circle on the horizon ring. Since the calendar is engraved concentrically with this zodiac circle, the date is indicated. Upon the release of a set-screw the globe can be moved within its mounting and used for another latitude. All drives are wound by means of a square at the South Pole; to wind the striking train one presses the square inward into the globe, and to wind the going train one pulls it one step outward. Duration is 28 hours. Originally the globe had a crowning figure with a compass upon which the eight principal winds were engraved. Signed on the meridian ring: ISAAC HABRECHT A STRASBURG AO MDCXLVI. Habrecht's original description and directions for use are in the Strasbourg archives. For Habrecht see Catalog No. 63.

Literature: Théodore Ungerer, "Une horloge à sphere tournante d'Isaac Habrecht III (1646)," *Archives Alsaciennes,* 1931, *10:* 150 ff. *An Inventory of the Navigation and Astronomy Collection in the National Maritime Museum, Greenwich* (Greenwich/London, 1970), p. 16.4. Maurice II, Fig. 263. Henry C. King, *Geared to the Stars* (Toronto, 1978), Fig. 5.24.

119 Armillary sphere with clockwork

Jost Bürgi. Kassel, c. 1600
Goldsmith work
by Anton Eisenhoit, Warburg
Stockholm, Nordiska Museet

Movement: brass plates and wheels
Case: fire-gilt bronze and brass
Dial: fire-gilt brass, cold enamel
Base: ebony
Height: 88.3 cm (34¾ in.)

Going, quarter-striking, and hour-striking trains. Going train with fusee and chain; the two striking trains are driven by a single spring (see Fig. 55). The way all the wheels are attached by screws to the pinion arbor is singular. The escapement is the original one; only the balance spring has been added. The clockwork movement is located in the base; it drives an astronomical chart (see figure at upper left), the hour and minute hands, and an annual index. Above the chart is a projection of the celestial houses (see Ch. 6). From the hour hand the drive originally went to the armillary sphere standing over the base. Inside the armillary sphere an image of the sun moves along the ecliptic. The movement of the sun, slowed in the aphelion and speeded in the perihelion, is achieved by means of an uneven distribution of teeth on the wheel which moves the sun forward in the ecliptic. Originally the armillary sphere could be equated by means of a telescopic arbor. Missing are the drive connection from the going train to the sphere and almost all mechanical parts within the sphere. On the sphere there were originally numerous stars attached by screw threads (most of them broken away). During restoration at the Bayerisches National-museum the principal repairs were replacement of the ecliptical coordinates on the Northern Hemisphere and restoration of the dragon hand, which had been broken into fragments. Engraved on the star chart are 49 constellations with their names (to the 48 classical constellations there was added in the 16th century Berenice's Hair, which had originally formed part of the constellation of Leo). The stars are represented according to six different magnitudes. Astro-logical tempers are assigned (by means of planetary symbols) to the constella-

tions and individual stars according to the systems devised by Ptolemy, Alfonso the Great, and Cardanus.

Checking on 20 engraved principal stars revealed that their marked positions did not deviate from their calculated positions by more than $\pm 0.4°$. From these positions of the stars one arrives at a date of 1600, ± 5 years, for the chart of the heavens. Before 1595 and after 1605 the engraved values, or else the computed values, differ from these. Since Eisenhoit died in 1603, the origin of this device narrows down to the years 1595–1603. Because of the hole bored through the middle of the celestial houses one cannot precisely reconstruct the latitude for which the apparatus was intended; projection gives about 52°. The Duke of Braunschweig gave the Emperor Rudolf II an apparatus devised by Jost Bürgi which also combined an armillary sphere and a chart of the stars (see Ch. 8). This circumstance, combined with the ingenious workmanship shown here, justifies ascribing this armillary sphere to Bürgi. It is possibly identical to the sphere delivered to Rudolf II. Since the device is set up for the latitude of Braunschweig, its utility at Prague would be only slightly diminished, for the armillary sphere could be elevated. (I thank Dr. Peter Förster, Prof. Dr. Paul Kunitsch, and Mrs. Elizabeth Zanzen, all of Munich, for assistance in this determination and computation.)

Literature: J. H. Leopold and K. Pechstein, *Der kleine Himmelsglobus 1594 von Jost Bürgi* (Lucerne, 1977), p. 27.

120 Heliocentric planetarium

Nikolaus Siebenhaar. Lübeck, 1651
Hillerød, Nationalhistoriske Museum
pa Frederiksborg

Movement: brass
Case: fire-gilt copper and brass
Base: polychrome wood
Height: 138 cm (4 ft 6⅜ in.)

An annual calendar with the saints' days is engraved around the edge of the bottom plate of the planetarium. Inside this two other circles with dials. On one of these, calendar and astronomical figures for 1651–1699 in columns, in the middle the number for the year and the arms of the owner, Duke Johann, Prince-Bishop of Lübeck (1606–1651), with the accompanying inscription: "Joan D.G. Epis. Lub. Her. Norv. Dux. Sclesv. Hols. Storm & Dith. Com. in Old. & Delm. 1651 F.C." On the other dial there is a three-dimensional representation of the Copernican concept of the universe. A terrestrial globe (map probably after W. J. Blaeu) revolves with its annual and diurnal motion around an image of the sun and is itself orbited by an image of the moon. By means of an inclined ring inside the model which corresponds to the course followed by the earth, the earth and the moon are driven up and down along the ecliptic over the course of a year. Inside the annual calendar are engraved allegorical figures of Day and Night facing each other and beneath the terres-

trial globe allegorical figures of the seasons of the year.

The commission for the model was issued in 1644, and it was completed in 1651. As early as 1696 Jacobaeus (*Museum Regium,* p. 66) wrote that it was believed the planetarium was the work of Nikolaus Siebenhaar, "rerum mechanicarum pertissimus." In 1634 Siebenhaar became a burgher at Lübeck, where there is record of him as late as 1668.

This instrument is one of the earliest surviving planetaria that represent the Copernican system of the universe.

Literature: Hans von Bertele, "Clockwork Globes and Orreries," *Horological Journal,* Dec. 1958, p. 805. Ernst Schlee, *Gemessene Zeit, Uhren in der Kulturgeschichte Schleswig Holsteins, Ausstellung Schloss Gottorf 1975,* p. 77. Maurice II, Fig. 267. Henry C. King, *Geared to the Stars* (Toronto, 1978), Fig. 6.12.

Acknowledgments

The exhibition *The Clockwork Universe* was conceived and planned jointly by the Bayerisches Nationalmuseum, Munich, and the National Museum of History and Technology, Washington, D.C. The suggestion for this collaborative venture had come from the Bayerisches Nationalmuseum's director, Lenz Kriss-Rettenbeck, and its curator of horology and metal crafts, Klaus Maurice.

The American version of the exhibition (Washington, D.C., November 7, 1980, to February 15, 1981) was prepared and carried out by the National Museum of History and Technology's Division of Mechanisms, consisting of Otto Mayr, curator, Carlene Stephens, museum specialist, Glenys Dyer and Elizabeth Reichelderfer, volunteers. Constantine Raitzky of Barto, Pennsylvania, designed the exhibit, and the staff of the Exhibits Department produced and installed it. The project could not have been carried out, however, without the energetic and generous help of innumerable other staff members of the National Museum of History and Technology and the Smithsonian Institution. Particular acknowledgment is due to Martha Morris Shannon for organizing and managing the complex loan arrangements; to Thomas K. Bush for overseeing the packing and shipping of the exhibit; to Alice Bryan for advice on insurance matters; to W. David Todd for designing and constructing five models of clock escapements; to Josiah Hatch and Geraldine Sanderson for publicity; to Glen Ruh for coordinating the publication of the exhibit catalog and to Christian Hohenlohe for expert financial assistance with the catalog.

The American version of the exhibit also benefited from a wide variety of generously given help from its Bavarian partner: the Bayerisches Nationalmuseum conducted all loan negotiations; restored a considerable number of the objects exhibited; provided various exhibition accessories, notably the models of two astronomical mechanisms made by Karl Jagemann, GmbH, and Peter Friess; and shared its rich collection of graphic materials. Other instances of help, hospitality, and friendship extended to the National Museum of History and Technology by the Bayerisches Nationalmuseum and its staff are too numerous to list here.

The production of the exhibition *The Clockwork Universe* and the publication of this catalog were made possible by a grant from the NCR Corporation, Dayton, Ohio. A grant from the Institut für Auslandsbeziehungen, Stuttgart, covered the cost of transporting the exhibit objects from Europe to America and back. The means of escort travel were provided by the Bayerische Landesstiftung, Munich. The archival researches of Eva Groiss were made possible by a grant from the Stiftung Volkswagenwerk, Hannover. Photography of objects published in this catalog was supported by a grant from the Deutsche Forschungsgemeinschaft, Bonn-Bad Godesberg. The Embassy of the Federal Republic of Germany, Washington, D.C., also supported the exhibition project in a variety of ways.

The clocks and automata exhibited in *The Clockwork Universe* were contributed by the following lenders:

Amsterdam, Rijksmuseum
Augsburg, Städtische Kunstsammlungen
Baltimore, Walters Art Gallery
Bamberg, Sammlung Textor
Basel, Historisches Museum
Boston, Museum of Fine Arts
Braunschweig, Herzog Anton Ulrich-Museum
Brussels, Musées Royaux d'Art et d'Histoire
Chicago, Adler Planetarium

Copenhagen, Danske Kongers Kronologiske Samling på Rosenborg
Copenhagen, Nationalmuseet
Darmstadt, Hessisches Landesmuseum
*Dresden, Grünes Gewölbe
*Dresden, Staatlicher Mathematisch-Physikalischer Salon
Furtwangen, Fachhochschule, Historische Uhrensammlung
Hillerød, Nationalhistoriske Museum på Frederiksborg
Innsbruck, Kunsthistorisches Museum, Sammlungen Schloss Ambras
Innsbruck, Tiroler Landesmuseum Ferdinandeum
Karlsruhe, Badisches Landesmuseum
Kassel, Staatliche Kunstsammlungen, Astronomisch-Physikalisches Kabinett
Kremsmünster, Kremsmünster Abbey
*Leningrad, Hermitage
London, National Maritime Museum
London, Bruno L. Schroder
London, Society of Antiquaries
London, Victoria and Albert Museum
Lübeck, Museum für Kunst und Kulturgeschichte
Milan, Museo Poldi Pezzoli
*Moscow, Kremlin Museum
Munich, Bayerisches Nationalmuseum
Munich, Bayerische Verwaltung der Staatlichen Schlösser, Gärten und Seen.
  Schatzkammer der Residenz
Munich, Wittelsbacher Ausgleichsfonds
Newark, Newark Museum
New York, Metropolitan Museum of Art
Rockford, Illinois, The Time Museum
Schwäbisch Gmünd, Städtisches Museum
Stockholm, Kungl. Husgerådskammaren
Stockholm, Nordiska Museet
Stuttgart, Württembergisches Landesmuseum
Ulm, Museum der Stadt
Vienna, Kunsthistorisches Museum
Washington, D.C., National Museum of History and Technology
Winterthur, Museum Lindengut
Winterthur, Uhrensammlung K. Kellenberger
Wuppertal, Historisches Uhren-Museum
Zurich, Museum der Zeitmessung Beyer
Unnamed private collections

The exhibition benefited from advice and help from the following individuals:
Jurgen Abeler, Wuppertal; Martin Angerer, Munich; Michael F. Ascher, Munich; Seth G. Atwood, Rockford; Franz Baumer, Munich; J. A. Bennett, London; Hans von Bertele, Vienna; Theodor Beyer, Zurich; A. L. den Blaauwen, Amsterdam; Claude Blair, London; Friedrich Blendinger, Augsburg; Ruth Blumka, New York; Luitpold von Braun, Munich; Albert Bruckmayer, Kremsmünster; Giuseppe Brusa, Milan; Bruno Bushart, Augsburg; Bruce Chandler, New York; Vera von Claer-Crodel, Hamburg; Rudolf Distelberger, Vienna; Walter Dürr, Schwäbisch Gmünd; Wolfgang Eckhardt, Hamburg; Erich Egg, Innsbruck; Povl Eller, Hillerød; Richard B. and Erna Flagg, Milwaukee; Stig Fogelmarck, Stockholm; J. T. Fraser, Westport; Johann Michael Fritz, Karlsruhe; Werner Ganz, Winterthur; Helmut Grötzsch, Dresden; Peter D. Guggenheim, New York; Tjark Hausmann, Berlin; John Hayward, London; Bodo Hedergott, Braunschweig; Carl Benno Heller, Darmstadt; Volker Himmelein, Stuttgart; Gerhard Hojer, Munich;

*The loan contributions from Moscow, Leningrad, and Dresden were unfortunately cancelled.

B. Hutchinson, London; Cedric Jagger, London; Lothar Jennes, Cologne; Jürgen Kalkbrenner, Bonn; Anne-Marie Kappeler, Langwiesen; H. C. King, Toronto; Friedrich Klemm, Munich; Björn R. Kommer, Lübeck; Rudolf Koppenhöfer, Bonn; Konrad Kraske, Bonn; Hans Lanz, Basel; Manfred Leithe-Jasper, Vienna; Arnold Lühning, Schleswig; Viktor Luthiger, Walchwil; Ludolf von Mackensen, Kassel; Peter Mediger, Munich; Michael Meier, Munich; Peter von Miller, Munich; Samuel C. Miller, Newark; Alessandra Mottola Molfino, Milan; Richard Mühe, Furtwangen; Ruth Mühlich, Munich; Kurt Müller, Bonn; Josy Muller, Brussels; Rudolf Neumeister, Munich; Margret Nida-Rümelin, Munich; Bengt Nyström, Stockholm; B. B. Piotrovski, Leningrad; Emmanuel Poulle, Paris; Richard H. Randall Jr., Baltimore; Haide Russell, Washington, D.C.; Elisabeth Scheicher, Innsbruck-Ambras; H. D. Schepelern, Copenhagen; K. Schillinger, Dresden; Ivo Schneider, Munich; Bruno L. Schroder, London; Helmut Seling, Munich; Hans Stiesdal, Copenhagen; Jürgen Teichmann, Munich; Susanne Thesing, Munich; F. H. Thompson, London; Erwin Treu, Ulm; Clare Vincent, New York; Madge and Roderick S. Webster, Chicago; Ruedi Wehrli, Winterthur; Helmuth Zebhauser, Munich.

Glossary

Prepared by Diana Menkes, Carlene Stephens, and W. David Todd. All words appearing in **boldface** are defined in the glossary.

alidade
: The rotatable indicator of an **astrolabe** used to sight a given star and determine its altitude above the horizon by reference to a scale of degrees inscribed on the instrument. Essentially replaced by gearworks in clocks with astrolabe faces. See pp. 51, 56.

arbor (*Welle*)
: Shaft, staff, stem, or axle that carries wheels and **pinions.** See p. 91.

armillary sphere
: A skeletonized **celestial globe**, sometimes driven by clockwork, consisting of rings or bands that represent the motions of the principal heavenly bodies around the earth, which is placed in the center of the sphere according to the Ptolemaic system of cosmology. Armillary spheres, which appeared in the 16th century, thus provided a three-dimensional representation of the movements indicated by dials on **planetary clocks.** See p. 100, Catalog Nos. 114, 119. Small spheres were also used decoratively on clocks, e.g., Catalog Nos. 30, 57.

aspect
: The angular relationship between the **zodiacal** positions of any two celestial bodies, important for certain astrological determinations. A diagram of aspects can be seen on many of the clocks illustrated, e.g., Catalog Nos. 9, 17, 24. See also p. 56, Fig. 26.

astrarium
: Synonymous with **planetarium**, this was the name given by Giovanni de 'Dondi to his complicated astronomical clock completed in 1364. See p. 22.

astrolabe
: A portable instrument used to measure the coordinates of celestial bodies, indicate their relative positions and movements, and thereby determine time and latitude. It often also incorporates functions for solving other astronomical, mathematical, and astrological problems. As discussed at length in Ch. 7, clocks were commonly fitted with astrolabe faces, e.g., in Catalog Nos. 13, 30, 32, and 49. See also Figs. 16, 17, 20–25.

Augsburg pine cone (*Pyr*)
: A stylized pine cone (an ancient Roman motif) which is the hallmark of the city of Augsburg. This mark was struck or stamped on clocks (in the 17th century obligatory for clocks with bases of true ebony, which also carried the stamp **EBEN**). See p. 124, Figs. 27, 70, Catalog Nos. 29, 43, 57, 98.

automaton
: During the time covered by the exhibit, automata were considered to be mechanisms capable of acting under their own power and direction; thus clocks were examples of automata. In the present context the term covers both **figure clocks** (animated mechanical figures which have a time-telling capability) and automata in the strict sense, i.e., mechanical figures without time indication. See pp. 15–16, 33–34, 160, 234–235, Fig. 5, Catalog Nos. 61–112.

azimuth
: The horizontal angle between an observer's meridian and the meridian of a celestial body being observed. See p. 52, Fig. 20.

balance
: An oscillating wheel which serves to regularize the motion of a clock or watch. First used as an alternative to the **foliot** and later used with the spiral **balance spring.** See Catalog No. 116.

balance spring (*Spiralfeder*)
: A spiral spring used to control the rate of oscillation of the **balance** of a portable clock or watch, thereby controlling more accurately the **going** rate of the clock. See pp. 24, 25, Catalog No. 50.

Bohemian hours
: See under **hours.**

bracket clock (*Konsoluhr*)
: Originally a clock made to be set on a bracket fixed to the wall. Now a generic term for shelf, mantel, and table clocks of various kinds. See Catalog Nos. 2, 3, 8.

calendar ring (*Kalendarscheibe*)
: A ring or disk engraved with the days of the month and often with the days of a quarter, half, or whole year. The more elaborate rings included the **saints' days** and **fixed feasts.** See Catalog Nos. 49, 52, 56, 115.

canonical hours
: The periods into which daily monastic life was divided in medieval times (matins, prime, tierce, sext, nones, vespers, complin). The length of these periods varied

constantly as they were based on the primitive form of sundial with perpendicular gnomon. See pp. 131, 146–147.

celestial globe (*Himmelsglobus*)
An imaginary sphere with the earth at its center according to the Ptolemaic system, and the heavenly bodies, including constellations of the **fixed stars** projected onto the sphere's surface. To the observer on earth the sphere appears to rotate about the celestial poles. Elaborate mechanical globes, driven by clockwork, were made in the 16th and 17th centuries. See pp. 88–89, 100–101, 113, Figs. 29, 40, 41, Catalog Nos. 113, 115–118.

chapter ring (*Stundenkreis*)
The circle of hour numerals on a clock dial; often a separate ring pinned to the dial plate. See Catalog Nos. 14, 20, 112.

chronometer
A precision timepiece that has been granted an official rating certificate and which, by virtue of its construction, can be relied on to keep that rate. Commonly used to determine longitude by ships at sea. See p. 25.

count wheel (*Schlossscheibe*)
Also "locking plate." A device in the form of a wheel or ring which is driven by the clock and determines the number of bell strokes for each hour or part thereof. In most instances the circumference of the wheel or ring has a given number of notches spaced at increasingly wider intervals to permit the clock to strike in the correct sequence. At the appropriate time a lever is lifted from a given notch. This lever raises another which releases the **striking train.** After the requisite number of strokes on the bell, the first lever drops into the next notch of the count wheel, allowing the second lever to drop into the path of the arresting pin or hoop and "lock" the striking mechanism. See Catalog Nos. 8, 56, 86.

cross-beat escapment
See under **escapement.**

crossings
The spokes of the wheel in a clock; hence "crossed-out wheel," Catalog No. 21.

crown wheel (*Kronrad*)
A wheel with fang-shaped teeth set at right angles to the plane of the wheel. Usually employed as the **escape wheel** in a **verge escapement.** See p. 92, Catalog Nos. 57, 117.

declination
An angular distance to a point on the **celestial sphere,** measured in degrees, from the celestial equator to a celestial body north or south of it. The equivalent of terrestrial latitude. See Catalog No. 115.

dominical letter (*Sonntagsbuchstaben*)
Also "Sunday letter." One of the letters A through G used to denote the Sundays in a particular calendar year. To determine the letter for a given year, the first seven days of January are assigned the corresponding first letters of the alphabet, and the letter of the first Sunday is the dominical letter for the year. Used to calculate the date for Easter, the dominical letters were one of the ecclesiastical **indications** on clocks, e.g., Catalog Nos. 21, 25, 49, 117.

dragon hand, dragon
(*Drachenzeiger, Drachen*)
In clocks of the Renaissance and later, the hand indicating the position of the lunar **nodes** was in the form of a dragon, the head indicating the ascending node, the tail the descending node. This hand shows the limits within which an eclipse can occur. The dragon shape reflects the ancient belief that eclipses were caused by a dragon swallowing the sun or moon. See pp. 55–56, Catalog Nos. 21, 33, 49.

drum clock
(*Trommelförmige Dosenuhr*)
A general term for a clock contained in a cylindrical case with the dial on the uppermost flat surface. Also called "table clock," or *Tischuhr* (see the richly decorated cylindrical clocks of Fig. 57, Catalog Nos. 9, 38–41); the latter term is also used in the catalog to include clocks with vertical dials.

EBEN
The mark used in Augsburg beginning in 1625 to indicate clock bases of true ebony as distinguished from stained woods such as pearwood. Augsburg was the main supplier of ebony bases. See pp. 124, 173, Fig. 70, Catalog Nos. 29, 54.

ecliptic
The apparent annual path of the sun in the **celestial sphere;** the intersection plane of the earth's solar orbit with the celestial sphere. See pp. 53–56. A sun hand (sometimes bearing an image of the sun) which showed the sun's movement around the ecliptic was one of the astronomical **indications** on the clocks illustrated, e.g., in Catalog Nos. 32, 113, 116.

epact
The age of the calendar moon at the beginning of a new year, from which its age at any subsequent date during the year can be determined. Used to calculate the date of Easter (the first Sunday after the fourteenth day of the calendar moon following the vernal **equinox**). See Catalog Nos. 21, 33.

epicycle

In the Ptolemaic system of cosmology, a small circle whose center moves on the circumference of a larger circle centered on the earth. Ptolemy introduced epicycles in order to describe the apparent motion of the planets around the earth, thus retaining circles—the "perfect figures" of Aristotelian philosophy—while explaining more plausibly the retrograde motion of such planets as Mars and Jupiter. See p. 212.

epicyclic train

A train of wheels having a fixed central wheel **arbor** with which other wheels and **pinions** mesh and about which they rotate. See Catalog Nos. 21, 22, 115.

equation of time (*Zeitgleichung*)

The difference between apparent **solar time** and mean solar time. Apparent solar days, as indicated on a sundial, are of unequal duration because of the sun's nonuniform motion along the **ecliptic** and the fact that the ecliptic is inclined to the equator. Mean solar time, as indicated on a clock, is divided into uniform days. The equation of time, actually an interval from zero (four times a year) to a maximum of about 16 minutes, is expressed as a graph or table of differences showing by how much a clock is fast or slow compared with the sun at any specific time.

equinox

Either of the two points on the **ecliptic** when the sun has zero **declination** and rises and sets due east and west. At these two points in the spring and fall (the vernal and autumnal equinoxes) the ecliptic coincides with the celestial equator, thus giving days and nights of equal length.

escapement (*Hemmung*)

In various forms, a mechanism that alternately checks and releases the driving force of a timepiece. The mechanism includes the **escape wheel**, which provides impulse to the **balance** or **pendulum**, usually by means of one or more **pallets**.

anchor escapement (*Ankerhemmung*)

Attributed to William Clement (1638–1704) and so called because its **pallets** resemble the flukes of an anchor. Generally in combination with a **pendulum** to regulate its motion more accurately than with the **verge escapement**. See Catalog No. 7.

cross-beat escapement (*Kreuzschlaghemmung*)

A form of escapement introduced into clockwork by Jost Bürgi in which two **foliot** arms in the vertical plane, each having one **pallet** upon its **arbor**, are driven by an **escape wheel** with fang-shaped teeth (as in the **anchor escapement**). This device kept more precise time than either the foliot or balance-wheel verge escapements. See also **Prague clock**. See pp. 91–93, 101, Figs. 30, 34, Catalog Nos. 56, 60.

crown-wheel escapement

See **verge escapement**.

dead-beat escapement (*Ruhende Hemmung*)

A modification of the **anchor escapement** in which the **escape wheel** remains stationary during impulses. Introduced in the 18th century by George Graham, this long remained a highly successful escapement.

pinwheel escapement

A form of escapement in which the teeth of the **escape wheel** are pins mounted at right angles to the wheel. The arms of the anchor engage with the pins. Invented by Galileo, it was reinvented by L. Amant and improved by J. A. Lepaute in 18th-century France. See p. 24.

verge escapement (*Spindelhemmung*)

Also "crown-wheel escapement." The oldest known escapement. Characterized by two **pallets**, mounted on an **arbor** at right angles to each other, which engage with an **escape wheel**, preventing the unchecked descent of the clock's driving weight and thus controlling the **going** rate of the clock. Used with **foliot, balance,** and **pendulum**. See p. 91, Catalog Nos. 7, 50, 52.

escape wheel

The rotating notched wheel periodically engaged and disengaged by the **pallets** in an **escapement**.

figure clock (*Figurenuhr*)

Also called "automaton clock." A clock with figures, animal or human, which have moving parts mechanically controlled by the **going train** or the **striking train** or both. Examples are Catalog No. 62, where the hours are indicated by the rotating crown of a Madonna, and No. 90, a lion whose eyes move with the going train and whose jaw and paw move when the hour is struck. Some figure clocks have a separate drive train which enables the entire clock to move across a table, like the jewelled, silver centaur and Diana group of Catalog No. 105. See pp. 121, 160, 234–235, Catalog Nos. 61–101, 105–109, 112.

fixed feast	A holy day in the liturgical year that is observed on the same date each year. Cf. **movable feast.**
fixed stars	The constellations of distant stars that form the backdrop of the **celestial sphere.** Contrasted to the "wanderers," the planets. The positions of these stars were determined and catalogs compiled and improved over the years, e.g., by Landgrave Wilhelm IV of Hesse in the 16th century. See pp. 88, 291, Catalog Nos. 113–120.
foliot	A horizontal bar fitted with adjustable weights and fixed at a right angle to the **verge** in a **verge escapement** clock. The period of oscillation of the foliot controls the regulation of the clock. See Catalog No. 9.
foreman (*Vorgeher*)	A master selected from among the individual crafts to head the combined guild of smiths. See p. 60, Catalog Nos. 19, 27, 50.
fusee (*Schnecke*)	A cone-shaped, grooved pulley used in a **spring-driven** movement to maintain a constant torque on the drive train as the mainspring unwinds and loses power. A chain or gut line (the best were made from sheep's intestines) connects the spring barrel to the fusee. The gut or chain is wound onto the fusee when the clock is wound up and is pulled back onto the barrel as the spring unwinds. See Ch. 10, Figs. 51, 67–69, Catalog Nos. 10, 21, 49, 55.
Geneva stopwork (*Malteser Kreus*)	Also "Maltese cross system." A mechanism that prevents overwinding of the mainspring. Composed of a wheel shaped like a Maltese cross and a finger piece on the winding square that engages with the wheel. See Catalog No. 48.
going train (*Gehwerk*)	The wheels and **pinions** that transmit driving power to the **escape wheel** of a clock; i.e., the timekeeping train of a clock. See pp. 159–160, Figs. 38, 51–53, 76, 79.
Golden Number (*Goldene Zahl*)	The number of a given year in its current 19-year Metonic cycle, or **lunar cycle.** The Greek astronomer Meton observed in the 5th century B.C. that after a period of 19 years (235 lunar months) the phases of the moon recur on the same days as in the preceding period. The ancient Greeks used the cycle to determine days for religious festivals and supposedly inscribed such dates in gold in public places. Dials for reading the Golden Number are part of the **indications** of the clocks illustrated, e.g., in Catalog Nos. 21, 25, 28.
Gothic clock	A weight-driven hanging wall clock with posted frame movement usually dating from the 15th and 16th centuries. Typical examples are clocks by the Liechti family. The Gothic style is shown in Catalog No. 2, the Liechti family mentioned in No. 6.
grande sonnerie	A system of striking not only the quarter hours but also the preceding hour at each quarter. See Catalog No. 86.
Great Clock	See under **hours.**
houses	See zodiac.
hours	In the 16th and 17th centuries various regional methods of reckoning the hours coexisted on the European continent. Comparisons and conversions of the different times were made by means of tables (see Fig. 75) and by dials on clocks (Catalog No. 7).
Bohemian hours or clock (*Böhmische Uhr*)	Divisions of the entire day/night period into 24 hours (1–24), beginning at sunrise. See p. 146, Catalog No. 7.
equal (or equinoctial) hours	See **temporal hours.**
Great Clock	See **Nuremberg hours.**
Italian hours or clock (*Italienische Uhr*)	Division into 24 hours (1–24), beginning at half an hour after sunset (although in practice the half hour was often ignored on clocks and sundials). See p. 146, Catalog Nos. 7, 40.
Nuremberg hours or clock (*Nürnberger Uhr*), or Great Clock (*Grosse Uhr*)	Division into day hours and night hours, the numbering of the day hours beginning with sunrise, the night with sunset. The numbers of hours in each segment vary seasonally (maximum and minimum being 1–16 and 1–8). Used specifically in Nuremberg and a few other imperial cites such as Regensburg. See pp. 110, 146, 148, 172, Catalog Nos. 1, 3, 41.

Small Clock (*Kleine Uhr*)	The conventional modern division into 24 hours numbered in two 12-hour periods (2 × 1–12), commencing at midnight. See pp. 146, 148, 172, Catalog Nos. 14, 24, 84.
temporal hours	Also "temporary hours," "seasonal hours," "unequal hours." Division of each of the periods sunrise to sunset and sunset to sunrise into 12 equal parts. In the 14th century this was replaced by the modern system of 2 × 12 equal hours. See Catalog Nos. 40, 49.
Whole Clock (*Ganz Uhr*)	Division into 24 hours (1–24), commencing at midnight. See pp. 110, 146, 172, Fig. 61, Catalog Nos. 7, 84.
indications	In addition to a dial which showed the time, German clocks of the 16th and 17th centuries were likely to have assorted auxiliary dials and other devices which revealed information of a practical, astronomical, astrological, or ecclesiastical nature. Among the dials included on the clocks in *The Clockwork Universe* are the following: setting dials for an alarm, checking dials for hour or quarter-hour striking, dials indicating the number of times a **remontoire** has functioned or the number of day-by-day windings; indications of the age and phase of the moon, the lunar **nodes**, the path of the moon through the **zodiac**, the path of the sun through the zodiac and along the **ecliptic, solar** and **lunar cycles,** days of the week and the month, the **ruling planets** for the week and sometimes for the hour, the **saints for the days, dominical letters,** the **Golden Number, epact,** dates for Easter, the number of days between Christmas and Shrove Tuesday, the rising and setting of the sun and moon, the duration of daylight, darkness, and occasionally of moonlight; comparative scales of lengths of days in different cities, comparative dials for various systems of hours (such as the **Whole Clock,** the **Italian hours,** the **Bohemian clock**), conversion dials for telling time by moonlight. Clocks were given **aspect** diagrams, Julian, Gregorian, and **perpetual calendars, astrolabes, armillary spheres,** compasses, **quadrants,** horizontal and vertical sundials, and **tympana** set for particular latitudes. Sun hands carried effigies of the sun, moon hands might be spheres which themselves rotated to show phases, the lunar **node** hand was shaped like a **dragon,** and other indicators took such forms as serpent's tongues and revolving crowns. Clocks which display a generous amount of indications are seen in Figs. 6, 60–63, Catalog Nos. 21, 32, 41, 48, 49.
inspection master, or sworn inspection master (*geschworener Geschaumeister*)	A master elected from the individual crafts to judge the quality of **masterpieces** in contested cases. See pp. 60–61, 66–68, Catalog Nos. 44, 57, 90, 110.
isochronism	Occurring in equal intervals of time. A **pendulum** is said to be isochronous when it completes oscillations of varying amplitude in identical periods of time—a phenomenon that was first described by Galileo. See p. 23.
Italian hours	See under **hours.**
jack (*Schlagfigur*)	Model of a human figure that strikes or appears to strike the time on a bell. Probably derived from the contraction *jacquemart = Jacques + marteau* (hammer). See Catalog No. 8.
lantern pinion	A form of **pinion** in which hollow pins are used instead of the usual solid leaves. Often used in **tower clocks,** as, e.g., in Catalog No. 6.
large clock (*grosse Uhr*)	General term for a public or **tower clock;** in guild regulations a designation covering all clocks with weights, regardless of size, and distinguished from **small clocks.** See pp. 67, 81–86.
locking plate	See **count wheel.**
lunar cycle	Also "Metonic cycle." A cycle of 19 years after which the phases of the moon recur on the same days of the year. See **Golden Number,** Catalog No. 21.
masterpiece (*Meisterstück*)	In the German clockmaking guilds of the 16th and 17th centuries a clock made according to regulations, required for admission to the guild and the designation of master. See pp. 59–61, 67–68, 78, 85–86, 88, 113, 172, Catalog Nos. 22, 39. For individual masterpieces see Fig. 33, Catalog Nos. 44, 45, 46, 54, 113.
monstrance clock (*Spiegeluhr*)	A clock having the form of a Roman Catholic monstrance, a receptacle for the consecrated bread of the Eucharist; also called "mirror clock" because of its shape. The monstrance clock was made chiefly in Augsburg (often as a masterpiece incorporating numerous astronomical **indications**), mainly during the period

1550–1650. See pp. 67, 172–173, Figs. 85–89, Catalog Nos. 42–46.

movable feast
A holy day in the liturgical year that is observed on a different date from year to year. Cf. **fixed feast.** See p. 110, Catalog No. 116.

nocturnal (*Nachtuhr*)
An instrument for determining the time at night by observing the positions of certain constellations. See p. 108, Fig. 65.

node (*Knoten*)
Either of two points on the **celestial sphere,** diametrically opposite, at which the path of a planet or the moon intersects the **ecliptic.** The ascending node is the intersection where the body crosses south to north; the descending node, from north to south. The alignment of the sun, moon, earth, and a lunar node determines the degree of eclipsing of the sun and moon. One of the astronomical **indications** on clocks was a hand in the form of a **dragon** which showed the lunar nodes. See pp. 55–56, Catalog No. 49.

Nuremberg hours
See under **hours.**

Orpheus clock
One of a group of ten south German table clocks made between 1560 and 1580 whose cases bear decoration in the form of scenes from the Orpheus legend. The clock movements, which vary considerably, are thought to have been made in four different workshops. See Catalog No. 40.

orrery
Generally a mechanical device, often run by clockwork, for representing the relative sizes, motions, and positions of bodies in the solar system. First made by George Graham, it was named after Charles Boyle (1676–1731), 4th Earl of Orrery, for whom one was made. Strictly defined, a device showing an orbiting and rotating earth with its orbiting moon.

pallet (*Anker*)
Usually occurring in pairs, the parts of a clock **escapement** that engage with the teeth of the **escape wheel** and receive impulses from them. They alternately hold and release each tooth of the escape wheel and so permit the **going train** to run at a controlled rate. See pp. 91–92, Catalog No. 117.

pendulum (*Pendel*)
A weight at the end of a light rod or cord suspended from a fixed support so that it swings freely under the force of gravity. In a timepiece, the oscillating device whose purpose is to impose a constant rate upon the **escapement.** The principle of the pendulum is acribed to Galileo, but it was not until Christiaan Huygens introduced the pendulum clock, first in 1657, that this device was regularly applied to the mechanical clock and thus revolutionized timekeeping. See pp. 23–25, 90–91.

perpetual calendar
A self-correcting calendar showing the day of the week, the date of the month, and the month of the year. Corrections for short months and leap years are automatically made by the mechanism. See p. 16.

pine cone
See **Augsburg pine cone.**

pinion
A small gear, usually with less than 20 teeth, or "leaves."

planetarium
Strictly defined, a device illustrating the orbits of heavenly bodies without representing the rotation of the earth. The earliest clockwork-driven planetarium was the 14th-century clock of Richard of Wallingford, followed by de 'Dondi's **astrarium,** which showed on seven dials the **epicyclic** motions of the sun, moon, and five planets. Clocks incorporating such astronomical indications are also referred to as astronomical or **planetary clocks.** See p. 22, Fig. 7, Catalog Nos. 53, 120.

planetary clock (*Planetenuhr*)
A clock that indicates the planetary signs ruling the hours, i.e., the **reigning planets.** Also, in more common usage, a clock that indicates the motions of the planets (including the sun and moon) around the earth, also called an astronomical clock or a **planetarium.** Most early planetary clocks showed the movements by dials; with the appearance of the **armillary sphere** in the 16th century these orbits were also shown by metal bands or rings. See pp. 16–18, 22, 97, 109–110, 212, Figs. 6, 51–54, 60–63, Catalog No. 53.

planets, reigning or ruling (*regierenden Planeten*)
Also references to "planetary gods" or "planetary deities." In astrological/astronomical usage of the 16 and 17th centuries seven planets—which from ancient times included the sun and moon in addition to Mercury, Venus, Mars, Jupiter, and Saturn—were considered to have influence over human affairs during each hour of the day. Dials indicating the planets which ruled the days of the week are

	seen, e.g., in Catalog Nos. 18, 43, 48. Less often the ruling planet is marked for each hour, either by a dial or table (see Catalog No. 21, Fig. 75).
Prague clock	Term used by Augsburg masters for a clock construction utilizing the **cross-beat escapement.** See p. 93, Fig. 30.
quadrant	An instrument in the shape of a quarter circle used basically for measuring altitudes, consisting of a 90° graduated arc with a movable radius for measuring angles. Quadrants were developed for different purposes, such as determining time (see the combined sundial quadrant and **nocturnal,** p. 108, Figs. 64, 65). A quadrant is incorporated into the **celestial globe** shown in Catalog No. 115.
regulating dial	A dial divided into equal increments, for indicating the amount of adjustment made to the regulating device of a timepiece in order to alter its rate. See Catalog Nos. 27, 33.
remontoire	A mechanical system, developed by Jost Bürgi, for transmitting a consistently equal force to the **escape wheel** of a clock. Power is transmitted from a clock's mainspring or driving weight, through a secondary spring or weight, to the **escapement.** The intermediate spring or weight is rewound at frequent intervals, thus exerting a constant force. See pp. 93–94, 97, 101, Fig. 52, Catalog Nos. 52, 53, 57.
rete	A rotatable openwork front plate of an **astrolabe** or an astrolabe face of a clock, on which the **zodiac** ring and a star map are engraved. See pp. 50–55, Figs. 16, 17, 21, 22, 24, 25, Catalog Nos. 7, 32, 40.
right ascension	An angular distance, measured eastward in units of time from the vernal **equinox** on the celestial equator to the celestial body under observation. The equivalent of terrestrial longitude. See Catalog No. 115.
rolling-ball clock (*Kugellaufuhr*)	A clock employing a ball rolling down an inclined track as a time standard replacing the **foliot** or **balance.** At the end of its run, the ball releases a mechanism that moves the hands. The concept is based on Galileo's observation that a ball rolling on an inclined plane takes the same time to cover the same distance at any point of the incline. See Catalog No. 54.
saints' days, or saint for the day (*Tagesheilige*)	Among the **indications** on the clocks were the names of the saints commemorated on each day of the year according to the Church. See also **calendar ring.** See Catalog Nos. 21, 32, 49, 120.
sidereal time	The standard time used by astronomers, the sidereal day is the time that elapses between two successive meridian transits of the vernal **equinox,** which is about 23 hours 56 minutes in terms of mean **solar time.** Since the sidereal day is about 4 minutes shorter than the mean solar day, there are 366 sidereal days in a civil year. The discrepancy between sidereal and solar time means that the star sphere makes one complete revolution in a year in addition to its diurnal motion.
small clock (*kleine Uhr*)	General term for a domestic clock, as used, e.g., in the clockmakers' guild regulations, as compared with a **large** or **tower clock.** See p. 67 for **masterpiece** specifications for small-clock makers, Augsburg, 1558. See Ch. 7, passim. (See also **Small Clock** under **hours.**)
smiths' eligibility (*Schmiedegerechtigkeit*)	A requirement, inherited or otherwise acquired, necessary for a master to qualify for his own workshop. See pp. 61–65, Fig. 28, Catalog Nos. 30, 43, 57, 75, 97, 109.
solar cycle	A period of 28 years, which is when the days of the month recur on the same days of the week. See Catalog No. 21.
solar time	The solar day is the time that elapses between two successive transits of the sun, some 4 minutes longer than a day measured in **sidereal time** because of the earth's orbital motion. Apparent solar time as measured by a sundial is nonuniform because of the sun's irregular motion along the **ecliptic** and the fact that the ecliptic is inclined to the equator. Mean solar time (or mean time or civil time) is based on a hypothetical sun which moves at a uniform rate, giving equal 24-hour days throughout the year. The difference between mean solar time and apparent solar time is the **equation of time.**
solstice (*Sonnenwende*)	One of two points on the **ecliptic** when the sun has maximum **declination.** The summer solstice, when the sun is farthest above the equator, is commonly known

as the longest day of the year, the winter solstice, when it is farthest below the equator, as the shortest day. On these two days the sun seems to stand still before it returns, as described by the German word, meaning "sun's turning point."

spring drive
The use of a spiral metal spring as motive power for a clock. Probably first used in portable clocks in the early 16th century, clocks before then being driven by weights.

stackfreed
A device consisting of a strip spring and cam that equalizes the torque of an unwinding mainspring by means of a gradually lessening opposing force. Used in early German timepieces, it was a less successful alternative to the **fusee.** See p. 114, Catalog Nos. 61, 65.

striking train (*Schlagwerk*)
The wheels and **pinions** that control the striking of a clock, either on the hour or quarter hour. See pp. 160–161, Figs. 42, 43, 55.

synodic month
Also "lunation." The period of time, 29 1/2 days, between two successive new moons. See p. 56, Catalog No. 6.

table clock
See **drum clock.**

temporal hours
See under **hours.**

touch marks
Knobs placed at the hours on a clock dial (a larger knob usually at 12 o'clock) to facilitate telling time in the dark. See Catalog Nos. 9, 17, 41.

tower clock (*Turmuhr*)
Also "turret clock." a **large clock,** usually of either posted frame or flat bed construction, installed in church towers, city halls, and similar structures as a public timekeeper. See pp. 19, 22, 68, Catalog Nos. 4–6.

train (*Räderwerk*)
A group of wheels and **pinions** interconnected to perform a particular function in a clock.

tropical year
Also "solar year." The period of time between two successive passages of the sun through the vernal **equinox** (365.24 days); the calendar year. See Catalog No. 6.

Turkish honorarium (*Türkenverehrung*)
As terms of a series of negotiated armistices, tributary gifts of money and costly products (including clocks) delivered annually in the 16th century to the Ottomon Porte in Constantinople by the court of the Holy Roman Empire. See Ch. 5, p. 75, Catalog No. 30.

tympanum
One or more plates underlying the **rete** of an **astrolabe** (or astrolabe clock face) that are engraved with representations of **declination** and **right ascension.** Each tympanum is calculated for use at a particular terrestrial latitude. See Catalog Nos. 21, 25, 28, 44.

verge (*Spindel*)
The **pallet arbor** of the **verge escapement.**

verge escapement
See under **escapement.**

volvelle
A set of scales and diagrams consisting of concentric disks made of parchment, paper, or metal for illustrating the relationships of heavenly bodies. See pp. 54–55, 109, Fig. 26.

weight drive
The use of a suspended weight as motive power for a clock. Mechanical clocks were all weight driven until the spiral spring was introduced, probably for the first time early in the 16th century. See Catalog Nos. 1–5, 7, 8, 12–14, 52, 56.

Whole Clock
See under **hours.**

year clock (*Jahresuhr*)
A clock constructed in such a way that it requires winding but once a year. See p. 25, Fig. 31, Catalog No. 57.

zodiac (*Tierkreis*)
A band in the heavens about 18° wide, centered on the **ecliptic,** which contains the orbits of the sun, moon, and major planets. Beginning eastward from the vernal **equinox,** it is divided into 12 equal 30° parts, known as the signs of the zodiac, named after constellations in the vicinity: Aries, Taurus, Gemini, Cancer, Leo, Virgo, Libra, Scorpio, Sagittarius, Capricorn, Aquarius, and Pisces, each represented by a pictorial figure and a glyph. In astrology, the signs are called the celestial or astrological houses (pp. 52–56, Fig. 25, Catalog Nos. 18, 48, 49). Dials with sun and moon hands indicating the positions of these bodies in the zodiac were often seen on clocks, e.g., in Catalog Nos. 9, 14, 18, 24, and 118. A zodiacal circle decorated with representations of the 12 signs is seen on the **armillary sphere,** Catalog No. 114.

Index of Names

Only those authors whose works were published before 1800 are included. The familial designations "the elder" and "the younger" are indicated by I and II. Names of clockmakers are in *italics*.

320

Photo Credits

Italic type refers to Catalog Nos., roman type to figures in the text, including the five introductory texts of the Catalog section of the book.

Augsburg, Städtische Kunstsammlungen   27
Berlin, Staatliche Museen Preussischer Kulturbesitz, Kupferstichkabinett   11
Cologne, Photo C. Hartzenbusch   *15, 85*
London, Colnaghi Ltd.   15
London, Photo P. J. Gates Ltd.   *115*
Munich, Bayerische Staatsbibliothek   8, 74, 82–84
Munich, Bayerisches Nationalmuseum (Walter Haberland)   6, 18–20, 23, 24, 26, 31, 32, 34–50, 54–56, 67–72, 75, 76, 78–81, 90, 91, 94–97, *6–8, 11, 14, 30, 34, 43, 49, 57, 58, 65, 70, 71, 75, 80, 91, 92, 94, 95*
Munich, Klaus Maurice   4, 21, 22, 25, 51–53, 66, 77, 85–89, *1, 4, 13, 17, 19, 21, 22, 26, 39, 41, 44, 45, 62, 66, 67, 73, 82, 83, 87, 90, 97, 98, 104*
Newark, Armen Photographers   *88, 110*
Paramus, Photo Taylor & Dull, Inc.   5
Vienna, Graphische Sammlung Albertina   93
Vienna, Kunsthistorisches Museum   *105, 111*
Vienna, Österreichische Nationalbibliothek   12
Vienna, Photo Meyer   *59, 105*
Winterthur, Photo H. R. Fellmann   33
Winterthur, Photo Michael Speich   *12*
All other photos are identified by the owners or lenders.

Library of Congress Cataloging in Publication Data

Main entry under title:

The Clockwork universe.

Catalog of an exhibition held in Munich, Germany, April 15 to Sept. 30, 1980, and in Washington, D.C., Nov. 7, 1980, to Feb. 15, 1981, and organized by the National Museum of History and Technology, Smithsonian Institution, Washington, D.C., and the Bayerisches Nationalmuseum, Munich.

Includes bibliographical references and index.
1. Clocks and watches—Exhibitions. 2. Automata—Exhibitions. I. Maurice, Klaus. II. Mayr, Otto. III. National Museum of History and Technology. IV. Munich. Bayerisches National-museum.

TS541.G32M858   681.1'13'09430740153   80-16780
ISBN 0-88202-188-5